The Mechanics of Soils and Foundations

Also available from Taylor & Francis

Introductory Geotechnical Engineering
Y.S. Fang et al.

Hb: ISBN 9780415304016
Pb: ISBN 9780415304023

Geotechnical Engineering 2nd ed
R. Lancellotta

Hb: ISBN 9780415420037
Pb: ISBN 9780415420044

Pile Design & Construction Practice 5th ed
M. Tomlinson et al.

Hb: ISBN 9780415385824

Piling Engineering
W.G.K. Fleming et al.

Hb: ISBN 9780415266468

Foundations of Geotechnical Engineering
I. Jefferson et al.

Hb: ISBN 9780415272407
Pb: ISBN 9780415272414

Decoding Eurocode 7
A. Bond et al.

Hb: ISBN 9780415409483

Geotechnical Physical Modelling
C. Leung et al.

Hb: ISBN 9780415420174

Information and ordering details

For price availability and ordering visit our website **www.tandfbuiltenvironment.com/**
Alternatively our books are available from all good bookshops.

The Mechanics of Soils and Foundations

Second Edition

John Atkinson

Taylor & Francis
Taylor & Francis Group

LONDON AND NEW YORK

First published 1993 by McGraw-Hill

Second edition published 2007 by Taylor & Francis
2 Park Square, Milton Park, Abingdon, Oxon, OX14 4RN

Simultaneously published in the USA and Canada
by Taylor & Francis
270 Madison Ave, New York NY 10016

Reprinted 2010 (twice)

Taylor & Francis is an imprint of the Taylor & Francis Group, an informa business

© 2007 John Atkinson

Typeset in Sabon by Keyword Group Ltd

Printed and bound in Great Britain by TJ International Ltd, Padstow, Cornwall

British Library Cataloguing in Publication Data
A catalogue record for this book is available from the British Library

Library of Congress Cataloging-in-Publication Data
Atkinson, J.H.
    The mechanics of soils and foundations/John Atkinson. – 2nd ed.
        p. cm.
    Includes bibliographical references and Index.
    ISBN 978-0-415-36256-6 (alk. paper) – ISBN 978-0-415-36255-9 (alk. paper)
1. Soil mechanics. 2. Foundations. 3. Slopes (Soil mechanics) I. Title.
    TA710.A786 2007
    624.1'5–dc22
    2006049709

ISBN 10: 0-415-36255-5 (hbk)
ISBN 10: 0-415-36256-3 (pbk)
ISBN 10: 0-203-01288-7 (ebk)

ISBN 13: 978-0-415-36255-9 (hbk)
ISBN 13: 978-0-415-36256-6 (pbk)
ISBN 13: 978-0-203-01288-8 (ebk)

Contents

Preface to the 1st edition

This book is about the behaviour of engineering soils and simple geotechnical structures such as foundations and slopes and it covers most of the theoretical geotechnical engineering content of a degree course in civil engineering. The book is aimed primarily at students taking first degree courses in civil engineering but it should also appeal to engineers, engineering geologists and postgraduate students wishing for a simple and straightforward introduction to the current theories of soil mechanics and geotechnical engineering. Although it deals specifically with soils and soil mechanics many of the theories and methods described apply also to rocks and rock mechanics.

The teaching and practice of geotechnical engineering has undergone significant changes in the past 25 years or so, both in the development of new theories and practices and in the standing of the subject within the civil engineering curriculum. Geotechnical engineering is now regarded as one of the major disciplines in civil engineering analysis (the others being hydraulics and structures). The most important development, however, has been the unification of shearing and volumetric effects in soil mechanics in the theories known generally as critical state soil mechanics and application of these theories in geotechnical analysis. In this book, unlike most of the other contemporary books on soil mechanics, the subject is developed using the unified theories right from the start, and theories for stability of foundations and slopes are developed through the upper and lower bound plasticity methods as well as the more commonly used limit equilibrium method. This is an up-to-date approach to soil mechanics and geotechnical engineering and it provides a simple and logical framework for teaching the basic principles of the subject.

The term 'critical state soil mechanics' means different things to different people. Some take critical state soil mechanics to include the complete mathematical model known as Cam Clay and they would say that this is too advanced for an undergraduate course. My view is much simpler, and by critical state soil mechanics I mean the combination of shear stress, normal stress and volume into a single unifying framework. In this way a much clearer idea emerges of the behaviour of normally consolidated and overconsolidated soils during drained and undrained loading up to, and including, the ultimate or critical states. It is the relationship between the initial states and the critical states that largely determines soil behaviour. This simple framework is extremely useful for teaching and learning about soil mechanics and it leads to a number of simple analyses for stability of slopes, walls and foundations.

This book is based on courses of lectures given to undergraduate students in civil engineering at City University. In the first year students take a course in geology and they also take a course in mechanics of materials within which there are six to eight lectures on soil mechanics and geotechnical engineering. These lectures cover the whole of the conventional syllabus (classification, seepage, strength, consolidation, bearing capacity and settlement, slope stability and earth pressure) but at lightning speed. The object is to introduce the students to the concepts and vocabulary of geotechnical engineering within the context of conventional mechanics of materials and structures and with reference to their everyday, childhood experiences of playing with sand, flour, plasticine and other soil-like materials so that, as the course develops in later years, they can relate particular topics into the whole scheme of civil engineering.

In the second year the students take a major course of lectures (with several laboratory sessions) in theoretical soil mechanics and geotechnical engineering. This is based on my earlier books – *The Mechanics of Soils* (with Peter Bransby) and *Foundations and Slopes*. This course depends entirely on the unification of shearing and volumetric effects which is introduced right from the start (and had been in the first year), although the phrase 'critical state soil mechanics' is rarely used. Theoretical soil mechanics is taken up to the development of a complete state boundary surface but stops short of the mathematical treatment of Cam clay. Stability problems are solved using upper and lower bound methods and these are then used to introduce limit equilibrium methods and standard tables and charts for bearing capacity, slope stability and earth pressure. In the third year the course covers practical aspects of geotechnical engineering through a series of lectures and projects on topics such as ground investigation, foundations, slopes, retaining walls and embankment designs.

This book covers the material in the second-year course (and also that summarized in the first year). It does not deal specifically with the practical aspects of geotechnical engineering which are introduced in the third year and are, in any case, generally better learned through working in practice with experienced engineers. This book should provide the basic text for an undergraduate course, but students will have to consult other books and publications to find more detailed coverage of particular topics such as laboratory testing, seepage, slope stability and foundation design.

The treatment of soil mechanics and geotechnical engineering in this book is simple, straightforward and largely idealized. I have tried to relate the behaviour of soils and geotechnical structures to everyday experiences, encouraging students to perform simple experiments themselves at home, on holiday and in a basic soil mechanics laboratory. I have described some simple tests which are designed to demonstrate the basic principles rather than generate highly accurate results. Only a few details are given of the apparatus and procedures since engineers should be trained to design and build simple equipment and work out how to make observations and analyse results themselves.

To illustrate the basic nature of soil strength and stiffness I have described the behaviour of soils in oedometer tests and in ideal shear tests in order to separate the effects of normal stress and compression from the effects of shearing and distortion. I have also described the behaviour of soils in triaxial tests, as these are the best tests to evaluate soil parameters. Readers will notice that I have not included data from tests on real soils or case histories of construction performance. This is quite deliberate and is common practice in undergraduate texts on structures, hydraulics, concrete and so on. As the book is intended primarily as an undergraduate teaching text it is kept simple and straightforward. The basic soil mechanics theories have been clearly demonstrated in earlier books from *Critical State Soil Mechanics* by Schofield and Wroth in 1968 to *Soil Behaviour and Critical State Soil Mechanics* by Muir Wood in 1991, and almost everything in this book follows from these well-established theories.

Throughout I have dealt with simple theories and idealizations for soil behaviour. I am very well aware that many natural soils behave in ways that differ from these idealizations and that there are a number of additional factors that may influence the design and analysis of geotechnical structures. Nevertheless, I am convinced that for the purposes of teaching the fundamental principles to students, it is better to maintain the simplicity of the idealized treatment, provided always that they appreciate that it is idealized. At many points in the text I have indicated where the behaviour of various natural soils may depart significantly from the idealized behaviour. I expect that individual lecturers will bring in other examples of the behaviour of natural soils drawn from their own experiences, but I hope that they would discuss these within the simple framework described in this book.

At the end of most chapters there is a short summary of the main points covered in the chapter and, in most cases, simple worked examples and exercises that illustrate the theories developed in the text. There is also a short selection of books and articles for further reading and a list of specific references quoted in the text.

The courses at City University which form the basis of this book were developed jointly with my colleagues Neil Taylor, Matthew Coop and John Evans and I am grateful to them for their contributions and for their comments and criticisms. I am grateful also for the very detailed comments that I received from many friends and colleagues, including Mark Allman, Eddie Bromhead, Peter Fookes, Charles Hird, Marcus Matthews, Sarah Stallebrass and Giulia Viggiani. The typing was shared between Anne-Christine Delalande and Robert Atkinson.

John Atkinson
City University
London

Preface to the 2nd edition

The first edition of *The Mechanics of Soils and Foundations* was published in 1993. Its objectives were set out in the Preface to the 1st edition and in the final chapter and they were, I think, largely achieved. It was not, and was not intended to be, a guide to ground engineering: it contained few design charts, few references to standards and codes and no case histories. It has been widely used as an undergraduate and post-graduate text in the more up-to-date university courses in soil mechanics and geotechnical engineering in UK and abroad. Teachers and students liked the very simple treatments of the basic theories of soil mechanics and their applications in analyses of slopes, foundations and retaining walls. Most of the engineers' offices I have been in have a copy.

When Tony Moore of Taylor & Francis approached me to prepare a second edition my initial response was say that I was very happy with the book as it was and couldn't he simply re-print it. It dealt with basic principles, little had changed in the time since the first edition was published and it was not therefore outdated. However, as I read it again in detail I saw that there were bits missing and there were places where I had not explained things as clearly as I would have liked. So this 2nd edition contains some new material and it includes extensive re-working and clarifications of the original descriptions and explanations; I hope to make them clearer.

I have added three new chapters. In the chapter on design parameters I have discussed which strengths, peak, critical state or residual, should be used with different analyses and with what factors. I have distinguished more clearly than before between a factor of safety which is intended to ensure that the structure is safe and a load factor which is intended to limit movements. I have related the basic strength and stiffness parameters of a soil to its classification. I have added a short chapter on unsaturated soils which describes the relationships between soil, water and air and aspects of the behaviour of unsaturated soils. I have added a chapter on soft ground tunnelling because this is an important part of modern ground engineering and the current theories and design methods illustrate the development of practice from a number of different methods of research.

As in the 1st edition, I have given a few charts of design parameters for slopes and foundations to illustrate the more common ones. This is a teaching text-book and it describes the basic theories upon which these charts are based. It is not a design manual and I have deliberately made these charts small so you will not be able to obtain reliable values from them. If you are designing a slope, a foundation or a retaining wall you should go to the original design charts or calculate the values you need.

I have described some simple experiments which you can do yourself at home or at the beach. These investigate the nature and state of soils and the behaviour of simple slopes and foundations and there are links to the parts of the book which deal with the relevant theories. It is important always to relate the behaviour of these simple models to the theories you are taught at university and to the designs you will do in practice.

I am grateful to all those friends and colleagues who have discussed with me soil mechanics and geotechnical engineering and how it should be taught to students. My belief is that university students should know and understand the basic principles which are described in my book. Applications of these theories through standards and codes should be taught only when sufficient time has been given to establishing the basic principles; practice is better learned by practice.

I am grateful for the very detailed comments and criticisms that I received from Eddie Bromhead, Federica Coteccia, Andrew McNamara, Sarah Stallebrass, Neil Taylor and Dave White who introduced me to the problem of the car going up hill which I have used in Chapter 22. I am grateful to my colleagues at City University for allowing me the time to complete this new edition.

John Atkinson
City University
London

A note on units

The SI system of units has been used: the basic units of measurement are:

Length   m
Time   s
Force   N   multiples   kiloNewton   1 kN $= 10^3$ N
                           megaNewton   1 MN $= 10^6$ N

Some useful derived units are:

Velocity   m/s
Acceleration   m/s^2
Stress (pressure) kN/m^2 = kiloPascal = kPa
Unit weight   kN/m^3

Unit force (1 N) gives unit mass (1 kg) unit acceleration (1 m/s^2). The acceleration due to the Earth's gravity is $g = 9.81$ m/s^2; hence the force due to a mass of 1 kg at rest on Earth is 9.81 N. (*Note*: there are about 10 apples in 1 kg: hence a stationary apple gives rise to a force of about 1 N acting vertically downwards.)

Greek alphabet

As in most branches of science and engineering, geotechnical engineering uses mathematics and symbols to develop general theories. Because the English alphabet has a limited number of characters use is made of the Greek alphabet:

A	α	alpha
B	β	beta
Γ	γ	gamma
Δ	δ	delta
E	ε	epsilon
Z	ζ	zeta
H	η	eta
Θ	θ	theta
I	ι	iota
K	κ	kappa
Λ	λ	lambda
M	μ	mu
N	ν	nu
Ξ	ξ	xi
O	o	omicron
Π	π	pi
P	ρ	rho
Σ	σ	sigma
T	τ	tau
Y	υ	upsilon
Φ	ϕ	phi
X	χ	chi
Ψ	ψ	psi
Ω	ω	omega

Glossary of symbols

Stress and strain parameters

One-dimensional compression and shear tests:
τ' shear stress
σ' normal stress
γ shear strain
ε_v volumetric strain = normal strain

Triaxial tests:
$q' = (\sigma_a' - \sigma_r')$ deviatoric stress
$p' = \frac{1}{3}(\sigma_a' + 2\sigma_r')$ mean normal stress
$\varepsilon_s = \frac{2}{3}(\varepsilon_a - \varepsilon_r)$ shear strain
$\varepsilon_v = \varepsilon_a + 2\varepsilon_r$ volumetric strain

Superscripts for strains
e   elastic
p   plastic

Subscripts for states
0   initial state   (i.e. q_0', p_0', v_0)
f   critical state   (i.e. q_f', p_f', v_f)
p   peak state   (i.e. q_p', p_p', v_p)
y   yield   (i.e. p_y')
r   residual

Subscripts for axes
z, h   vertical and horizontal
a, r   axial and radial

Normalizing parameters
$\ln p_c' = (\Gamma - v)/\lambda$
$v_\lambda = v + \lambda \ln p'$
$\log \sigma_c' = (e_\Gamma - e)/C_c$
$e_\lambda = e + C_c \log \sigma'$

A	area
A	activity
B	breadth or width
B	parameter for peak state power law envelope
B	pore pressure parameter
C	compliance
C	cover (above tunnel)
C_c	slope of the normal compression line
C_s	slope of a swelling and recompression line
D	depth
D_r	relative density
E	work done by external loads
E	Young's modulus (E' for effective stress; E_u for undrained loading)
F	force
F_a	axial force
F_n	normal force
F_s	shear force
F_s	factor of safety
G	shear modulus (G' for effective stress; G_u for undrained loading)
G_p, H_p	parameters for peak strength in triaxial tests
G_o'	shear modulus at very small strain
G_s	specific gravity of soil grains
H	height or thickness
H	maximum drainage path
H	horizontal load (on a foundation)
H_c	critical height (of a slope)
I_l	liquidity index
I_P	plasticity index
I_σ	influence coefficient for stress
I_ρ	influence coefficient for settlement
J	stiffness modulus coupling shear and volumetric parameters
K'	bulk modulus
K_0	coefficient of earth pressure at rest
K_a	coefficient of active earth pressure
K_p	coefficient of passive earth pressure
L	length
L_f	load factor
M'	one dimensional modulus
N	normal force
N_c, N_γ, N_q	bearing capacity factors
N_d	number of equipotential drops (in a flownet)
N_f	number of flow channels (in a flownet)
N_s	stability number (for undrained slopes)
P	length of tunnel heading
P	potential
P	force on retaining wall
P_a	force due to active pressure

P_p	force due to passive pressure
P_w	force due to free water
Q	flow (volume)
Q	pile load
Q_b	pile base resistance
Q_s	pile shaft resistance
R	radius
R	rigidity
R_o	overconsolidation ratio for one-dimensional compression and swelling
R_p	overconsolidation ratio for isotropic compression and swelling
S	stiffness
S	settlement (above a tunnel)
S_r	degree of saturation
S_s	stress state parameter
S_v	volume state parameter
T	surface tension force
T	torque (on a shear vane)
T	shear force
T_c	tunnel stability number
T_r	time factor for radial consolidation
T_v	time factor for one-dimensional consolidation
U	force due to pore pressures
U_t	average degree of consolidation after time t
V	vertical load (on a foundation)
V	velocity (of seepage)
V	volume
V_a	allowable load
V_c	collapse load
V_s	safe load
V_s	volume of soil grains
V_w	volume of water
W	work dissipated by interval stresses
W	weight
W_s	weight of soil grains
W_w	weight of water
Y_o	yield stress ratio for one-dimensional stresses
Y_p	yield stress ratio for isotropic stresses
a	acceleration
b	thickness or width
c'	cohesion intercept in Mohr–Coulomb failure criterion
c_r	coefficient of consolidation for radial consolidation
c_s	coefficient of consolidation for spherical consolidation
c_v	coefficient of consolidation for one-dimensional consolidation
d_s	grain diameter
d_v	voids diameter
e	voids ratio

e_0	voids ratio of normally consolidated soil at $\sigma' = 1.0$ kPa
e_κ	voids ratio of overconsolidated soil at $\sigma' = 1.0$ kPa
e_Γ	voids ratio of soil on the critical state line at $\sigma' = 1.0$ kPa
g	shear modulus for states inside the state boundary surface
h_w	height of water in standpipe
i	width of settlement trough (above tunnel)
i	slope angle
i_c	critical slope angle
i	hydraulic gradient
i_c	critical hydraulic gradient
k	coefficient of permeability
m, n	slope stability numbers (for drained slopes)
m_v	coefficient of compressibility for one-dimensional compression
n	scale factor (for modelling)
n_l	degree of non-linearity
p	surface stress
p'_m	maximum past stress
q	surcharge pressure
q	rate of seepage
q	bearing pressure
q_c	bearing capacity
q_n	net bearing pressure
q_a	allowable bearing pressure
q_s	safe bearing pressure
r	radius
r_u	pore pressure coefficient
s	length along a flowline
s_u	undrained strength
s_w	shear stress (on a retaining wall)
t	time
u	pore pressure
u_0	initial steady state pore pressure
u_a	pore air pressure
u_w	pore water pressure
u_∞	long term steady state pore pressure
$\bar{u}$	excess pore pressure
v	volume loss ratio (for tunnelling)
v	specific volume
v_κ	specific volume of overconsolidated soil at $p' = 1.0$ kPa
w	water content
w_l	liquid limit
w_p	plastic limit
Γ	specific volume of soil on the critical state line at $p' = 1.0$ kPa
Δ	large increment of
M	slope of CSL projected to $q':p'$ plane
N	specific volume of normally consolidated soil at $p' = 1.0$ kPa
Σ	sum of

α factor for undrained shear stress on pile shaft
γ unit weight
γ_d dry unit weight
γ_w unit weight of water ($= 9.81 \text{ kN/m}^3$)
δ small increment of
δ' angle of friction between structure and soil
η q'/p'
κ slope of swelling and recompression line
λ slope of normal compression and critical state lines
ν Poisson's ratio (ν' for drained loading, $\nu_u = \frac{1}{2}$ for undrained loading)
ρ density
ρ_a allowable settlement
ρ_d dry density
ϕ'_m mobilized angle of friction
ϕ'_s safe angle of friction
ϕ'_r residual angle of friction
σ_t tunnel support pressure
τ shear stress on pile shaft
ρ settlement
ρ_c consolidation settlement
ρ_i initial settlement
ρ_t settlement at time t
ρ_∞ final consolidation settlement
ϕ' angle of friction
ϕ'_a allowable angle of friction
ϕ'_c critical state angle of friction
ϕ'_p peak angle of friction
ψ angle of dilation

Some simple experiments to illustrate geotechnical engineering

1 Introduction

I have suggested a set of simple experiments which you can do at home or on the beach without any special apparatus. These illustrate some of the fundamental theories of soil mechanics and geotechnical engineering covered in the book. For each experiment I have indicated the sections in the book where you can find the relevant theories and explanations.

2 Soil grains

For this you will need a magnifying glass and a clear bottle. Find some different soils and inspect the grains closely. If you are near a beach you will be able to find sand and gravel. At home, get some soil from the garden and get some sand from a builder. Dry your samples in an oven at about 100°C (gas mark $\frac{1}{4}$) and, if the grains are sticking together, separate them with a pestle and mortar or with a small hammer.

Separate the grains into piles with grains in the sizes:

<0.01 mm
0.01 mm to 0.1 mm
0.1 mm to 1 mm
1 mm to 10 mm
>10 mm

You can see grains larger than about 0.1 mm with your naked eye and grains larger than about 0.01 mm with a magnifying glass. Estimate the proportions by weight in each pile and draw a grading curve. Grading curves are described in Sec. 5.3.

Put some of the soil into a bottle of water: the water should fill the bottle and the soil sample should fill the bottle about $\frac{3}{4}$ full. Shake well, stand the bottle upright and wait until the water at the top becomes clear or for a maximum of 24 hours. The largest grains will sediment first and they will be at the bottom: the grains become finer upwards. If the water at the top is still cloudy after 24 hours the soil there is clay. Estimate the proportions of the different sizes and plot a grading curve. If you have done your experiments on the same soil you should have the same grading curves.

Examine the grains larger than 1 mm and describe their shape and surface texture: are they rounded or angular, are they elongated or flaky, are they rough or smooth?

You have now done the basic descriptions of the nature of soils described in Chapter 5.

3   Soil state

For this you will need some clean sand and a straight-sided glass. The sand should be about 1 mm size and rounded sand is best. You can get some from a beach or use coarse soft sand from a builder.

Measure the internal diameter of the glass and weigh enough dry sand to fill the glass so it is about $\frac{3}{4}$ full. Stand the empty glass in a saucer and fill it to the brim with water. Slowly pour the sand into the glass so the water overflows. Collect all the water which has been displaced and weigh it. This is the volume occupied by the sand grains and you can calculate the specific gravity of the sand grains. (The answer should be about $G_s = 2.6$ to 2.7.)

The sand is loosely packed. Measure its height in the glass and calculate the volume it occupies. You can now calculate the voids ratio and specific volume of the sand, its water content and its unit weight. These calculations are described in Sec. 5.5. Calculate the total vertical stress, the pore water pressure and the effective stress at the bottom of the glass. These calculations are described in Sec. 6.2.

Vibrate the glass: you can do this by gently tapping its side with a rubber hammer or a block of soft wood but don't break the glass. The top surface of the sand will settle as the grains pack together more closely. The sand is now dense. Measure the new height and re-calculate all the parameters you calculated for its loose state. Some should be the same and some should be different.

4   Slopes in sand

For this you will need some clean sand, a plastic cup, a washing-up bowl, a bottle and a rubber tube. The sand should be about 1 mm size and rounded sand is best. You can get some from a beach or use coarse soft sand from a builder.

Pour dry sand from the plastic cup onto a rough flat surface and measure the angle of the slope of the cone. This is the critical state friction angle. The critical state friction angle is defined in Sec. 9.3 and the relationship between this and a slope angle is given in Sec. 21.6.

Half fill the bottle with dry sand and roll it along a table so there is a continuously failing slope. The angle of the slope is the critical state friction angle. A better version of this experiment is shown in Fig. 9.14.

Half fill the washing-up bowl with dry sand, shake it or vibrate it to make it dense and leave the surface flat. Tip the bowl up until the slope starts to move and measure the angle of the slope just before it moves. This is the peak friction angle. The peak strength of soil is discussed in Sec. 10.4. When the slope has come to rest measure the new slope angle. This is the critical state friction angle.

Pour the sand slowly, totally under water without causing currents in the water. (This is quite difficult to do and you will need to do it in deep water in a large container.) The angle of the slope will be the same as the angle of the slope in dry sand. This shows that the critical state friction angle is the same in dry and saturated sand.

Using the rubber tube slowly syphon the water from the container without causing currents in the water. As water drains from the sand the slope becomes flatter. This demonstrates the effect of seepage on slope stability discussed in Sec. 21.7.

5   Unconfined compressive strength

For this you will need some clean sand with grain size about 1 mm, some clay, some plastic cups and some kitchen scales. You can get clay from your garden or modelling clay from an art shop.

Add some water to the sand and make a sandcastle using a plastic cup. Notice how much water you added to make a good sandcastle.

Measure the strength and water content of your sandcastle. To measure its strength put it on the kitchen scales and note its weight. With your hand press down on the top of the sandcastle until it fails and note the maximum reading on the scales. If you divide the force required to fail the sandcastle (less its weight) by its area you have measured its unconfined compressive strength. This is one of the strengths defined in Sec. 3.3. Put all your sandcastle in the oven at 100°C to dry it and then weigh the dry sand. You can now calculate the water content of your sandcastle (see Sec. 5.5). Repeat the test with larger and smaller water contents and plot a graph of unconfined compressive strength against water content. What happens at large and small water contents?

Get some of your clay and adjust its water content by adding water or drying it until you can mould it like plasticine. Press lumps into a plastic cup with your fingers and then cut the plastic cup away to make a 'claycastle'. Measure its strength and water content. Repeat the test with more and less water and plot a graph of unconfined compressive strength against water content. What happens at large and small water contents?

Compare the results from your sandcastles and your claycastles. What are these results telling you about soil strength and the pressures of water in the pores of the soil? Why is the 'claycastle' much stronger than the sandcastle?

The strength of both the 'claycastle' and the sandcastle arise from a combination of friction and effective stress as described in Sec. 9.3. The effective stress depends on the total stress and the pore pressure as described in Sec. 6.5. The radial total stresses are zero, the pore pressures are negative (i.e. they are suctions) so the effective stresses are positive. The suctions depend on the grain size as described in Sec. 6.4. In the 'claycastle' the grains are very small so the suctions and the strength are relatively large: in the sandcastle the grains are larger so the suction and the strength are smaller.

6   Undrained loading of sand

Go to the seaside and find some smooth sand near the sea. Notice that as you walk across the sand your footprints look dry for a little while. Put your foot on the ground and push it sideways. The sand around your foot will look dry but, if you keep pushing, it will return to its original appearance. If you then take your foot away the sand will be wet and there may be free water.

The sand is dense. When you push sideways quickly it is sheared undrained and the pore pressures decrease, as discussed in Sec. 11.2 and shown in Fig. 11.1. The negative pore pressures make the surface of the sand appear dry; you are looking at meniscuses. If you wait the excess (negative) pore pressures dissipate and as the water flows towards your footprint the surface returns to its original appearance. When you remove your foot the water appears at the surface.

7   Excavations

Go to the seaside and dig holes in the beach in different places. (Do not dig your holes deeper than your waist because if the sides collapse you could be killed.) Dig one hole well above the high tide mark and another near the water.

Observe whether it is difficult or easy to dig the hole; what do the sides of the hole look like; what does the sand which you have dug out look like when you dump it on the ground; what happens when you reach water?

You will observe the behaviour of a simple excavation described in Sec. 21.10 and illustrated in Fig. 21.20. When you reach the water table, water will flow into the hole and you will not be able to dig it any further. If the sand is wet (i.e. it has a high water content) it is difficult to dig out because the rapid excavation is undrained and there will be suctions in the pore water. If the sand is nearly dry (i.e. the degree of saturation is small) the suctions will be small and the sand will be relatively easy to excavate.

Carefully inspect the sides of your holes. Is the sand all one size or are there layers of sand and gravel? Where does the soil look wet and where does it look dry? Poke your finger into the sides of your hole: is this easy or difficult to do?

You have done a simple ground investigation, constructed a test pit as described in Sec. 17.4, and done some simple probing tests as described in Sec. 17.5. You could take some samples home and determine their gradings, friction angles, water contents and unconfined compressive strengths.

8   Shallow foundations

For this you will need a washing-up bowl, some sand, some cups or glasses of different sizes and a piece of tube. Fill a washing-up bowl with dry sand with the rubber tube placed down the inside to the bottom of the bowl. Put a container such as a tall glass, which you have weighed, on the surface and measure the settlement. Slowly fill the glass with sand, measuring the weight of the sand in it and measure its settlement at stages of loading. Measure the diameter of the glass and calculate the stress applied to the soil surface. Plot the applied stress against the settlement divided by the diameter of the glass.

You should obtain a plot of the bearing pressure against the settlement similar to that shown in Fig. 22.3. Has the foundation reached its bearing capacity?

With the glass full of sand so the foundation is loaded, slowly pour water through the tube down the inside of the washing-up bowl so you are pouring water into the bottom of the sand. (It helps if you put a layer of gravel 10 to 15 mm thick below the sand.) Observe what happens to the loaded foundation as water rises in the sand. As the water table rises the effective stresses in the sand decrease and the bearing capacity of the foundation reduces: there will be additional settlements. The bearing capacity of shallow foundations is discussed in Sec. 22.5. The influence of the water table on bearing capacity is shown in Eq. 22.10.

If you are able to do so, drill a hole in the bottom of the washing-up bowl so the water drains out. What happens to the sand and to the foundation? This is like the experiment shown in Fig. 6.7.

As water drains down the pore pressures decrease and the effective stresses increase. The soil will not dry completely. The suctions will depend on the degree of saturation and these are related by a water retention curve like those shown in Fig. 26.5. The soil will be stiffer and stronger than when it was dry. You can add load to the foundation and the settlements will be relatively small. If you now block the hole in the bottom of the washing-up bowl and pour more water into the tube the foundation will probably fail.

Fill the washing-up bowl with polystyrene peas. You can get these from any large department store: they are used to fill pillows and cushions. Put your empty cup or glass on the surface and it immediately sinks to the bottom. Why can sand support a foundation but polystyrene peas cannot? Look at Eq. 22.10. The polystyrene soil is dry so $\gamma_w = 0$ and the polystyrene is very light so $\gamma \approx 0$ and the bearing capacity is negligible.

You are driving a car off-road up a slight slope. If the wheels start to spin should the passengers get out or should you put more people in the car? Does it matter if the ground is sand or clay? What you should do is described in Sec. 22.7.

9   Piled foundations

For this you will need some sand and some clay, some plastic cups or tin mugs and a pencil. Put marks on the pencil at intervals of 5 mm. Put some dry sand into a cup or mug so it is loose. Put it on the kitchen scales and note the weight. Push the pencil slowly and continuously into the sand and as each 5 mm mark passes the surface record the force on the scales. Now pull the pencil out slowly and continuously and record the force as each mark emerges from the surface.

Put fresh sand into the cup or mug and vibrate it so it becomes dense. Repeat the penetration and extraction experiment.

Adjust the water content of your clay until it is like plasticine. Mould this into a cup or mug and repeat the penetration experiment. Measure the water content of your clay. Repeat the penetration experiment with your clay with different water contents.

For each test, plot the load against the displacement of the pencil during penetration and extraction and plot the results from all tests on the same graph. Look at the ways in which the load builds up with penetration. Compare the behaviour between sand and clay, between penetration and extraction and between dense and loose sand and as the water content of the clay changes.

Assuming that the shear stress between the side of the pencil and the sand is the same for penetration and extraction, calculate the contribution to the total load arising from the resistance at the tip of the pencil and arising from shear stress along its sides and plot these against the depth of penetration for each test.

You have now done pile load tests and the behaviour should be something like that shown in Fig. 23.2(b).

10   Failure of earth dams

Go to the seaside and find a place where a little stream of water is flowing across the beach towards the sea. Build a small dam across the stream to make a pond. Watch what happens. If the pond overtops the dam the flow of water will quickly erode the downstream slope, the dam will fail and there will be a large breach in the dam.

If the dam is not overtopped, water will seep through the dam because sand is highly permeable and it will emerge near the toe of the dam where the slope will be relatively flat. The seepage conditions near the toe are like those described in Sec. 21.5 and shown in Fig. 21.7(b). Water emerging from the downstream slope will erode soil from the toe and this will progress back into the dam which will fail. This is hydraulic erosion described in Sec. 14.6.

Get some clay from your garden. Adjust its water content by adding water or drying it until you can mould it like plasticine. Mould a dam across the middle of your washing-up bowl and fill one side with water. The dam does not leak and it remains stable. The clay dam does not leak because clay has a relatively very low permeability. The permeabilities of sand and clay are described in Sec. 6.10.

11   Summary

I hope that you will do some of these simple experiments yourselves and use the theories given later in the book to understand what is happening. Many of the fundamental principles of geotechnical engineering are revealed by these experiments.

As part of your course you will probably do further laboratory experiments on soil samples and on model foundations, slopes and walls. You should observe what happens and relate your observations to the simple theories described later.

Chapter 1

Introduction to geotechnical engineering

1.1 What is geotechnical engineering?

The use of engineering soils and rocks in construction is older than history and no other materials, except timber, were used until about 200 years ago when an iron bridge was built by Abraham Darby in Coalbrookdale. Soils and rocks are still one of the most important construction materials used either in their natural state in foundations or excavations or recompacted in dams and embankments.

Engineering soils are mostly just broken up rock, which is sometimes decomposed into clay, so they are simply collections of particles. Dry sand will pour like water but it will form a cone, and you can make a sandcastle and measure its compressive strength as you would a concrete cylinder. Clay behaves more like plasticine or butter. If the clay has a high water content it squashes like warm butter, but if it has a low water content it is brittle like cold butter and it will fracture and crack. The mechanics that govern the stability of a small excavation or a small slope and the bearing capacity of boots in soft mud are exactly the same as for large excavations and foundations.

Many engineers were first introduced to civil engineering as children building structures with Meccano or Lego or with sticks and string. They also discovered the behaviour of water and soil. They built sandcastles and they found it was impossible to dig a hole in the beach below the water table. At home they played with sand and plasticine. Many of these childhood experiences provide the experimental evidence for theories and practices in structures, hydraulics and soil mechanics. I have suggested some simple experiments which you can try at home. These will illustrate the basic behaviour of soils and how foundations and excavations work. As you work through the book I will explain your observations and use these to illustrate some important geotechnical engineering theories and analyses.

In the ground soils are usually saturated so the void spaces between the grains are filled with water. Rocks are really strongly cemented soils but they are often cracked and jointed so they are like soil in which the grains fit very closely together. Natural soils and rocks appear in other disciplines such as agriculture and mining, but in these cases their biological and chemical properties are more important than their mechanical properties. Soils are granular materials and principles of soil mechanics are relevant to storage and transportation of other granular materials such as mineral ores and grain.

Figure 1.1 illustrates a range of geotechnical structures. Except for the foundations, the retaining walls and the tunnel lining all are made from natural geological materials. In slopes and retaining walls the soils apply the loads as well as provide strength

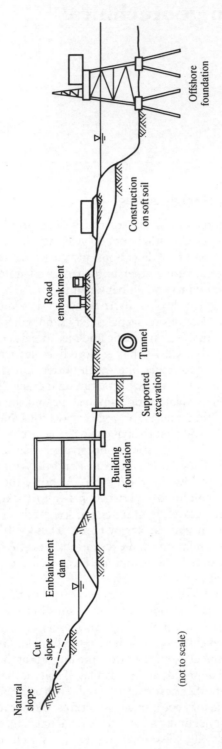

Figure 1.1 Examples of geotechnical engineering construction.

Natural slope

Cut slope

Embankment dam

Building foundation

Supported excavation

Tunnel

Road embankment

Construction on soft soil

Offshore foundation

(not to scale)

and stiffness. Geotechnical engineering is simply the branch of engineering that deals with structures built of, or in, natural soils and rocks. The subject requires knowledge of strength and stiffness of soils and rocks, methods of analyses of structures and hydraulics of groundwater flow.

Use of natural soil and rock makes geotechnical engineering different from many other branches of engineering and more interesting. The distinction is that most engineers can select and specify the materials they use, but geotechnical engineers must use the materials that exist in the ground and they have only very limited possibilities for improving their properties. This means that an essential part of geotechnical engineering is a ground investigation to determine what materials are present and what their properties are. Since soils and rocks were formed by natural geological processes, knowledge of geology is essential for geotechnical engineering.

1.2   Principles of engineering

Engineers design a very wide variety of systems, machines and structures from car engines to very large bridges. A car engine has many moving parts and a number of mechanisms, such as the pistons, connecting rods and crankshaft or the camshaft and valves, while a bridge should not move very much and it certainly should not form a mechanism. Other branches of engineering are concerned with the production and supply of energy, the manufacture of washing machines and personal computers, the supply, removal and cleaning of water, moving vehicles and goods and so on.

Within civil engineering the major technical divisions are structural (bridges and buildings), hydraulic (moving water) and geotechnical (foundations and excavations). These are all broadly similar in the sense that a material, such as steel, water or soil, in a structure, such as a bridge, river or foundation, is loaded and moves about. The fundamental principles of structural, hydraulic and geotechnical engineering are also broadly similar and follow the same fundamental laws of mechanics. It is a pity that these subjects are often taught separately so that the essential links between them are lost.

In each case materials are used to make systems or structures or machines and engineers use theories and do calculations that demonstrate that these will work properly; bridges must not fall down, slopes or foundations must not fail and nor must they move very much. These theories must say something about the strength, stiffness and flow of the materials and the way the whole structure works. They will investigate ultimate limit states to demonstrate that the structure does not fall down and they will also investigate serviceability limit states to show that the movements are acceptable.

Notice that engineers do not themselves build or repair things; they design them and supervise their construction by workers. There is a common popular misconception about the role of engineers. The general public often believes that engineers build things. They do not; engineers design things and workmen build them under the direction of engineers. Engineers are really applied scientists, and very skilled and inventive ones.

1.3   Fundamentals of mechanics

In any body, framework or mechanism, changes of loads cause movements; for example a rubber band stretches if you pull it, a tall building sways in the wind and

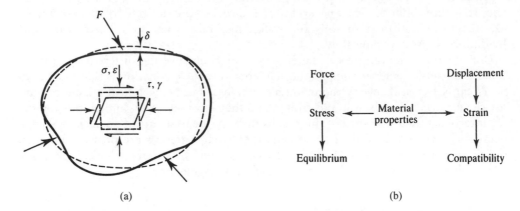

Figure 1.2 Principles of mechanics.

pedalling a bicycle turns the wheels. The basic feature of any system of forces and displacements and stresses and strains are illustrated in Fig. 1.2. Forces give rise to stresses and these must be in equilibrium or the body will accelerate. Displacements give rise to strains which must be compatible so the material does not tear or overlap. (Relationships between forces and stresses and between displacements and strains are given in Chapter 2.) These two separate requirements (of equilibrium and compatibility) are quite simple and they apply universally to everything. The relationships between stresses and strains (or between forces and displacements) are governed by the characteristics of the material.

There are a number of branches or subdivisions of mechanics which depend on the material, the type of problem and any assumptions made. Obviously soil mechanics is the mechanics of structures made of soils and there are also rock mechanics for rocks and fluid mechanics for fluids. Some important branches of mechanics are illustrated in Fig. 1.3; all of these are used in soil mechanics and appear later in this book.

Rigid body mechanics deals with mechanisms, such as car engines, in which all the moving parts are assumed to be rigid and do not deform. Structural mechanics is for framed structures where deformations arise largely from bending of beams and columns. Fluid mechanics is concerned with the flow of fluids through pipes and channels or past wings, and there are various branches depending on whether the fluid is compressible or not. Continuum mechanics deals with stresses and strains throughout a deforming body made up of material that is continuous (i.e. it does not have any cracks or joints or identifiable features), while particulate mechanics synthesizes the overall behaviour of a particulate material from the response of the individual grains. You might think that particulate mechanics would be relevant to soils but most of current soil mechanics and geotechnical engineering is continuum mechanics or rigid body mechanics.

1.4 Material behaviour

The link between stresses and strains is governed by the properties of the material. If the material is rigid then strains are zero and movements can only occur if there

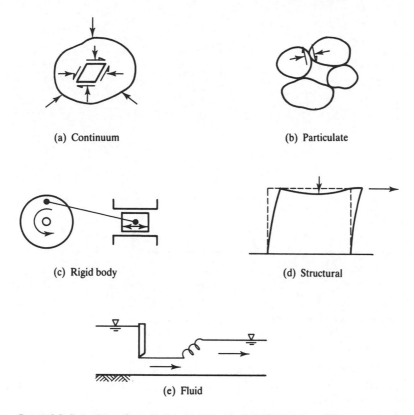

(a) Continuum

(b) Particulate

(c) Rigid body

(d) Structural

(e) Fluid

Figure 1.3 Branches of mechanics used in geotechnical engineering.

is a mechanism. Otherwise materials may compress (or swell) or distort, as shown in Fig. 1.4. Figure 1.4(a) shows a block of material subjected to a confining pressure σ and Fig. 1.4(c) shows a relationship between the pressure and the change of volume; the gradient is the bulk modulus K. The stress can be raised more or less indefinitely and the material continues to compress in a stable manner and does not fail; K continues to increase with stress and strain.

Figure 1.4(b) shows a block of material subjected to shearing stresses τ so that it distorts in shear. Notice that compression in Fig. 1.4(a) involves a change of size while shear distortion involves a change of shape; in a general loading, compression and distortion occur simultaneously. Figure 1.4(d) shows a simple relationship between shear stress and shear strain; the gradient is the shear modulus G and this reduces with stress and strain. The material fails when no more shear stress can be added and then it continues to strain at constant shear stress τ_f; this is the shear strength of the material.

Figure 1.4 illustrates the two most important aspects of material behaviour: stiffness and strength. Stiffness relates changes of stress and changes of strain by

$$K = \frac{\mathrm{d}\sigma}{\mathrm{d}\varepsilon_v} \quad G = \frac{\mathrm{d}\tau}{\mathrm{d}\gamma} \tag{1.1}$$

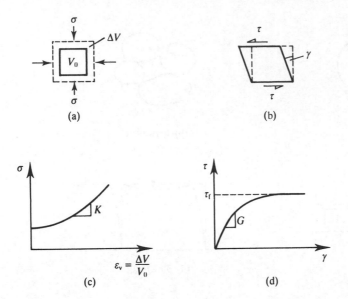

Figure 1.4 Compression and distortion.

where $\varepsilon_v = \Delta V / V_0$ is the volumetric strain and γ is the shear strain. The simplest theory for stiffness is the theory of elasticity in which K and G are constants and apply equally to loading and unloading.

Strength is the limiting shear stress that the material can sustain as it suffers large shear strains. The two most common theories for strength are to say that the material is cohesive and the limiting shear stress is a constant for the material given by

$$\tau_f = s \tag{1.2}$$

or to say that the material is frictional so that the strength is proportional to the confining pressure given by

$$\tau_f = \sigma \mu = \sigma \tan \phi \tag{1.3}$$

where μ is a coefficient of friction and ϕ is a friction angle. Later we will find that both of these theories apply to soils, but in different circumstances.

Values for the stiffness parameters K and G and the strength parameters s and μ (or ϕ) will obviously depend on the material, but they may also depend on other things such as temperature and rate of loading. For example, if the strength depends on the rate of strain the material is said to be viscous. The first part of this book, up to Chapter 15, deals largely with the basic theories for the strength and stiffness of soils and other granular materials.

1.5   Basic characteristics of soils

At first sight soils appear to behave rather strangely. For example, you can pour dry sand like water and you can pour saturated sand under water in the same way, yet you can make sandcastles from damp sand that will support loads. Clays can be squeezed and moulded like plasticine and appear to behave very differently from sands, but very old slopes in clay have angles comparable to those in sands.

The essential features of soil behaviour which we will examine later in this book are as follows:

1. External loads and water pressures interact with each other to produce a stress that is effective in controlling soil behaviour.
2. Soil is compressible; volume changes occur as the grains rearrange themselves and the void space changes.
3. Soil shearing is basically frictional so that strength increases with normal stress, and with depth in the ground. We will find that soil stiffness also increases with normal stress and depth.
4. Combining these basic features of soil behaviour leads to the observation that soil strength and stiffness decrease with increasing water pressure and with increasing water content.
5. Soil compression and distortion are generally not fully recoverable on unloading, so soil is essentially inelastic. This is a consequence of the mechanism of compression by rearrangement of the grains; they do not 'un-rearrange' on unloading.

We will see later that there is no real distinction between sands and clays and that the apparent differences arise from the influence of pore pressures and seepage of water in the void spaces between the grains.

1.6   Basic forms of geotechnical structure

The four basic types of geotechnical structure are illustrated in Fig. 1.5; most other cases are variations or combinations of these. Foundations (Fig. 1.5(a)) transmit loads to the ground and the basic criterion for design is that the settlements should be relatively small. The variables of a foundation are the load V, the size of the base B and the depth D. Foundations may support loads that are relatively small, such as car wheels, or relatively large, such as a power station. Slopes (Fig. 1.5(b)) may be formed naturally by erosion or built by excavation or filling. The basic variables are the slope angle i and the depth H, and the design requirement is that the slope should not fail by landsliding.

Slopes that are too deep and too steep to stand unsupported can be supported by a retaining wall (Fig. 1.5(c)). The basic variables are the height of the wall H and its depth of burial D, together with the strength and stiffness of the wall and the forces in any anchors or props. The basic requirements for design are complex and involve overall stability, restriction of ground movements and the bending and shearing resistance of the wall. In any structure where there are different levels of water, such as in a dam (Fig. 1.5(d)) or around a pumped well, there will be seepage of water. The seepage

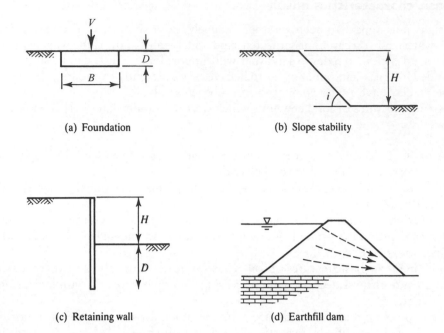

Figure 1.5 Geotechnical structures.

causes leakage through a dam and governs the yield of a well and it also governs the variation of pressure in the groundwater.

The structures in Fig. 1.5 clearly should not fail. There are, however, situations where the material must fail; these include excavation and flow of mineral ore or grain from a storage silo. Solutions to problems of this kind can be found using the theories of soil mechanics. Other problems in geotechnical engineering include movement of contaminants from waste repositories and techniques for ground improvement.

1.7   Factors of safety and load factors

All structural and geotechnical analyses contain uncertainties of one kind or another. These may involve uncertainties in prediction of maximum loads (particularly live loads due to wind, waves and earthquakes) approximations in the theories adopted for material behaviour and structural analysis, and uncertainties in the determination of strength and stiffness parameters. To take account of these approximations and uncertainties it is usual to apply a factor of safety in the design. These factors may be applied as partial factors to reflect the various uncertainties or as a single lumped value.

All applied sciences that analyse and predict natural events involve assumptions, approximations and simplifications because the real world is very complicated. Many people believe that the uncertainties in geotechnical engineering are very large because of the variability of natural soils in the ground and the apparent complexity of theoretical soil mechanics. It is true that geotechnical engineering is less exact than many applications of physics and chemistry, but it is probably less approximate than, say, sociology and economics. You can usually, but not always, improve a theory by making it more complicated and by adding more variables. For example, if material strength and stiffness parameters are allowed to vary with ambient temperature the theories will become more complex but possibly more realistic. In this book I shall be dealing with fairly simple theories of soil mechanics and geotechnical engineering which are suitable for most routine design problems.

Although it is always essential to consider the ultimate limit states of structures to demonstrate that they will not collapse, the principal design criterion for many structures, particularly foundations, is the need to limit ground movements and settlements. In practice this is often done by applying a factor to the collapse load. In my first job as a young engineer I was involved in the design of a very large earthfill dam, where the consequences of collapse would have been catastrophic and would certainly have meant major loss of life: the chief engineer required a factor of safety of about 1.25 against slope failure. In my second job I was asked to design the foundations for a small store shed which was part of a water treatment works: the chief engineer required a factor of 3.

I was puzzled by this inconsistency until I discovered that the large factor required for the foundations of the store shed was not really a factor of safety but was a factor to limit the settlement. The chief engineer knew that if the collapse load of a foundation was reduced by a factor of 3 the resulting settlements would be small. The point is illustrated in Fig. 1.6, which shows the settlement ρ of a foundation with an increasing vertical load V. In Fig. 1.6(b) there is a collapse load V_c and a safe load V_s that is about 80 per cent of V_c, corresponding to a factor of safety of about 1.25. There is also an allowable load F_a and for this load the settlements are small.

The safe load V_s is given by

$$V_s = \frac{1}{F_s} V_c \qquad (1.4)$$

where F_s is a factor of safety. Values of F_s in geotechnical engineering are normally in the range 1.25 to 1.5, depending on the consequences of failure and the uncertainties in the analyses and determination of the loads and the soil parameters.

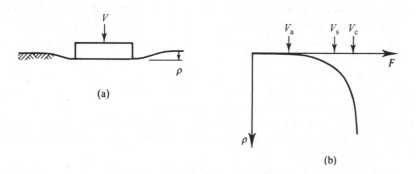

Figure 1.6 Factor of safety and load factor for a foundation.

The allowable load V_a is given by

$$V_a = L_f V_c \tag{1.5}$$

where L_f is a load factor. Values for L_f in geotechnical engineering are normally in the range 1/4 to 1/3, depending on the sensitivity of the structure to movements and the uncertainties in the analyses and determination of the loads and the soil parameters.

1.8   Standards and codes of practice

Construction engineering is regulated by standards and codes of practice. These are intended to ensure that structures are designed and built safely, economically and of good quality. These have evolved over many years and are based on theories and practices which may have become out-dated. They were drawn up by committees and often contain an amalgamation of current practices and interests.

In the UK, construction is currently regulated largely by British Standards. In this book I have referred to only three; BS1377 for soil testing, BS5930 for site investigations and BS8004 for foundations but there are, of course, many others. These will soon be replaced by Eurocodes and the relevant one for ground engineering is Eurocode 7: Geotechnical design; EN1997.

As a practising engineer you will often be required to deliver designs which meet standards and codes of practice. You should, however, ensure that your designs do not conflict with the basic theories for geotechnical engineering set out in this book.

In geotechnical engineering it is essential to distinguish between a factor of safety which is intended to ensure safety and a load factor which is intended to limit settlements and ground movements.

1.9   Summary

1. Geotechnical engineering is a branch of engineering and deals with the analysis and design of foundations, slopes and structures made from soils and rocks.
2. The basic theories of mechanics (equilibrium and compatibility) and of material behaviour (stiffness and strength) apply equally in geotechnical engineering.
3. The basic behaviour of soil is influenced both by the loads on the soil grains and the pressures in the water in the void spaces.
4. Soil mechanics describes the relationships between stresses and strains in soils. These will be dealt with in Chapters 8 to 15. We will find that soil behaviour is essentially frictional, compressible and largely inelastic.
5. Methods and theories for analysis and design of geotechnical structures, such as foundations, slopes and retaining walls, and for seepage of groundwater will be covered in Chapters 19 to 25.
6. In geotechnical design safe loads are found by applying factors of safety while movements are often restricted by applying a load factor.

Further reading

Chadwick, A. and J. Morfet (1998) *Hydraulics in Civil and Environmental Engineering*, E & FN Spon.

Gordon, J. E. (1991) *The New Science of Strong Materials*, Penguin, Harmondsworth.

Gordon, J. E. (1978) *Structures*, Penguin, Harmondsworth.

Gulvanessian, H. and R. Driscoll (2001) Eurocodes – the new environment for structural design. Proc. ICE, Civil Engineering, 144, pp. 3–7.

Montague, P. and R. Taylor (1989) *Structural Engineering*, McGraw-Hill, London.

Palmer, A. C. (1976) *Structural Mechanics*, Oxford University Press.

Spencer, A. J. M. (1980) *Continuum Mechanics*, Longman, London.

Chapter 2

Basic mechanics

This chapter, and the following one, cover the basic methods for the analysis of stress and strain using the Mohr circle constructions and the general features of material behaviour. These techniques are essential for understanding soil behaviour and for analysing soil structures and will be used extensively throughout the book. The topics should be covered in other courses on strength of materials, but here they are put into the context of soil mechanics. Readers are advised to skim through these two chapters and come back to them to work through the details as necessary.

2.1 Introduction

Mechanics is the study of forces and displacements, or stresses and strains, and there are a number of branches of mechanics associated with particular materials or with particular applications. The fundamental principles of mechanics are simply the application of equilibrium and compatibility. For any body that is not accelerating the forces and moments must be in equilibrium: this is simply Newton's first law. For any body, or system of bodies, that is distorting or moving around the strains and displacements must be compatible. This means that material does not vanish and gaps do not appear; this is simply common sense. What we can do is to analyse states of stress (or strain) so that we can calculate the stresses (or strains) on any plane at a point from the stresses (or strains) on any other pair of planes.

2.2 Stresses and strains

I shall assume that readers have been introduced to the basic ideas of stress and strain in other courses. A stress is basically an intensity of loading given by a force acting on a unit area, while a strain is basically an intensity of deformation given by a displacement over a unit gauge length. In geotechnical engineering there are two minor differences from the definitions of stress and strain usually adopted for metals and concrete, and these account for the particulate nature of soils. Firstly, the unit area or gauge length must be large enough to include a representative number of soil grains and, secondly, because uncemented soils cannot sustain tensile stresses compressive stresses are positive.

Figure 2.1 shows stresses and strains in a cube of soil subjected to normal and shear forces. The changes of normal stress $\delta\sigma$ and normal strain $\delta\varepsilon$ due to a change of normal

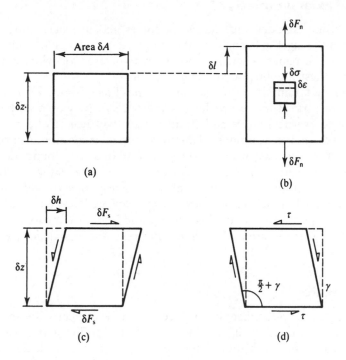

Figure 2.1 Stress and strain.

load δF_n are given by

$$\delta\sigma = -\frac{\delta F_n}{\delta A} \tag{2.1}$$

$$\delta\varepsilon = -\frac{\delta l}{\delta z} \tag{2.2}$$

(Notice that negative signs have been added so that compressive stresses and strains are positive quantities.) The changes of shear stress $\delta\tau$ and shear strain $\delta\gamma$ due to a change of load δF_s are given by

$$\delta\tau = -\frac{\delta F_s}{\delta A} \tag{2.3}$$

$$\delta\gamma = -\frac{\delta h}{\delta z} \tag{2.4}$$

(Notice that negative signs have been added so that positive shear stresses and shear strains are associated with increases in the angles in the positive quadrants of the element as shown.)

2.3   Plane strain and axial symmetry

In general we should consider stresses and strains, or forces and displacements, in three dimensions, but then the algebra becomes quite complicated and it is difficult to represent general states on flat paper. There are, however, two cases for which only two axes are required and these are illustrated in Fig. 2.2.

Figure 2.2(a) shows plane strain where the strains in one direction are zero and the stresses and strains are vertical (σ_z, ε_z) or horizontal (σ_h, ε_h). (It would be best to use v as the subscript for vertical stress and strain but we will need to keep the subscript v for volumes and volumetric strains.) This corresponds to conditions in the ground below a long structure, such as an embankment or wall or a strip foundation. Figure 2.2(b) shows axial symmetry where the radial stresses and strains (σ_r, ε_r) are equal and the other stresses and strains (σ_a, ε_a) are axial. This corresponds to conditions in the ground below a circular foundation or a circular excavation. Throughout this book I will consider only plane strain and axial symmetry and I will use the axes z, h (vertical and horizontal) for plane strain and the axes a, r (axial and radial) for axial symmetry.

2.4   Rigid body mechanics

When soils fail they often develop distinct slip surfaces; on a geological scale these appear as faults. Slip surfaces divide soil into blocks and the strains within each block may be neglected compared with the relative movements between blocks, so the principles of rigid body mechanics are applicable for failure of slopes and foundations. To demonstrate this examine how sandcastles and claycastles fail in unconfined compression tests.

Equilibrium is examined by resolution of forces in two directions (together with moments about one axis) and this is done most simply by construction of a polygon of forces: if the polygon of forces closes then the system of forces is in equilibrium. Figure 2.3(a) shows a set of forces acting on a triangular block. We will see later that this represents the conditions in soil behind a retaining wall at the point of failure. Figure 2.3(b) shows the corresponding polygon of forces where each line is in the same direction as the corresponding force and the length is proportional to the magnitude of the force. The forces are in equilibrium because the polygon of forces is closed.

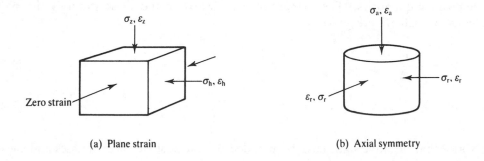

(a) Plane strain   (b) Axial symmetry

Figure 2.2 Common states of stress.

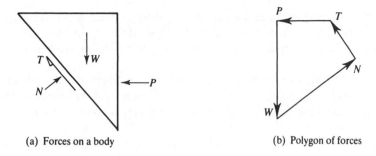

(a) Forces on a body

(b) Polygon of forces

Figure 2.3 Conditions of equilibrium.

You will study conditions for equilibrium of forces in other courses on mechanics or strength of materials and this is just the same.

Compatibility of displacement is examined most conveniently by construction of a displacement diagram. Figure 2.4(a) illustrates two triangular blocks moving as illustrated by the arrows; we will see later that this could represent the displacement of soil below a foundation at the point of failure. Each block is given an identifying letter and O represents stationary material. In Fig. 2.4(b) each arrow represents the direction and magnitude of the displacement of one of the rigid blocks and the displacement diagram closes. The letters on the arrows represent the relative displacements, thus oa is the displacement of A with respect to O and ab is the displacement of B with respect to A.

The relative movements of rigid bodies in mechanisms like that shown in Fig. 2.4 can be examined by making simple models from stiff card. (From a flat sheet of card cut a triangular recess, cut two triangular shapes like A and B and demonstrate that you have a compatible mechanism. To get them to move it is necessary to drill small holes at the corners.)

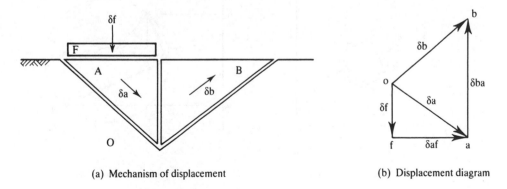

(a) Mechanism of displacement

(b) Displacement diagram

Figure 2.4 Conditions of compatibility.

2.5   Analysis of stress

Within a loaded body the stresses generally vary from point to point so, for example, the stresses below the edge and centre of a foundation are different. At any point the stresses are different on different planes and it is necessary to relate the stresses on the different planes.

The simplest form of analysis is through the Mohr circle construction which is covered in courses on strength of materials. The only difference for soil mechanics is that the sign convention is changed so that compressive stresses and counter-clockwise shear stresses are positive.

Figure 2.5(a) shows principal stresses σ_z and σ_h on the faces of an element of soil and Fig. 2.5(b) shows the corresponding Mohr circles of stress. The pole P of the Mohr circle is defined so that a line from P to σ_z gives the direction of the plane on which σ_z acts. In Fig. 2.5(a) there is an element rotated to an angle θ as shown and the stresses $(\tau_n, \sigma_n$ and $\tau_m, \sigma_m)$ on the faces of this element are at N and M in Fig. 2.5(b). From the geometry of the Mohr circle the angle 2θ subtended at the centre by the point representing the major principle plane and the point N is twice the angle between the planes on which these stresses act. From the geometry of the figure, $\tau_n = \tau_m$. Using Fig. 2.5(b) it is possible to calculate τ_n, σ_n and τ_m, σ_m from σ_z and σ_h or vice versa, and in order to construct the Mohr circle it is necessary to know the stresses on two (preferably orthogonal) planes.

2.6   Analysis of strain

Analysis of strains at a point using the Mohr circle of strain is similar to that for stress, but there are a few points to note about strains. Firstly, while it is possible to talk about a state of stress with respect to zero stress (taken as atmospheric pressure), there is no absolute zero for strain so we have to talk about changes, or increments, of strain. These may be small increments (denoted by $\delta\varepsilon$) or large increments (denoted by $\Delta\varepsilon$) and generally they occur as a result of corresponding large or small increments of stress.

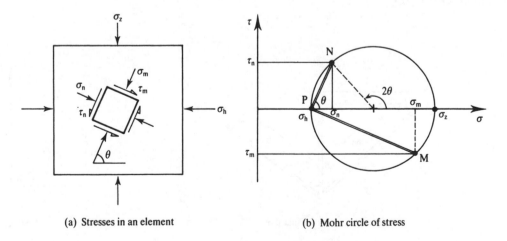

(a) Stresses in an element   (b) Mohr circle of stress

Figure 2.5 Analysis of stress using a Mohr circle.

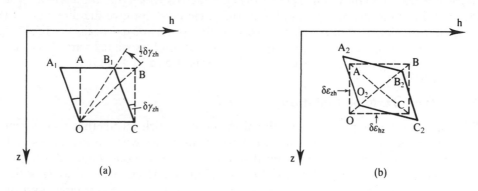

Figure 2.6 Shear strains in an element.

Secondly, while stresses in soils are almost always positive (particulate materials cannot usually sustain tensile stresses unless the grains are attached to one another), strains may be positive (compressive) or negative (tensile) and in an increment of strain there will usually be compressive and tensile strains in different directions. Thirdly, we must be careful to distinguish between pure shear strains and engineers' shear strains $\delta\gamma$ and take account of any displacements of the centre of area of distorted elements.

Figure 2.6(a) shows an element OABC strained by $\delta\gamma_{zh}$ to a new shape OA_1B_1C. It can be seen that the diagonal OB has rotated to OB_1 through $\frac{1}{2}\delta\gamma_{zh}$. Figure 2.6(b) shows the strained element rotated and translated to $O_2A_2B_2C_2$ so that the centre and the diagonals coincide and the edges have now all strained through the same angle $\delta\varepsilon_{zh} = \delta\varepsilon_{hz} = \frac{1}{2}\delta\gamma_{zh}$.

Figure 2.7(a) shows a plane element with principal strains $\delta\varepsilon_z$ and $\delta\varepsilon_h$ (which is negative), while Fig. 2.7(b) is the corresponding Mohr circle of strain. The pole is at P

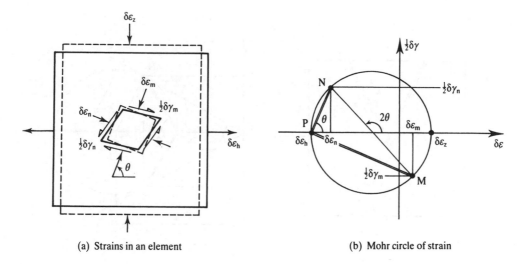

(a) Strains in an element

(b) Mohr circle of strain

Figure 2.7 Analysis of strain using a Mohr circle.

so the line from P to the point $\delta\varepsilon_z$ gives the plane across which the strain is $\delta\varepsilon_z$. (Notice that the line from the pole to a point on the circle does not give the direction of the strain but the direction of the plane perpendicular to the normal strain.) In Fig. 2.7(a) there is an element rotated at an angle θ and the strains associated with this element ($\frac{1}{2}\delta\gamma_n, \delta\varepsilon_n$ and $\frac{1}{2}\delta\gamma_m, \delta\varepsilon_m$) are given by the points N and M as shown.

2.7   Stress ratio and dilation

We will see later that soils are frictional materials, which means that their strength (i.e. the maximum shear stress they can sustain) increases with normal stress and so the stress ratio τ/σ is more important than the shear stress alone. Figure 2.8(a) shows a stressed element and Fig. 2.8(b) is the corresponding Mohr circle of stress with the pole at P. There are two lines ON which are tangents to the Mohr circle and these define the points on which the stress ratio is given by

$$\frac{\tau}{\sigma} = \tan\phi_m \tag{2.5}$$

where ϕ_m is the mobilized angle of shearing resistance. From the geometry of Fig. 2.8(b) $t = \frac{1}{2}(\sigma_z - \sigma_h)$ and $s = \frac{1}{2}(\sigma_z + \sigma_h)$ and

$$\frac{t}{s} = \sin\phi_m = \frac{\sigma_z - \sigma_h}{\sigma_z + \sigma_h} \tag{2.6}$$

or

$$\frac{\sigma_z}{\sigma_h} = \frac{1 + \sin\phi_m}{1 - \sin\phi_m} = \tan^2\left(45° + \tfrac{1}{2}\phi_m\right) \tag{2.7}$$

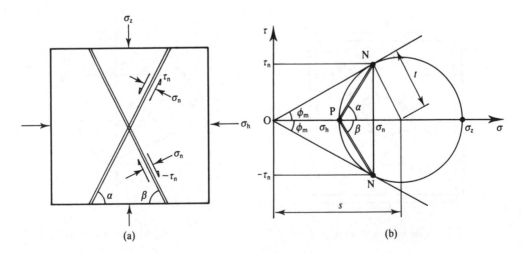

(a)   (b)

Figure 2.8 Limiting stress ratio and angle of shearing resistance.

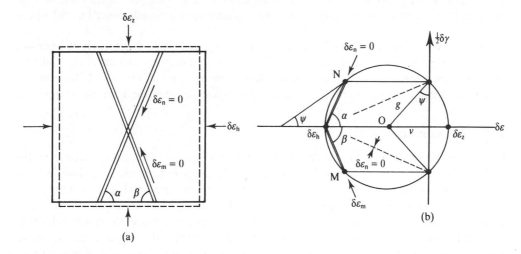

Figure 2.9 Angle of dilation and zero extension lines.

The planes, shown by double lines, on which this stress ratio occurs are at angles α and β as shown and, from the geometry of the figure,

$$\alpha = \beta = 45° + \tfrac{1}{2}\phi_{\mathrm{m}} \tag{2.8}$$

For frictional materials these correspond to the planes on which the most critical conditions occur and they should be the planes on which failure will occur.

When the major and minor principal strains have opposite signs the origin of the axes is inside the Mohr circle, as shown in Fig. 2.9(b). There are two planes, shown by broken lines in Fig. 2.9(b), across which the normal strains are zero, and so there are two directions, shown by double lines, at angles α and β along which the strains are zero as shown in Fig. 2.9(a). These planes are defined by an angle of dilation ψ. From Fig. 2.9(b), the lengths $v = \tfrac{1}{2}(\delta\varepsilon_z + \delta\varepsilon_h)$ and $g = \tfrac{1}{2}(\delta\varepsilon_z - \delta\varepsilon_h)$, and if the volumetric strain is $\delta\varepsilon_v = \delta\varepsilon_z + \delta\varepsilon_h$ then the angle of dilation is given by

$$\tan\psi = -\frac{\delta\varepsilon_v}{\delta\gamma} \tag{2.9}$$

or

$$\sin\psi = \frac{\delta\varepsilon_z + \delta\varepsilon_h}{\delta\varepsilon_z - \delta\varepsilon_h} \tag{2.10}$$

where $\delta\gamma$ is the increment of shear strain across the plane. (The negative signs are required in Eqs. (2.9) and (2.10) so positive angles ψ are associated with dilation or negative volumetric strains.) From the geometry of the figure,

$$\alpha = \beta = 45° + \tfrac{1}{2}\psi \tag{2.11}$$

Comparing Figs. 2.8 and 2.9, the angle of dilation ψ describing the strain ratio $\delta\varepsilon_v/\delta\gamma$ has similar properties to the angle of shearing resistance ϕ_m which describes the stress ratio τ/σ.

You can visualize how materials strain by drawing a circle with a felt-tipped pen on a sheet of thin rubber and stretching it. The circle will distort into an ellipse and its area may increase. You can probably see that there are two diameters of the original circle that remain the same length and these correspond to the directions of zero strain.

2.8   Slip surfaces

Figure 2.9 represents homogeneous straining where there are no discontinuities, or slip surfaces, like those that appear during rigid body deformation and the double lines show the directions of zero strain. Figure 2.10(a) shows material that is deforming by intense shearing in a very thin zone AB and Fig. 2.10(b) shows a detail of the slip zone. This thin zone of shearing material has a small but finite thickness which is usually too small to see; in soils it is probably of the order of ten grains thick. Shear zones usually appear to have no thickness and so they are called slip planes or slip surfaces.

Since the length of AB in Fig. 2.10(a) remains constant, because the material on either side is rigid, it is a zero extension line and its direction is given by $\alpha = 45° + \frac{1}{2}\psi$, as in Fig. 2.9. From Fig. 2.10(b),

$$\delta\gamma = \frac{\delta h}{H_0} \qquad \delta\varepsilon_v = \frac{\delta v}{H_0} \tag{2.12}$$

$$\tan\psi = \frac{\delta\varepsilon_v}{\delta\gamma} = \frac{\delta v}{\delta h} \tag{2.13}$$

so that the movement across the slip surface A $\to$ A$_1$ and B $\to$ B$_1$ is at angle ψ to the direction of the slip surface as shown.

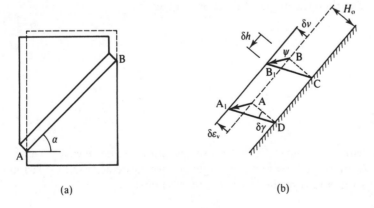

(a)   (b)

Figure 2.10 Discontinuous slipping and slip surfaces.

2.9 Summary

1. Forces and stresses in any body of material must be in equilibrium: this means that the polygon of forces acting on the body, or on an element inside the body closes.
2. Strains and displacements in any distorting body must be compatible: this means that the material does not tear or overlap and the displacement diagram closes.
3. States of stress or strain at a point can be analysed using the Mohr circle construction so that the stresses or strains on any plane can be calculated from the geometry of the circle.
4. If slip surfaces develop, their directions correspond to the directions of zero extension lines and the relative movement across a slip surface is at an angle ψ to its direction.

Worked examples

Example 2.1: Equilibrium of forces using a force polygon   Figure 2.11(a) shows forces acting on a rigid triangular block of soil with a slip surface; two of the forces are known to be $W = 160$ kN and $T = 60$ kN. Figure 2.11(b) shows the corresponding polygon of forces. Scaling from the diagram, or by calculation, $P = 75$ kN.

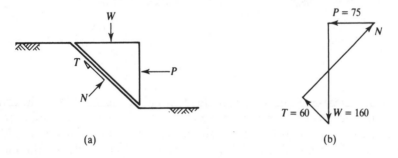

(a)   (b)

Figure 2.11

Example 2.2: Compatible displacements using a displacement diagram   Figure 2.12(a) shows two rigid blocks separated by slip surfaces where all the angles are 45° or 90°; the left-hand block moves with a vertical component of displacement 1 mm as

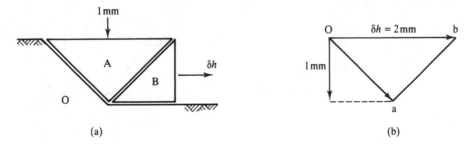

(a)   (b)

Figure 2.12

shown. Figure 2.12(b) shows the corresponding displacement diagram. Scaling from the diagram, or by calculation, $\delta h = 2$ mm.

Example 2.3: Stress analysis using a Mohr circle of stress   Figure 2.13(a) shows an element of soil behind a retaining wall; the effective vertical and horizontal stresses are $\sigma_z = 300$ kPa and $\sigma_h = 100$ kPa and these are principal stresses. Figure 2.13(b) shows the Mohr circle of stress. Scaling from the diagram, $\phi_m = 30°$, the angles of the critical planes are $\alpha = \beta = 60°$ and the stresses on these planes are $\sigma = 150$ kPa and $\tau = \pm 87$ kPa.

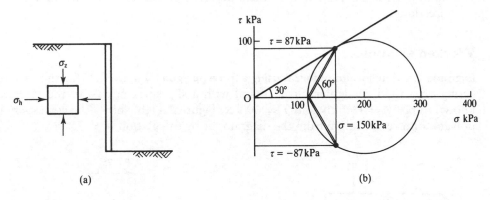

Figure 2.13

Example 2.4: Analysis of strain using a Mohr circle of strain   Figure 2.14(a) illustrates an increment of displacement of a retaining wall from the broken to the solid lines. The strains in an element of soil behind the wall are $\delta \varepsilon_z = 0.10\%$ and $\delta \varepsilon_h = -0.20\%$ and these are principal strains. Figure 2.14(b) shows the Mohr circle for the increment of strain. Scaling from the diagram, the angle of dilation is $\psi = 20°$. The zero

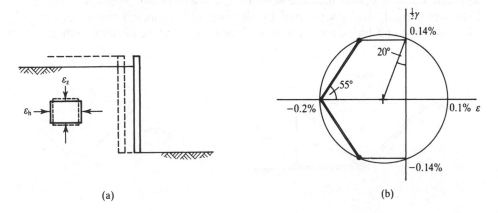

Figure 2.14

extension lines are at angles $\alpha = \beta = 55°$ and the shear strains across zero extension lines are given by $\frac{1}{2}\delta\gamma = \pm 0.14\%$.

Further reading

Atkinson, J. H. (1981) *Foundations and Slopes*, McGraw-Hill, London.
Case, J., A. H. Chilver and C. T. F. Ross (1999) *Strength of Materials and Structures*, Edward Arnold, London.

Chapter 3

Essentials of material behaviour

Before reading this chapter, read the note at the beginning of Chapter 2.

3.1 Stress–strain behaviour, stiffness and strength

In Chapter 2, I considered the states of stress and strain in loaded and deforming material. The analyses that were developed for stresses and strains, using Mohr circles, are not dependent on the material and they are equally applicable for steel, concrete or soil. In order to analyse any kind of structure, or any kind of solid or fluid continuum, it is necessary to have relationships between stresses and strains. These are called constitutive relationships and they take a number of different forms depending on the nature of the material and on the loading.

Figure 3.1 shows an idealized relationship between stress and strain and it is similar to the stress–strain curves for common engineering materials like metals, plastics, ceramics and engineering soils. For soils and other granular materials, it is necessary to deal with something called effective stress to take account of pore pressures in the fluid in the voids between the grains. (In simple terms effective stresses can be thought of as the stresses effective in the soil grains.) Effective stress will be covered in Chapter 6 where it will be shown that all soil behaviour, including stiffness and strength, is governed by an effective stress which is denoted by a prime (as in σ'). As this book is about soil I will use effective stresses from now on.

Stiffness is the gradient of the stress–strain line. If this is linear the gradient is easy to determine but, if it is curved, the stiffness at a point such as A in Fig. 3.2 may be quoted as a tangent or as a secant and given by

$$\text{tangent stiffness} = \frac{d\sigma'}{d\varepsilon} \tag{3.1}$$

$$\text{secant stiffness} = \frac{\Delta\sigma'}{\Delta\varepsilon} \tag{3.2}$$

The stiffness of a material largely determines the strains and displacements in structures, or in the ground, as they are loaded or unloaded. Another term often used in soil mechanics to describe the relationship between stress and strain is 'compressibility', but this is basically the reciprocal of stiffness. Often there is a marked change in the gradient of a stress–strain curve at a yield point, as shown in Fig. 3.1. This is associated

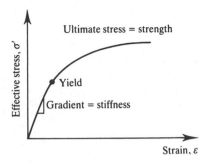

Figure 3.1 A typical stress–strain curve for soil.

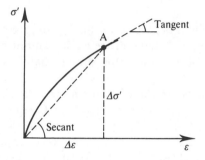

Figure 3.2 Tangent and secant stiffness moduli.

with a fundamental change in behaviour, often from elastic and recoverable straining to inelastic and irrecoverable straining.

In simple terms the strength of a material is the largest shear stress that the material can sustain and it is this which governs the stability or collapse of structures.

Stiffness and strength are quite different things: one governs displacements at working load and the other governs the maximum loads that a structure can sustain. Materials may be stiff (i.e. have high stiffness) or soft and they may be strong or weak and they may have any reasonable combination of stiffness and strength. Steel is stiff and strong while margarine is soft and weak; blackboard chalk is relatively stiff and weak while rubber is relatively soft and strong.

3.2 Choice of parameters for stress and strain

Figure 3.1 shows the characteristics of material behaviour, but axes of stress and strain are not carefully defined. The choice of axes will depend on the tests carried out to examine the material behaviour and the parameters required. For metals that are essentially elastic and then plastic the parameters required are Young's modulus E, Poisson's ratio ν and the yield and ultimate stresses, and these can be obtained from a simple uniaxial extension test. For concrete, the required parameters can be obtained from

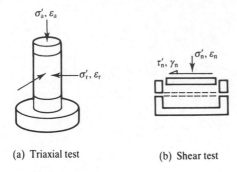

(a) Triaxial test   (b) Shear test

Figure 3.3 Common soil tests.

a uniaxial compression test. For soils, volume changes that occur during compression and shearing are very important and to describe soil behaviour we must examine separately shearing and volumetric strains and responses to shearing and normal loading and unloading.

The two tests commonly used in soil mechanics are the triaxial test and the shear test illustrated in Fig. 3.3. These will be considered in more detail in Chapter 7. The relationships between the stresses in the two tests can be obtained from the Mohr circle construction, as shown in Fig. 3.4. This illustrates that, within the triaxial specimen with stresses (σ_a', σ_r') there are elements with stresses (τ_n', σ_n') like those in a shear specimen and vice versa.

In a shear test the sample could be loaded or unloaded by increasing or decreasing the effective normal stress σ_n' with zero shear stress and it would compress or swell with normal strains ε_n. Alternatively it could be sheared to the left or to the right and there would be shear strains γ. The triaxial sample could be tested by increasing or decreasing either σ_a' or σ_r' and there would be axial and radial strains ε_a and ε_r.

It is convenient to define special stress and strain parameters for triaxial test samples. These describe shearing and normal or volumetric effects and they are defined as:

$$q' = \sigma_a' - \sigma_r' \tag{3.3}$$

$$p' = 1/3 \left(\sigma_a' + 2\sigma_r'\right) \tag{3.4}$$

$$\varepsilon_s = 2/3 \left(\varepsilon_a - \varepsilon_r\right) \tag{3.5}$$

$$\varepsilon_v = \varepsilon_a + 2\varepsilon_r \tag{3.6}$$

The parameter q' is the diameter of the Mohr circle and it is a measure of the maximum shear stress. The parameter p' is the average stress and it is approximately equal to the distance of the centre of the circle from the origin as shown in Fig. 3.4. The parameter ε_s is equivalent to the shear strain and ε_v is simply the volumetric strain.

The exact relationships between the pairs of parameters τ' and q' for shear stress; σ_n' and p' for effective normal stress; γ and ε_s for shear strains; ε_n and ε_v for volumetric

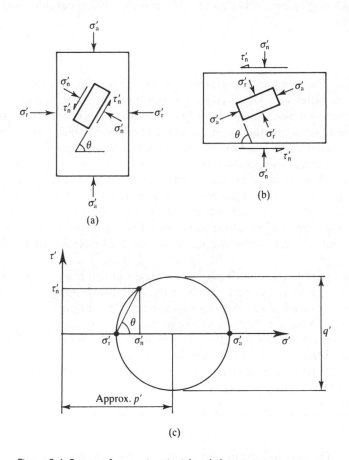

Figure 3.4 States of stress in triaxial and shear tests.

strains depend on a number of factors but principally on the angle θ between the directions of the planes in Figs. 3.4(a) and (b).

The volumetric strain ε_v describes the change in size of an element while the shear strain ε_s describes the change in its shape. The value $\frac{2}{3}$ in Eq. (3.5) is required for consistency. During an increment of straining the work done per unit volume of soil δW must be invariant (i.e. independent of the choice of parameters). In terms of axial and radial stresses and strains

$$\delta W = \sigma'_a \varepsilon_a + 2\sigma'_r \delta\varepsilon_r \tag{3.7}$$

and also

$$\delta W = q'\delta\varepsilon_s + p'\delta\varepsilon_v \tag{3.8}$$

You should substitute Eqs. (3.3) to (3.6) into Eq. (3.8) and demonstrate that this reduces to Eq. (3.7). When considering the behaviour of soils in triaxial tests I will generally use the parameters q', p', ε_s and ε_v and for shear tests I will generally use the parameters τ_n', σ_n', γ and ε_n.

3.3   Strength

The strength of a material describes the ultimate state of stress that it can sustain before it fails. (For soils that can suffer very large strains we will have to define failure very carefully, but this will be considered in detail later.) People talk about tensile strength, compressive strength, shear strength, and so on, as though they were all different, but these should really all be related to some fundamental characteristic strength.

The link between these different strengths is the maximum shear stress, or the size of the largest Mohr circle that the material can sustain. Figure 3.5(a) and (b) shows uniaxial tensile and compression tests and the corresponding Mohr circle of stress; the test samples fail when the Mohr circle reaches the limiting size given by the radius τ_f'. Figure 3.5(c) shows a vertical cut and the shear and normal effective stresses on some inclined planes are τ_n', and σ_n'; failure will occur when the Mohr circle reaches its limiting size.

We can say that materials that have strength can sustain shear stresses and the strength is the maximum shear stress that can be sustained. Only materials with strength can have slopes because shear stresses are required to maintain a slope. A material that cannot sustain a slope, like stationary water, has no strength, there are no shear stresses in it and the Mohr circle reduces to a point as shown in Fig. 3.5(d).

There are two fundamentally different failure criteria to consider. The first illustrated, in Fig. 3.6(a), is called cohesion and it says that the material will fail when the Mohr circle of stress touches an envelope given by

$$\tau' = c' \tag{3.9}$$

where c' is the cohesion. The second, illustrated in Fig. 3.6(b), is called friction and it says that the material will fail when the Mohr circle of effective stress touches an envelope given by

$$\tau' = \sigma'\mu = \sigma'\tan\phi' \tag{3.10}$$

where μ is the coefficient of friction and ϕ' is the angle of friction. You probably did experiments at school to measure the coefficients of friction of different materials.

There is a third criterion of failure called the Mohr–Coulomb criterion and it is simply the sum of cohesion and friction. It is illustrated in Fig. 3.6(c) and it says that the material will fail when the Mohr circle touches a line given by

$$\tau' = c' + \sigma'\tan\phi' \tag{3.11}$$

The development of the Mohr–Coulomb criterion from Coulomb's original research is described by Heyman (1972). All three criteria are used to describe the strength of soils under different conditions of drainage and strain and these will be discussed in later chapters.

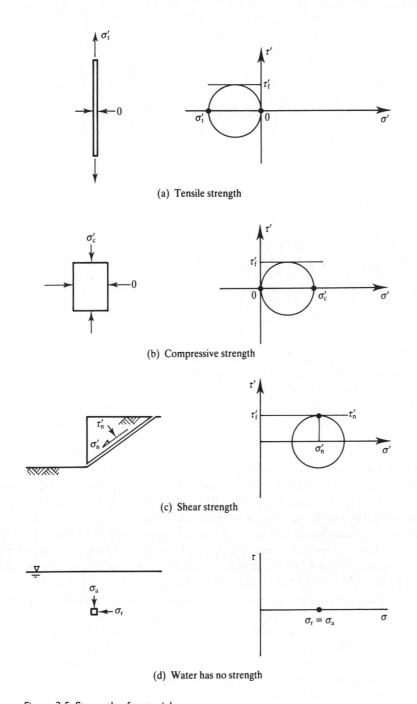

(a) Tensile strength

(b) Compressive strength

(c) Shear strength

(d) Water has no strength

Figure 3.5 Strength of materials.

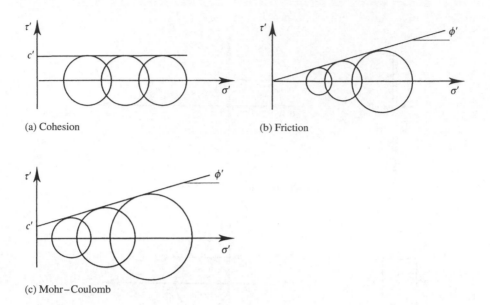

(a) Cohesion

(b) Friction

(c) Mohr–Coulomb

Figure 3.6 Failure criteria.

Dry sugar is a frictional material and its strength is given by Eq. (3.10) while butter is cohesive and its strength is given by Eq. (3.9); the strength will depend on the temperature of the butter. Damp sugar will have a strength given by the Mohr–Coulomb criterion; there is a cohesion because the sugar grains are stuck together.

3.4   Brittle and ductile

Strength defines the stress conditions at which failure occurs but it is important to think about how the material fails. If you bend a biscuit it will not deform very much but it will suddenly snap. This is brittle behaviour. If you load warm butter by pressing it with your hand it will gradually deform. This is ductile behaviour.

Concrete is brittle: in compression tests there are relatively small strains and it fails with a bang. Many metals such as steel, copper and aluminium (but not cast iron) are ductile. In compression tests they gradually deform and continue to deform at relatively large strains. (In tension tests metal samples snap but that is because the area decreases as the sample necks.)

Soils and rocks may be either brittle or ductile. Soft clay is ductile when it has a relatively high water content but, if it is highly compressed, stiff clay becomes brittle. Rocks are generally brittle when they are near the ground surface but they become ductile if they are at great depth. We will see later that soils and rocks behave in essentially the same way and what is important is the state, which is the combination of confining stress and water content.

Structures made of brittle materials are inherently unsafe. They fail catastrophically after very little deformation. Brick and masonry buildings fail in this way particularly during earthquakes. Structures made of ductile materials such as steel and timber are inherently much safer. They suffer very large deformations and so give plenty of warning before they fail. Tall steel-framed buildings wave around with wind or earthquake loadings but they do not often fall down.

3.5   Stiffness

As discussed earlier in Sec. 3.1 and illustrated in Figs. 3.1 and 3.2 stiffness is the relationship between stress and strain. The stiffness modulus, which is the gradient of the stress–strain curve, may be a tangent or a secant: if the material is linear these are the same.

For isotropic loading for which q' remains constant we can define a bulk modulus K' and for triaxial loading with p' constant we can define a shear modulus G' as:

$$K' = \frac{\mathrm{d}p'}{\mathrm{d}\varepsilon_\mathrm{v}} \tag{3.12}$$

$$3G' = \frac{\mathrm{d}q'}{\mathrm{d}\varepsilon_\mathrm{s}} \tag{3.13}$$

For loading in a shear test illustrated in Fig. 3.3(b) with zero or constant shear stress we can define a one-dimensional modulus M' and for shearing with constant normal stress the shear modulus is G' where

$$M' = \frac{\mathrm{d}\sigma'_\mathrm{n}}{\mathrm{d}\varepsilon_\mathrm{v}} \tag{3.14}$$

$$G' = \frac{\mathrm{d}\tau'}{\mathrm{d}\gamma} \tag{3.15}$$

Alternative stiffness parameters are Young's modulus E' and Poisson's ratio v'. These are obtained directly from a uniaxial compression or extension test in which the radial stress σ'_r is held constant (or zero) and are given by

$$E' = \frac{\mathrm{d}\sigma'_\mathrm{a}}{\mathrm{d}\varepsilon_\mathrm{a}} \tag{3.16}$$

$$v' = -\frac{\mathrm{d}\varepsilon_\mathrm{r}}{\mathrm{d}\varepsilon_\mathrm{a}} \tag{3.17}$$

In soil mechanics the shear and bulk moduli G' and K' are often used instead of Young's modulus and Poisson's ratio because it is important to consider shearing and change of shape separately or decoupled from compression and change of size.

These parameters are often called elastic parameters because they are usually derived in text books for elastic materials. In Eqs. 3.12 to 3.17 they have been defined as tangent moduli and they are simply the gradients of the appropriate stress–strain curves.

3.6   Strength, stiffness and rigidity

The strength of a material is the maximum shear stress which it can sustain and its stiffness is the ratio of change of stress to the resulting strain. These were discussed in Secs. 3.3 and 3.5.

A material may be relatively strong or relatively weak: it may be relatively stiff or relatively soft. Concrete (in compression) and rubber have similar strengths but concrete is much stiffer than rubber. Aluminium and glass have similar stiffnesses but glass is much stronger than aluminium. There is no requirement for any relationship between strength and stiffness.

The ratio of stiffness to strength is called rigidity R and this is commonly defined as

$$R = \frac{E'}{q'_f} \tag{3.18}$$

where E' is Young's modulus given by Eq. (3.16) and q'_f is strength expressed as the diameter of the Mohr circle at failure. Table 3.1 gives typical values for stiffness, strength and rigidity for some common materials, including soft and stiff clays, showing that values of rigidity vary over several orders of magnitude.

Figure 3.7(a) shows the stress strain response of a material which is linear with Young's modulus E' and which first fails at a stress q'_f when the strain is ε_f. Figure 3.7(b) is the corresponding relationship between stiffness and strain. Hence, from the geometry of Fig. 3.7(a), the rigidity R is

$$R = \frac{E'}{q'_f} = \frac{1}{\varepsilon_f} \tag{3.19}$$

From Fig. 3.7(a)

$$q'_f = E' \varepsilon_f \tag{3.20}$$

Table 3.1 Typical values for stiffness, strength and rigidity of some common materials

Material	Young's modulus E' MPa	Strength q'_f MPa	Rigidity R
Concrete	28,000	40	700
Glass	70,000	1000	70
Mild steel	210,000	430	500
Copper	120,000	200	600
Aluminium	70,000	100	700
Rubber	10	20	0.5
Timber	10,000	20	500
Soft clay	100	0.05	2000
Stiff clay	300	0.3	1000

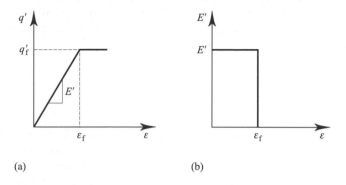

Figure 3.7 Relationship between strength, stiffness and rigidity.

where $E'\varepsilon_f$ is the area beneath the stiffness–strain curve. This result, that the strength of a linear material is equal to the area beneath the stiffness strain curve is an important one. In fact, with a few restrictions, this result holds for all materials and for states before failure.

3.7 Constitutive equations

During a general loading in the ground both shear and normal stresses are likely to change simultaneously so there will be shearing and volumetric straining together. For soils it turns out that shearing and volumetric effects are coupled so that shearing stresses cause volumetric strains and normal stresses cause shear strains. This is quite surprising and we will see later how the particulate nature of soils gives rise to shear and volumetric coupling.

A simple constitutive equation relating shearing and volumetric stress–strain behaviour can be written as

$$\begin{Bmatrix} \delta q' \\ \delta p' \end{Bmatrix} = \begin{bmatrix} S_{11} & S_{12} \\ S_{21} & S_{22} \end{bmatrix} \begin{Bmatrix} \delta \varepsilon_s \\ \delta \varepsilon_v \end{Bmatrix} \tag{3.21}$$

where $[S]$ is a stiffness matrix containing stiffness moduli. The components of $[S]$ are

$$S_{11} = \frac{\partial q'}{\partial \varepsilon_s} = 3G' \tag{3.22}$$

$$S_{22} = \frac{\partial p'}{\partial \varepsilon_v} = K' \tag{3.23}$$

$$S_{12} = \frac{\partial q'}{\partial \varepsilon_v} = J'_1 \tag{3.24}$$

$$S_{21} = \frac{\partial p'}{\partial \varepsilon_s} = J'_2 \tag{3.25}$$

For materials that are isotropic and elastic and perfectly plastic (see Sec. 3.9), $J_1' = J_2'$, and the stiffness matrix is symmetric, while for materials that are isotropic and elastic, $J_1' = J_2' = 0$ (see Sec. 3.8) so that shearing and volumetric effects are decoupled. Alternatively, a constitutive equation can be written as

$$\begin{Bmatrix} \delta\varepsilon_s \\ \delta\varepsilon_v \end{Bmatrix} = \begin{bmatrix} C_{11} & C_{12} \\ C_{21} & C_{22} \end{bmatrix} \begin{Bmatrix} \delta q' \\ \delta p' \end{Bmatrix} \tag{3.26}$$

where $[C]$ is a compliance matrix containing compliance parameters. Comparing Eqs. (3.21) and (3.26), $[C]$ is the inverse of $[S]$ and, in general, there are no simple relationships between the stiffness parameters in $[S]$ and the compliance parameters in $[C]$. However, for materials that are isotropic and elastic, shear and volumetric effects are decoupled so that $C_{12} = C_{21} = 0$ and in this case $C_{11} = 1/S_{11} = 1/3G'$ and $C_{22} = 1/S_{22} = 1/K'$.

Since the stress–strain behaviour of soil is largely non-linear stiffness and compliance parameters will not be constants, but will vary with strain. They also depend on the current stresses and on the history of loading and unloading.

3.8   Elasticity

Materials that are elastic are conservative so that all of the work done by the external stresses during an increment of deformation is stored and is recovered on unloading: this means that all the strains that occur during an increment of loading are recovered if the increment is removed. An important feature of isotropic and elastic materials is that shear and volumetric effects are decoupled so that the stiffness parameters J_1' and J_2' are both zero and Eq. (3.21) becomes

$$\begin{Bmatrix} \delta q' \\ \delta p' \end{Bmatrix} = \begin{bmatrix} 3G' & 0 \\ 0 & K' \end{bmatrix} \begin{Bmatrix} \delta\varepsilon_s^e \\ \delta\varepsilon_v^e \end{Bmatrix} \tag{3.27}$$

(where the superscripts e denote elastic strains) and the complete behaviour is as shown in Fig. 3.8. For materials that are elastic but anisotropic the coupling moduli J_1' and J_2' are equal, so that the matrix in Eq. (3.21) is symmetric. Elastic materials can be non-linear, in which case all the elastic moduli vary with changing stress or strain. (Stretching and relaxing a rubber band is an example of non-linear and recoverable elastic behaviour.)

The more usual elastic parameters are Young's modulus E' and Poisson's ratio v'. These are obtained directly from the results of uniaxial compression (or extension) tests with the radial stress held constant (or zero), and are given by

$$E' = \frac{d\sigma_a'}{d\varepsilon_a^e} \tag{3.28}$$

$$v' = -\frac{d\varepsilon_r^e}{d\varepsilon_a^e} \tag{3.29}$$

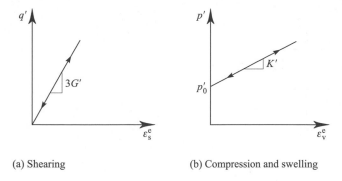

(a) Shearing   (b) Compression and swelling

Figure 3.8 Behaviour of ideal linear elastic material.

Most texts on the strength of materials give the basic relationships between the various elastic parameters and, for isotropic materials, these are

$$G' = \frac{E'}{2(1 + v')} \tag{3.30}$$

$$K' = \frac{E'}{3(1 - 2v')} \tag{3.31}$$

3.9   Perfect plasticity

When the loading has passed the yield point in Fig. 3.1 simultaneous elastic and plastic strains occur and the stiffness decreases. During an increment of plastic deformation the work done is dissipated and so plastic strains are not recovered on unloading. (Bending a paper clip so it remains permanently out of shape is an example of plastic deformation.)

At the ultimate state there are no further changes of stress (because the stress–strain curve is horizontal) and so all the strains at failure are irrecoverable. The plastic strains at failure in Fig. 3.1 are indeterminate; they can go on more or less for ever and so we can talk about plastic flow. Although it is impossible to determine the magnitudes of the plastic strains at failure, it is possible to say something about the relative rates of different strains such as shear and volumetric strains.

Figure 3.9(a) illustrates an element of material loaded to failure with different combinations of some arbitrary stresses, σ'_x and σ'_y. The combinations of stress that cause failure and plastic flow are illustrated in Fig. 3.9(b) and are represented by a failure envelope. At any point on the envelope the vector of the failure stress is σ'_f and Fig. 3.9(c) shows the corresponding plastic strains. Since the stresses remain constant the strains accumulate with time and so the origin is arbitrary. The direction of the vector of an increment of the plastic straining is given by $\delta\varepsilon^p_x / \delta\varepsilon^p_y$. The relationship between the failure envelope and the direction of the vector of plastic strain is called a flow rule.

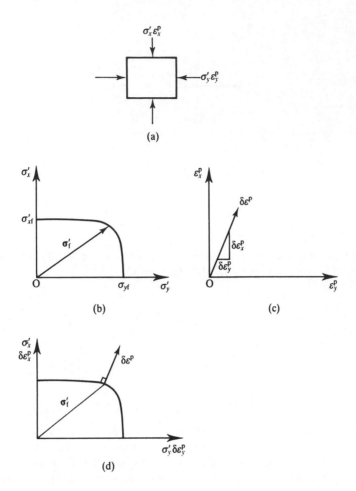

Figure 3.9 Behaviour of ideal perfectly plastic materials.

Figure 3.9(d) contains the same information as Figs. 3.9(b) and (c) with the axes superimposed and the origin for plastic strains placed at the end of the appropriate vector of failure stress. For a perfectly plastic material the vector of plastic strain is normal to the failure envelope, and this is known as the normality condition of perfect plasticity.

Another common way of describing the flow rule for plastic straining is to define a plastic potential envelope that is orthogonal to all the vectors of plastic straining, as shown in Fig. 3.10. Then the material is perfectly plastic if the plastic potential is the same as the failure envelope. This is called an associated flow rule as the plastic potential is associated with the failure envelope. Of course the normality condition and an associated flow rule are different ways of saying the same thing.

An important feature of plastic straining is that the strains depend on the state of stress and do not depend on the small change of stress that causes the failure. This is

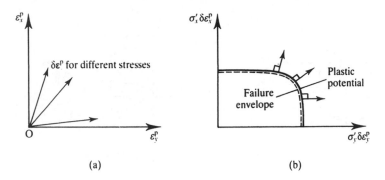

(a)   (b)

Figure 3.10 Plastic potential.

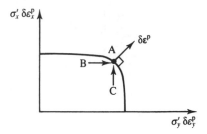

Figure 3.11 Vector of plastic straining for different loadings.

in contrast to elastic straining where the strains depend on the increments of stress as given by Eq. (3.27). Figure 3.11 shows two different loadings B → A and C → A, both of which cause failure at A. The plastic strains are the same for both loading paths; they are governed by the gradient of the failure envelope at A and not by the loading path.

The behaviour of an ideal elastic–perfectly plastic material can be represented by the behaviour of the simple model illustrated in Fig. 3.12(a). This consists of a soft rubber block with a frictional sandpaper base and a rigid platen bonded to the top. A constant normal force F_n and variable horizontal forces F_x and F_y are applied to the platen. If the horizontal forces are less than required to cause frictional sliding of the sandpaper over the table all deformations of the platen are due to elastic deformation of the rubber block. Thus increments of force $\pm\delta F_x$ cause displacements $\pm\delta x^e$ in the direction of the force as shown in Fig. 3.12(b). If, however, there is frictional sliding then the direction of plastic (irrecoverable) displacement δ^p is in the direction of the resultant force F and is independent of the increment of load δF_x or δF_y, as shown in Fig. 3.12(c).

3.10 Combined elasto–plastic behaviour

With reference to Fig. 3.1, the stress–strain behaviour is elastic up to the yield point and is perfectly plastic at the ultimate state. Between the first yield and failure there are simultaneous elastic and plastic components of strain.

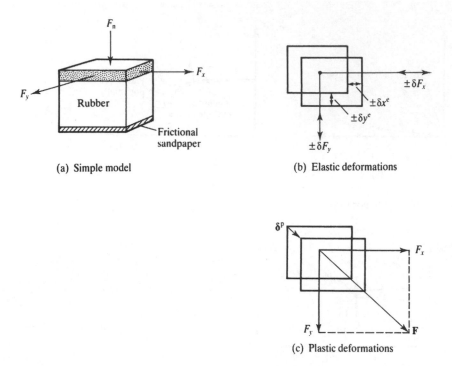

(a) Simple model   (b) Elastic deformations

(c) Plastic deformations

Figure 3.12 A physical model for elastic and plastic behaviour.

In Fig. 3.13 material is loaded from O_1 and is elastic until yielding occurs at Y_1, where the yield stress is σ'_{x1}. It is then strained further and unloaded to O_2 where there are irrecoverable plastic strains $\delta\varepsilon^p_{x1}$. When the material is reloaded from O_2 it is elastic until yielding occurs at Y_2, where the yield stress is σ'_{x2}. If the material is then strained further and unloaded to O_3, on reloading it will have a new yield stress σ'_{x3} and so on. Thus the principal consequences of straining from Y_1 to Y_2 (or from Y_2 to Y_3) are to cause irrecoverable plastic strains and to raise the yield point from σ'_{x1} to σ'_{x2} (or from σ'_{x2} to σ'_{x3}). This increase of the yield point due to plastic straining is called hardening and the relationship between the increase in the yield stress $\delta\sigma'_x$ and the plastic straining $\delta\varepsilon^p_x$ is known as a hardening law. In Fig. 3.13 there is a broken line to the left of the first yield point, which suggests that there could be even lower yield points for previous loadings; this simply demonstrates that the origin of strains O_1 was arbitrarily chosen.

Yielding and plastic straining may cause hardening (i.e. an increase in the yield stress), as shown in Fig. 3.14(a), or softening (i.e. a decrease in the yield stress), as shown in Fig. 3.14(b). In the latter case the state has reached, and passed, a peak in the stress–strain curve, and this is a feature commonly found in the behaviour of soils. In each case the total strains are the sum of the elastic and plastic components and the plastic strains are related to the change of the yield stress by a hardening law.

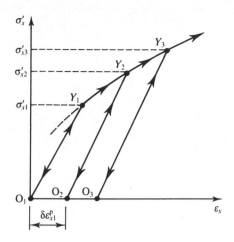

Figure 3.13 Material behaviour during load cycling.

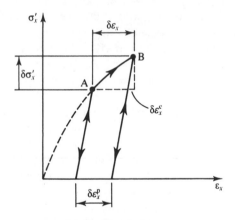

(a) Strain hardening

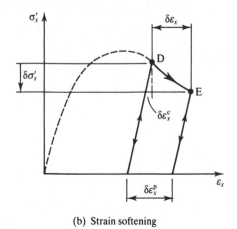

(b) Strain softening

Figure 3.14 Yielding and plastic straining.

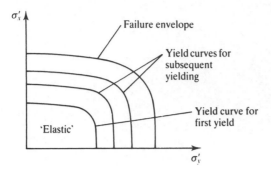

Figure 3.15 Examples of simple yield curves.

Yielding under combined stresses may be represented by a set of yield curves which are similar to the failure envelope, as illustrated in Fig. 3.15. This shows a yield curve for the first yield, two yield curves for subsequent yielding and a failure envelope. For states inside the first yield curve the behaviour is elastic. The state cannot reach the region outside the failure envelope. If the plastic strains are perfect then the vectors of plastic strain are normal to the yield curves. Thus, for the loading path A → B in Fig. 3.16 which crosses successive yield surfaces the vectors of plastic strain are normal to the yield surface.

Since each yield curve in Fig. 3.16 is associated with a particular plastic strain we can use the plastic strain as a third axis to develop a yield surface, as shown in Fig. 3.17. For any state on the yield surface there are plastic strains that are normal to the appropriate yield curve and are given by the movement of the stress point across the surface. For any state inside the surface, during first loading or due to unloading, the behaviour is elastic. Thus, for the loading and unloading O → A → B → C in Fig. 3.17 the behaviour is elastic for the paths O → A and B → C. For the path A → B there are simultaneous elastic and plastic strains.

It is now possible to assemble the flow rule, the hardening law and the elastic stress–strain equations into an explicit constitutive equation for the complete range of loading

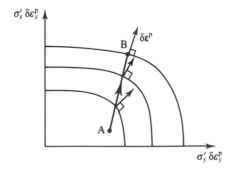

Figure 3.16 Plastic straining for loading on a yield surface.

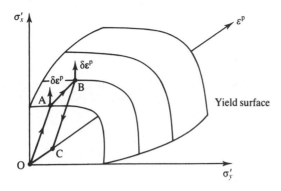

Figure 3.17 Behaviour during a cycle of loading on and under yield surface.

up to failure. We will develop such a constitutive equation for soil in Chapter 12 when we have obtained equations for the yield surface and for the successive yield curves for soil.

3.11   Time and rate effects

In developing constitutive equations for materials we have, so far, considered only relationships between changes of effective stress and changes of strain. This means that no strains occur at constant load (except at failure). In addition it was assumed that the relationships between stresses and strains were independent of the rate of loading or the rate of straining. In soils there are a number of time and rate effects mainly due to drainage of water and, to a limited extent, due to creep and viscosity in the soil skeleton.

Time-dependent straining due to drainage of water is known as consolidation and it is a coupling of deformations due to effective stress with seepage. Theories for consolidation will be considered in Chapter 15.

The theory of viscosity relates stresses in moving materials (usually fluids) to the velocity of flow, so that the shear stresses in water flowing in a pipe are related to the velocity of the flow. In solid materials such as steel, concrete or soil, the strength or stiffness may be governed by the rate of loading or by the rate of straining. It turns out that the important mechanical properties of most soils, except peats and organic soils, are not significantly influenced by the rate of loading, and usually we will not have to worry about viscous effects in soil mechanics.

Materials under constant stress generally continue to strain, but at a rate that diminishes with time; this is known as creep. The basic relationship for creep is

$$\delta\varepsilon^c = C_\alpha \ln(t/t_0) \tag{3.32}$$

where C_α is a creep parameter that depends on a number of factors, including the magnitudes of the (constant) stresses, and t_0 is time from which the creep strains are measured. Equation (3.32) can be differentiated to give the creep strain rate as

$$\frac{d\varepsilon^c}{dt} = \frac{C_\alpha}{t} \tag{3.33}$$

showing that the creep strain rate decreases with time. Many soils, particularly soft clays and peats, show significant creep strains. These can also influence the subsequent behaviour, as we will discuss later.

3.12   Summary

1. The basic mechanical properties of structural materials are stiffness and strength. Stiffness relates changes of stress to changes of strain and this governs deformations and ground movements. Strength is the largest shear stress that a material can sustain before it fails and this governs the ultimate states of collapse of structures.
2. Strength can be described as cohesion with $\tau' = c'$ (where c' is the cohesion) or as friction with $\tau' = \sigma' \tan \phi'$ (where ϕ' is the angle of friction) or by the Mohr–Coulomb criterion which combines cohesion and friction.
3. Purely elastic strains are recovered on unloading. In metals the elastic stress–strain line is approximately linear so the elastic parameters G' and K' are approximately constants.
4. A perfectly plastic material continues to strain with constant stresses and the vector of plastic strain, which relates the rates of plastic shear and volumetric strains, is normal to the current yield curve.
5. Theories for elasto-plastic straining can be obtained by adding the elastic and plastic components of strain.
6. Most time and rate effects in soils are due to coupling of stiffness with seepage of pore water. Creep and viscous effects are usually neglected except in peats and other organic soils.

Worked examples

Example 3.1: Stress and strain in a triaxial test   In a triaxial compression test on a sample of soil the pore pressure is zero so total and effective stresses are equal. The radial stress is held constant at $\sigma'_r = 200$ kPa and the axial stress is changed from $\sigma'_a = 350$ kPa to 360 kPa. The strains for this increment were $\delta\varepsilon_a = 0.05\%$ and $\delta\varepsilon_r = -0.01\%$.

At the start of the increment,

$$q' = \sigma'_a - \sigma'_r = 350 - 200 = 150 \text{ kPa}$$

$$p' = \tfrac{1}{3}(\sigma'_a + 2\sigma'_r) = \tfrac{1}{3}(350 + 400) = 250 \text{ kPa}$$

During the increment $\delta\sigma_a' = 10\,\text{kPa}$, $\delta\sigma_r' = 0$ and, from Eqs. (3.3) to (3.6),

$$\delta q' = (\delta_a' - \delta\sigma_r') = 10\,\text{kPa}$$

$$\delta p' = \tfrac{1}{3}(\delta\sigma_a' + 2\sigma_r') = \tfrac{1}{3} \times 10 = 3.3\,\text{kPa}$$

$$\delta\varepsilon_s = \tfrac{2}{3}(\delta\varepsilon_a - \delta\varepsilon_r) = \tfrac{2}{3}(0.05 + 0.01) = 0.04\%$$

$$\delta\varepsilon_v = \delta\varepsilon_a + 2\delta\varepsilon_r = 0.05 - 0.02 = 0.03\%$$

Example 3.2: Calculation of shear and bulk modulus The soil in Example 3.1 is isotropic and elastic (i.e. shearing and volumetric effects are decoupled). For the increment,

$$\text{Shear modulus } G' = \frac{\delta q'}{3\delta\varepsilon_s} = \frac{10}{3 \times 0.04/100 \times 1000} = 8.3\,\text{MPa}$$

$$\text{bulk modulus } K' = \frac{\delta p'}{\delta\varepsilon_v} = \frac{3.3}{0.03/100 \times 1000} = 11.1\,\text{MPa}$$

$$\text{Young's modulus } E' = \frac{\delta\sigma_a'}{\delta\varepsilon_a} = \frac{10}{0.05/100 \times 1000} = 20\,\text{MPa}$$

$$\text{Poisson's ratio } \nu' = -\frac{\delta\varepsilon_r}{\delta\varepsilon_a} = -\frac{-0.01}{0.05} = 0.2$$

From Eqs. (3.24) and (3.25), substituting for E' and ν',

$$G' = \frac{E'}{2(1 + \nu')} = \frac{20}{2(1 + 0.2)} = 8.3\,\text{MPa}$$

$$K' = \frac{E'}{3(1 - 2\nu')} = \frac{20}{3(1 - 0.4)} = 11.1\,\text{MPa}$$

Further reading

Atkinson, J. H. (1981) *Foundations and Slopes*, McGraw-Hill, London.

Calladine, C. R. (1969) *Engineering Plasticity*, Pergamon Press, London.

Case, J., A. H. Chilver and C. T. F. Ross (1999) *Strength of Materials and Structures*, Edward Arnold, London.

Heyman, J. (1972) *Coulomb's Memoir on Statics*, Cambridge University Press.

Jaeger, J. C. (1969) *Elasticity, Fracture and Flow*, Methuen, London.

Palmer, A. C. (1976) *Structural Mechanics*, Oxford University Press.

Timoshenko, S. P. and J. N. Goodier (1969) *Theory of Elasticity*, McGraw-Hill, New York.

Chapter 4

The structure of the Earth

4.1 Introduction

Soils occur very near the surface of the Earth and are essentially the products of the action of the weather and the climate on rocks. Weathering of rock *in situ* leads to the formation of residual soils. These may be eroded, transported and laid down as deposited soils. The engineering properties of soils and how they occur in the ground depend to a great extent on their geological origins and so geotechnical engineers need to know something about geology.

In this one chapter I cannot possibly cover the whole of geology, or even all the parts related to engineering. You will find a number of simple and easy-to-read books on geology for engineers and on engineering geology and you will probably attend lectures on the subject. What I want to do here is set down what I consider to be the most interesting and important aspects of geology related to geotechnical engineering in soils. This is my personal list and other geotechnical engineers and geologists will probably want you to know about other things. This does not really matter because if you want to be a good geotechnical engineer you will need to study geology in some detail.

4.2 The Earth's crust

The Earth has a radius of about 8000 km and a crust of soils and rocks about 25 to 50 km thick (see Fig. 4.1(a)). The ratio of the thickness of the crust to its radius of curvature is about the same as that of an eggshell. Below the crust is a mantle of hot plastic material and plates of crust move about on the mantle. This drift of the continental crust accounts for mountain building, earthquakes and volcanic activity at boundaries between the plates. It also accounts for evidence of glacial deposits in Australia and tropical soils in Antarctica. In a single core of rock taken almost anywhere on Earth, there will be rocks deposited in conditions that were like all the known present-day environments.

The surface of the crust (the land surface and the sea bed) has altitudes and depths above and below mean sea level of the order of 8 km (see Fig. 4.1(b)). Materials near the surface are soils and rocks although there is not a very clear distinction between the two; at low stresses soils fracture like rocks while at high stresses rocks will deform plastically like soils. For engineering purposes soils rarely occur below a depth of about 300 m (see Fig. 4.1(c)). Geologically old soils (older than about 2 millions years)

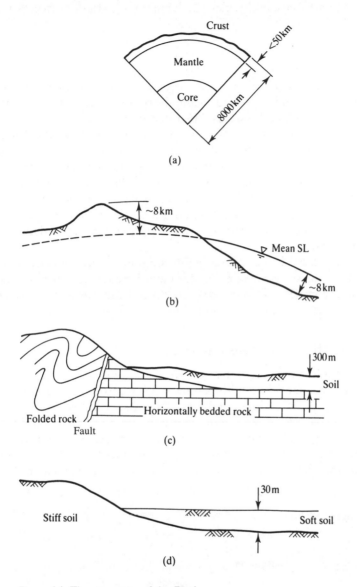

Figure 4.1 The structure of the Earth.

are usually relatively stiff and strong while young soils (Glacial and Post-Glacial) are usually relatively soft and weak, but these are rarely deeper than about 30 m (see Fig. 4.1(d)). Notice that the slope of the land reflects the strength of the underlying material; in rocks mountain slopes can be steep and high while in soils the slope angles are much more gentle and the heights are much less. Spread out over most of the land surface is a layer of soil of variable thickness, but usually less than 1 m, that supports plant life. This is called topsoil; it is of great interest to farmers and gardeners but not to engineers, except to save and replace as landscaping after construction is complete.

4.3   Geological processes

Soils and rocks close to the surface of the Earth are exposed to the atmosphere and are weathered, eroded, transported and deposited, while deep burial converts soils back to rocks. The general cycle of soils and rocks is illustrated in Fig. 4.2.

Collisions between drifting continental plates raise mountain chains like the Andes, the Rockies and the Himalayas. Rain, snow and sunshine weather rocks and soils; this may consist simply of mechanical breakdown of a rock mass into chunks of material that have the same composition as the parent rock or chemical alteration to new clay minerals. Water, ice and wind transport these weathered rock fragments and, at the same time, degrade, polish and sort them into different sizes. When these transporting agents slow down, the soil particles are deposited and as further material is deposited above they become compressed.

The cycle illustrated in Fig. 4.2 is, of course, highly simplified and there are many additional influences and processes. For example, rocks weathered *in situ* form residual soils while tectonic activity recycles molten material from below the crust to the surface, sometimes causing rocks and soils to metamorphose in the process.

4.4   Stratigraphy and the age of soils and rocks

A borehole drilled down into the crust will pass through strata, or layers, of soils and rocks which generally become older with depth. Stratigraphy is the study of the sequence of strata that represent geological history. At a particular location there will have been periods of volcanic activity, mountain building and erosion and so a single borehole will not reveal the complete sequence of Earth's history. A break in the stratigraphic column in a borehole record is known as an unconformity; often the materials at an unconformity have been eroded before deposition of new material.

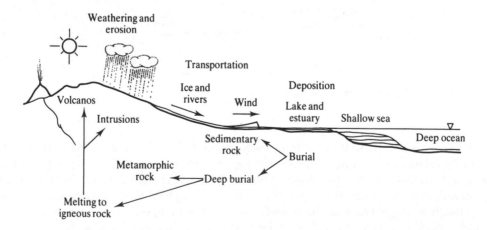

Figure 4.2 Simplified representation of the cycling of rocks and soils.

A highly simplified version of the stratigraphic column is shown in Table 4.1; this gives the name and approximate age of the major divisions, the general nature of the deposits and typical examples from the United Kingdom. In other parts of the world, the major divisions and their ages are the same but the nature of the deposits may well be different: for example, the Cretaceous Chalk in South East England is the same age as the Deccan Lavas in India.

In order to describe the chronological history of the Earth geologists classify major strata according to their age, not what they are. Notice that the initial letters of geological names are capitals (e.g. Old Red Sandstone, London Clay, etc.) whereas the engineering descriptions (e.g. overconsolidated clay) have lower-case initial letters. For example, the deposit called London Clay is of Eocene age and was deposited 40 to 60 million years ago. The deposit is found in South East England and is also found in Belgium, where it is called Boom Clay. In the London region it is largely a marine clay but to the west of London, in the Hampshire Basin, it is mostly silt and fine sand with very little clay. Old Red Sandstone is of Devonian age and was deposited 350 to 400 million years ago. It is generally red in colour, unlike the Carboniferous rocks above and the Silurian rocks below, which are both grey, but it is not all sandstone and it contains thicknesses of mudstones and siltstones.

Generally soils and rocks become stiffer and stronger with age: London Clay is obviously stronger than the soils found in the English Fens and the slates in North Wales are stronger still. As a very rough guide, materials of Cenozoic age are generally regarded as soils for engineering purposes; materials of Mesozoic age are generally regarded as soft rocks and materials of Palaeozoic age are regarded as hard rocks. The soils and rocks in the stratigraphic column contain fossils which are the most important indicators of their age and provide a record of evolution on Earth. Cambrian and Ordovician rocks contain mollusc shells and corals; land plants occur in the Devonian, reptiles in the Carboniferous, amphibians in the Permian, dinosaurs in the Triassic and birds in the Jurassic; the dinosaurs became extinct in the Cretaceous. Mammals, fishes, insects and birds had evolved by the Eocene, but modern man did not evolve until the middle of the Pleistocene, about 1 million years ago.

Since the engineering properties of sands, silts and clays and of sandstones, siltstones and mudstones are likely to be different, the standard geological age-based classifications will only be of limited use in geotechnical engineering. Much better schemes for engineering classifications of soils and rocks are based on the nature of the grains and on the state of stress and water content. These are described in Chapter 5.

4.5   Depositional environments

The nature of the weathering and the mode of transport largely determine the nature of a soil (i.e. the size and shape of the grains, the distribution of grain sizes and their mineralogy). The environment into which it is deposited and the subsequent geological events largely determine the state of the soil (i.e. the denseness or looseness of the packing of the grains) and its structure (i.e. the presence of features such as fissuring, bedding, bonding and so on).

As you move about the world you can see weathering, erosion, transportation and deposition taking place. In the present day in the United Kingdom most of the transportation is by water (rivers look dirty because they are carrying soil particles) and

Table 4.1 Simplified stratigraphical column in Britain

Name of geological group or era	Name of geological system or period (approx. age in millions of years)		General nature of deposits Major orogenies (mountain building) and igneous activity (in italics)	Examples in the United Kingdom
CENOZOIC (= recent life)	Quaternary	Recent	Alluvium, blown sand, glacial deposits, etc.	Extensive drift cover to most of the bedrock
		Pleistocene (2)	At least five major ice ages separated by warmer periods	
	Tertiary	Pliocene	Sand, clays and shell beds	East Anglian Crags
		Miocene	*Alpine orogeny*	
		Oligocene		
		Eocene (65)	*Igneous activity in west Scotland and Northern Ireland*	London Clay
MESOZOIC (= middle life)		Cretaceous	Sand, clays and chalk	Chalk, Gault Clay, Weald Clay
		Jurassic	Clays, limestones, some sands	Lias Clay, Portland Stone
		Triassic (250)	Desert marls, sandstone marls	Keuper Marl, Bunter Sandstone
PALAEOZOIC (= ancient life)	Newer	Permian	Breccias, marls, dolomitic limestone *Hercynian orogeny* *Igneous activity*	Magnesian Limestone
		Carboniferous	Limestones, shales, coals and sandstones	Coal Measures and Mountain Limestone
		Devonian (and Old Red Sandstone) (400)	Siltstones, sandstones and limestones (Lacustrine sandstones and marlstones) *Igneous activity* *Caledonian orogeny*	Red beds in Hereford and S. Wales
	Older	Silurian	Thick shallow-water limestones, shales and sandstones	Rocks of Wales and Lake District
		Ordovician	*Volcanic activity in the Ordovician* Slates, siltstones, sandstone and greywackes	Rocks of North Wales, Central and North Scotland
		Cambrian (600)		
PRE-CAMBRIAN		Torridonian	Schists Sandstones	Rocks of North West Scotland
		Uriconian	Sandstones Lavas and tuffs (Shropshire)	
		Lewisian	*Pre-Cambrian orogenies*	

most of the deposition is in lakes, estuaries and in the near-shore region of the sea bed. In the past there have been many different climates and environments, because what is now the United Kingdom moved about the Earth on a drifting continental plate. Today, in cold regions and at high altitudes, you can see transportation by glaciers and in deserts by wind, while in the tropical regions there are deep deposits of residual soils being formed *in situ*. The study of depositional environments is a fascinating subject and is the key to the understanding and interpretation of engineering ground investigations. The basic principles are that all soils and rocks were deposited in one of a relatively small number of depositional environments, all of which can be found somewhere in the world today; if you know what the depositional environment was then you can infer much about the likely nature and properties of a deposit.

Figure 4.3 illustrates three typical depositional environments. Figure 4.3(a) shows the end of a moving glacier transporting eroded soil and rock. It deposits a basal till and a terminal moraine; the soils in both these deposits are well graded (i.e. they contain a wide variety of particle sizes from clays to boulders and they are often called boulder clay). Water from the melting glacier transports material away from the glacier but sorts the sizes, depositing first gravels, then sands and moving clays considerable distances. Figure 4.3(b) shows deposition into lakes or estuaries or into the oceans. Slow flowing rivers can only carry fine-grained soils, so the deposits will be largely silts and clays. Still water deposits tend to be layered horizontally while delta

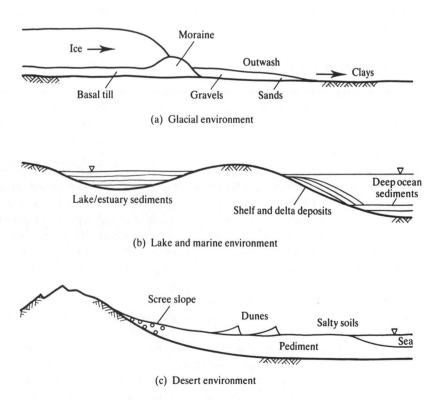

Figure 4.3 Characteristic depositional environments.

and moving water deposits are built in steps. Figure 4.3(c) shows a desert environment. Hot days and cold nights cause thermal weathering of rock mountains which produces scree slopes. Rare flash floods transport material across the desert floor (or pediment), depositing coarse material first and fine material later, probably in fans and layers. Winds cause migration of sand dunes. Large daily temperature variations with occasional rainfall cause physical alteration of the soils in the pediment.

These are only three typical depositional environments. They are discussed in more detail, together with other examples, by Fookes and Vaughan (1986). Much of the United Kingdom north of a line from the Thames to the Severn estuaries is covered with a veneer of glacial deposits. Most natural and man-made lakes are currently collecting layered silt and clay deposits. Large rivers (e.g. Nile, Ganges, Mississippi) are currently building delta deposits. Modern deserts occur widely throughout Asia, Australia, Africa, North and South America and the Middle East. Glacial environments occur in high latitudes (e.g. Greenland, Antarctica) and at high altitudes.

These typical depositional environments can be recognized in ancient rocks. For example, the London Clay was deposited in a shallow sea; the Chalk is calcium carbonate deposited in a warm sea; the New Red Sandstone in the Triassic and the Old Red Sandstone in the Devonian are ancient desert deposits. The important point to make here is that you should study present-day depositional environments as an aid to interpretation of ground investigations; if a geologist can tell you the environment into which a soil or rock was deposited you have a very good idea of what to expect.

4.6   Recent geological events

Although the depositional environment has a major influence on the formation of soils and rocks, they are altered by later geological events such as further deposition or erosion, folding and faulting and volcanic activity. For soils and soil mechanics the most significant recent geological events are rising or falling land and sea levels which lead to continuing deposition or erosion.

Land and sea levels rise and fall relative to one another for a variety of reasons, including plate movements and mountain building. One of the most important causes of changes of sea level is temperature change. During an ice age the sea cools and contracts and ice remains on the land as glaciers; the weight of ice depresses the land which rebounds as the ice melts. At the end of the last ice age, about 20 000 years ago, the sea level was about 100 m lower than it is now, so the UK coastline was west of Ireland and you could have walked to France (if you could cross the large river flowing through the Straits of Dover).

During a period of rising sea levels (e.g. at the end of an ice age) soils are deposited around the coasts. If the sea level remains stationary for some time vegetation grows, which is submerged and decays to peat as the sea level rises again. In the United Kingdom extensive deposits of this kind are found in the Wash and in the Somerset Levels. Continuing sea level rise and deposition leads to deposits of normally consolidated soils (see Chapter 15) which are soft and weak near the surface but become stronger with depth. During a period of falling sea level (e.g. at the beginning of an ice age) the land becomes exposed and subject to weathering, erosion and transportation.

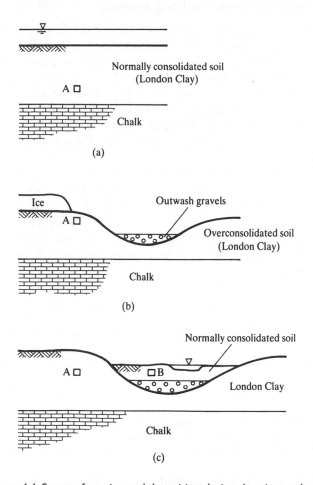

Figure 4.4 Stages of erosion and deposition during changing sea levels.

As the ground is eroded the soils become overconsolidated (see Chapters 16 and 18) due to unloading, but they do not recover their original state. Overconsolidated soils have stiffnesses and strengths which are more or less uniform with depth and which are larger than those of normally consolidated deposits at the same depth.

Figure 4.4 illustrates a sequence of falling and rising sea levels; this is a highly simplified model of the recent geology of the Thames estuary north east of London. In Fig. 4.4(a) London Clay is deposited in a shallow sea; notice an element at A just above the Chalk. Figure 4.4(b) shows a glaciation with a very low sea level and a nearby glacier. Much of the London Clay has by now been eroded so the element A is nearer the surface. Meltwater from the glacier has eroded a river channel which has been partly filled with outwash gravels. Figure 4.4(c) shows the present day; rising sea levels have led to deposition of soft soils in the valley. The soil at B is about the same depth as that at A, but it is normally consolidated and so is relatively soft and weak.

4.7   Importance of geology in geotechnical engineering

It is obvious that an understanding of the geology of a location will aid the interpretation of ground investigations. All soils were deposited or formed *in situ* in one of only a few characteristic environments. These environments, together with later geological events, determine the nature and state of soils and rocks. Very nearly all the environments that have occurred on Earth can be found somewhere in the world today. What you have to do is find a geologist who can identify the geological environment of a deposit for you. Do not ask the geologist what the soil or rock is as you can usually see what it is yourself; instead, ask how did it get there (i.e. how was it deposited) and what has happened to it since.

The most important thing to remember is that natural soils and rocks arrived where they are and have the properties they have through natural geological processes; there was no divine intervention mixing things up and changing them around. For every site there is a geological story to be told which explains how all the natural geological materials got there and why they are as they are. If you cannot tell the geological story for your site you do not understand it properly and the chances are you will make a mistake designing the groundwork.

Further reading

Blyth, F. G. H. and M. H. De Freitas (1984) *A Geology for Engineers*, Butterworth-Heinemann Ltd.

Fookes, P. G. and P. R. Vaughan (1986) *A Handbook of Engineering Geomorphology*, Surrey University Press.

Gass, I. G., P. J. Smith and R. C. L. Wilson (1971) *Understanding the Earth*, Open University Press, Sussex.

Holmes, A. H. (1993) *Principles of Physical Geology*, Nelson, London.

ICE (1976) *Manual of Applied Geology for Engineers*, ICE, London.

McLean, A. C. and C. D. Gribble (1985) *Geology for Civil Engineers*, Chapman & Hall, London.

Read, H. H. and J. Watson (1970) *Beginning Geology*, Macmillan, London.

Waltham, A. C. (2001) *Foundations of Engineering Geology*, Spon.

Chapter 5

Classification of soils

5.1 Description and classification

Soils consist of grains, usually rock fragments or clay particles, with water and gas, usually air or water vapour, in the void spaces between the grains. If there is no gas present the soil is saturated and if there is no water it is dry, while if there is both water and gas in the voids the soil is unsaturated. The mechanics of unsaturated soils is very complicated and in this book I will normally consider saturated or dry soils. Fortunately, in civil engineering applications soils are mostly saturated, except in hot dry environments or when compacted.

The mechanical properties of a soil (i.e. its strength and stiffness) depend principally on the nature of the grains (i.e. what they are) and the state of the soil (i.e. how the grains are packed together). You can dig up a sample of soil from your garden or from the beach and describe what you see. You can describe its colour, the size and shape of the grains (if you can see them) and some aspects of the behaviour, such as its response to moulding in your fingers. To be useful, however, you will need a scheme of classification that separates groups of soils with markedly different behaviour. Any useful scheme of soil classification should be based on relatively simple tests and observations.

It is important to distinguish between soil description and soil classification. Description is simply what you see and how the soil responds to simple tests; you may want to describe only a single soil sample or a soil profile exposed in a cliff face, in an excavation or from a number of samples from a borehole. A classification is a scheme for separating soils into broad groups, each with broadly similar behaviour. There are various classification schemes for different purposes: there are agricultural classifications based on how soils support crops and geological classifications based on the age of the deposit or the nature of the grains. For civil engineering purposes soil classifications should be based mainly on mechanical behaviour.

5.2 Description of soils

Soil description is essentially a catalogue of what the soil is and it is helpful to have a simple scheme to describe the essential features. There are several such schemes published in National Standards and to some extent these reflect the characteristics of the most common soils in the region; you should look up the relevant standard for the region you will work in. In the United Kingdom these are the British Standards for site investigations (BS 5930:1999) and for soil testing (BS 1377:1990) but slightly different schemes are used in other regions. A simple and universal scheme for soil description is as follows:

1. The nature of the grains. The most important features of soil grains are their size and the grading (i.e. the proportions of different sizes), together with the shape and surface texture of the grains and their mineralogy.
2. The current state of the soil. The important indicators of the state of a soil are the current stresses, the current water content and the history of loading and unloading: these are reflected by the relative strengths and stiffnesses of samples of the soil.
3. The structure of the soil. This consists of fabric and bonding. Natural soils are rarely uniform and they contain fabric features, such as layers, which are seen in small samples and in large exposures. In some natural soils the grains are weakly bonded together. (If the grains are strongly bonded the material has become a rock.) Soil structure will be discussed further in Chapter 16.
4. The formation of the soil. Soils are formed in different ways. They may be deposited naturally from water, ice or wind; they may be the residual products of rock weathering; they may be compacted by machines into embankments and fills.

A more detailed scheme for description of soils is given in BS 5930:1999. This is similar to the scheme described above but is more detailed and gives helpful quantitative values for a number of visual observations.

The nature of a soil does not usually change during normal civil engineering works; occasionally weak and brittle soil grains may fracture during loading so the grading changes. On the other hand, the state of a soil does change as soils near foundations and excavations are loaded or unloaded and compress or swell.

The manner of formation of a soil will influence both its nature, its initial state and its structure (i.e. layering, fissuring and bonding). In this book I will be examining the basic behaviour of soils observed in remoulded and reconstituted samples where any fabric and bonding has been removed by the preparation of the sample. Since most natural soils have some structure it is important always to test some intact samples, but their behaviour should be examined within the basic framework established for reconstituted samples.

5.3   Soil particle sizes, shapes and gradings

The range of particle sizes in soils is very large and ranges from clay grains that are smaller than 2 μm (0.002 mm) to boulders that are larger than 200 mm. A particular range of particle sizes is given a name, as in Fig. 5.l, so that, for example, in UK practice medium sand is 0.2 to 0.6 mm. As a general guide, individual sand-sized and coarser particles are visible to the naked eye while individual silt-sized particles are visible using a $\times$ 10 hand lens. If you can wash fine grained soil off your boots it is probably silt, but if you have to scrape it off it is probably clay; similarly, if silt dries

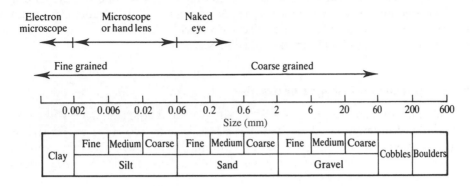

Figure 5.1 Soil particle sizes.

on your hands it will dust off while dry clay will leave your hands dirty and will have
to be washed off.

Soil particle shapes also differ considerably. Clay grains are usually plate-like while
silt, sand and gravel grains are more rotund.

Words such as sand, silt and clay are used both to classify a particular grain size and
to describe a soil which may contain lesser quantities of other sizes. The distribution
of particle sizes in a soil is represented by a grading curve on a particle size chart, as
shown in Fig. 5.2. If the grading curve is flat the soil contains a wide variety of different
particle sizes and is known to engineers as well graded; if the curve is steep and one size

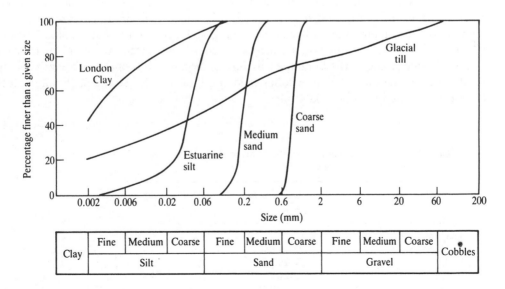

Figure 5.2 Grading curves plotted on a particle size distribution chart. (After BS 1377:1990.)

predominates the soil is poorly graded. The grading of a soil often reflects its origin. Soils deposited by rivers or wind tend to be poorly graded while boulder clays and tills deposited from ice tend to be well graded with a wide distribution of sizes. Tests to determine the grading of soils are described in Sec. 7.3.

5.4 Properties of fine-grained soils

The behaviour of a coarse-grained soil (i.e. silt-sized and coarser), is very like that of an assembly of different sized marbles, but clays differ in two respects. Firstly, some clay grains themselves may show significant volume changes as the loading and water content changes; this accounts for clays tending to crack as they dry. Secondly, particle surface effects become significant.

The surface of a soil grain carries a small electrical charge which depends on the soil mineral and may be modified by an electrolyte in the pore water. These charges give rise to forces between soil grains in addition to their self-weight. The magnitudes of the interparticle forces are proportional to the surface areas of the grains, while self-weight forces are proportional to the volumes of the grains. As particle sizes decrease the surface forces diminish with the square of the effective diameter, whereas the self-weight forces diminish with the cube; consequently the effects of surface forces are relatively more important in fine-grained than in coarse-grained soils.

The relative importance of the surface and self-weight forces may be described by the specific surface. This is defined as the total surface area of all grains in unit mass. Table 5.1 lists typical values for the specific surface of three common clay minerals and of clean sand; the differences in the values of specific surface for sand and clay are very large.

In coarse-grained soils such as silt, sand and gravel, particle surface forces are negligible compared to their self-weight forces, so that dry sand will run through an hour-glass and form a cone at the base. Dry fine-grained materials, such as kitchen flour, behave differently and if you squash a handful of flour in your hand it will form a coherent lump. This is because, as the grains become densely packed and the number of contacts in unit volume increases, the slight surface forces give rise to a small cohesive strength; the lump is easily broken because the cohesive strength is very small. We will see later that true cohesive strength in soils is usually negligible unless they are cemented by other materials.

Table 5.1 Approximate values for the specific surface of some common soil grains

Soil grain	Specific surface (m^2/g)	Activity
Clay minerals		
Montmorillonite	Up to 840	>5
Illite	65–200	≈0.9
Kaolinite	10–20	≈0.4
Clean sand	2×10^{-4}	—

5.5   Specific volume, water content and unit weight

Many important mechanical properties of soil depend on the closeness of the packing of the grains, so that loose soils (i.e. where there is a high proportion of voids) will be weaker and more compressible than dense soils. The state of a soil can be described by the specific volume v given by

$$v = \frac{V}{V_s} \tag{5.1}$$

where V is the volume of a sample containing a volume V_s of soil grains. Sometimes the voids ratio e is used instead of specific volume, where

$$e = \frac{V_w}{V_s} \tag{5.2}$$

and V_w is the volume of the voids which, in saturated soil, are filled with water. Since $V = V_w + V_s$,

$$v = 1 + e \tag{5.3}$$

For coarse-grained soils, where surface forces are negligible, the grains pack together like spheres. The maximum specific volume of a loose assembly of uniform spheres is 1.92 and the minimum specific volume of a dense assembly is 1.35; common sands and gravels have specific volumes in the range $v = 1.3$ to 2.0. For fine-grained clay soils surface effects may be significant, especially at low stresses, and the maximum specific volume of a recently sedimented clay will depend on the clay mineral and any electrolyte in the pore water. Montmorillonite clays with large specific surfaces may exist with specific volumes in excess of 10, while kaolinite clays which have smaller specific surfaces have a maximum specific volume around 3. Under large loads the specific volumes of clay soils may be reduced to as little as $v = 1.2$ as the flat clay plates become nearly parallel.

   Specific volume cannot be measured directly but it can be calculated from other easily measured parameters. The most convenient is water content w, defined as

$$w = \frac{W_w}{W_s} \tag{5.4}$$

and unit weight γ defined as

$$\gamma = \frac{W}{V} \tag{5.5}$$

where W_w is the weight of water evaporated by heating soil to $105°C$, W_s is the weight of dry soil, $W = W_w + W_s$ is the weight of a sample with volume V. Standard tests to measure water content and unit weight are described in Sec. 7.3. For a typical clay soil the water content might be in the range 0.20 to 0.70 (i.e. 20% to 70%) and the unit

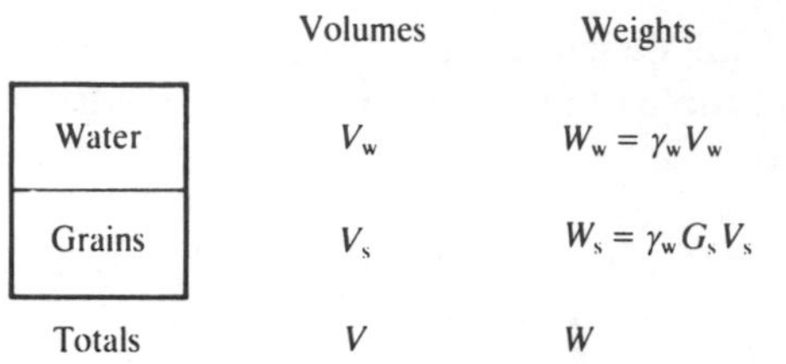

	Volumes	Weights
Water	V_w	$W_w = \gamma_w V_w$
Grains	V_s	$W_s = \gamma_w G_s V_s$
Totals	V	W

Figure 5.3 Volumes and weights of grains and water in saturated soils.

weight might be 18 to 22 kN/m^3 (i.e. about twice that of water: $\gamma_w =$ approximately 10 kN/m^3).

Relationships between these and specific volume can be obtained from Fig. 5.3 together with Eqs. (5.1) to (5.5) as

$$v = 1 + wG_s = 1 + e \tag{5.6}$$

$$v = \frac{G_s - 1}{\left(\dfrac{\gamma}{\gamma_w}\right) - 1} \tag{5.7}$$

where G_s is the specific gravity of the soil grains which, for many soils, is approximately $G_s = 2.65$.

5.6   Limits of consistency

As the water content and specific volume of a soil are increased it will soften and weaken; this is well known to farmers and football players. If the water content is very large we just get muddy water and if it is very small we get a material that is hard and brittle like rock. Obviously there are limits to the water content within which a soil has the consistency of soil rather than the consistency of a liquid or a brittle rock. Tests to determine the precise water contents at which soil behaviour becomes liquid or brittle are the Atterberg limits tests described in Sec. 7.3; these determine the liquid limit (w_L) where the soil starts to flow like a liquid and the plastic limit (w_P) where it ceases to be plastic and becomes brittle.

The Atterberg limits apply to fine-grained soils. (Soils for which it is possible to determine the Atterberg limits are often called plastic, but this term must not be confused with the strict meaning of plastic as a type of constitutive relationship, discussed in Sec. 3.9.) For coarse-grained sands and gravels the appropriate limits are the minimum density of a very loosely poured sample and the maximum density of a vibrated and heavily loaded sample (Kolbuszewski, 1948). Thus the minimum density of a sand is equivalent to the liquid limit of a clay, while the maximum density is equivalent to the plastic limit. The relationships between the Atterberg limits and the maximum and minimum densities are illustrated in Fig. 5.4.

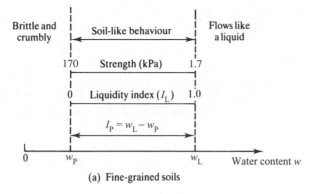

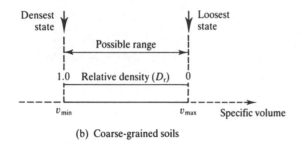

(a) Fine-grained soils

(b) Coarse-grained soils

Figure 5.4 Limits of consistency of soils.

An important parameter for clay soils is the plasticity index (I_P), defined as

$$I_P = w_L - w_P \qquad (5.8)$$

This defines the range of water content for a soil and is related to the maximum volume change (or compressibility) of the soil. Similarly, the difference between the maximum and minimum densities is related to the relative compressibility. These limits depend on the grading and on the mineralogy, shape and surface texture of the grains, so they describe the nature of the soil. The Atterberg limits are measured on the fraction of a soil sample smaller than 425 μm so they describe only part of a well-graded soil. The activity A is defined as

$$A = \frac{I_P}{\% \text{ by weight of clay}} \qquad (5.9)$$

This is closely related to the specific surface and to the mineralogy of the clay. Typical values for common clay minerals are given in Table 5.1.

5.7   Current state

Because soil is both frictional and relatively highly compressible its stiffness, strength and specific volume all depend on the current stresses and history of loading and unloading during deposition and erosion. In Fig. 5.5(a) the soil at a shallow depth z is lightly loaded by the small vertical stress σ_z due to the weight of soil above and it is loose. After deposition of a substantial depth of soil z_1 as in Fig. 5.5(b), the same soil is heavily loaded and has become dense. After erosion back to the original ground level, as in Fig. 5.5(c), the same soil is again lightly loaded but remains relatively dense. Thus the current water content or density of a soil will depend on the current stress and on the history of loading and unloading.

The current state of a soil can be related to the limiting states. For fine-grained clay soils the liquidity index (I_L) is defined as

$$I_L = \frac{w - w_P}{w_L - w_P} \tag{5.10}$$

where w is the current water content and for coarse-grained soils the relative density (D_r) is defined as

$$D_r = \frac{v_{max} - v}{v_{max} - v_{min}} \tag{5.11}$$

where v is the current specific volume. These relationships are illustrated in Fig. 5.4. Notice that a liquidity index of 1.0 (corresponding to the loosest or wettest state) corresponds to a relative density of zero.

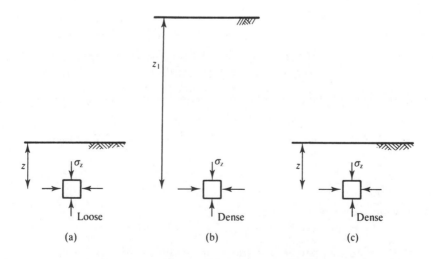

Figure 5.5 Changes of state during deposition and erosion.

Table 5.2 Strength of clay soils estimated from observations in hand samples

Consistency	Identification	Undrained strength s_u kPa
Very soft	Extrudes between fingers	<20
Soft	Easily moulded in fingers	20–40
Firm	Moulded by strong finger pressure	40–75
Stiff	Cannot be moulded in fingers	75–100
Very stiff	Can be indented by thumb nail	>150

Another measure of the consistency of a clay soil is its immediate strength. We will see later that the (undrained) strength of a clay is related to the liquidity index, as illustrated in Fig. 5.4(a). When the water content of a clay soil is at its liquid limit the strength is close to 1.7 kPa and when the water content is at the plastic limit the strength is close to 170 kPa. Rapid estimates of the strength of clays can be made in hand samples using the criteria in Table 5.2.

5.8   Origins of soils

The mechanical behaviour of a soil is determined principally by its nature and its current state, but these are governed, to some extent, by the manner of formation of the soil which may be deposited, residual or compacted by machines. Detailed discussions of the influence of the manner of formation of soils on their nature and state are beyond the scope of this book and are contained in books on engineering geology, but there are a few simple observations to be made:

1. Deposited soils. Soils may be deposited from water, ice or wind and the grading and particle shape and texture are governed largely by the transporting agent. Soils deposited from water or air are poorly graded because the ability of rivers or wind to move different sizes depends on the velocity, while soils deposited from ice (i.e. boulder clays) are well graded because ice can move all particle sizes equally. Abrasion in moving water or air produces rounded and polished grains while soil grains transported by ice generally retain their original angular shape and rough texture. The mineralogy of transported soils is simply that of the parent material, which may be rock fragments or weathered and eroded clay. The fabric of deposited soil is usually bedded and layered, reflecting changes in the depositional environment.
2. Residual soils. These are the products of weathering of rocks, or soils, *in situ*. Their grading and mineralogy depend in part on the parent material but principally on the depth and type of weathering and on details of the drainage conditions. Residual soils usually have low water contents and liquidity indices (or high relative density) and may be unsaturated. The fabric of immature residual soils often reflects the fabric of the parent rock.
3. Compacted soils. Soils may be compacted into fills by rolling, vibration or impact. They are usually unsaturated initially but may later become saturated. Often soils are compacted in layers and may show horizontal structure.

5.9   Summary

Classification of soils requires a careful and detailed description of the soil *in situ* and in samples together with some simple classification tests. The important characteristics required for description of soils are:

1. The nature of the grains including the grading (i.e. the distribution of particle sizes) and the mineralogy, particularly of clay soils. The Atterberg limits give indications of clay mineralogy.
2. The state of a soil is given by the stresses, the history of deposition and erosion and the water content. Important indicators of soil state are the liquidity index of fine-grained soils or the relative density of coarse-grained soils.
3. Structure including bedding, layering, fissuring, jointing and bonding.
4. The method of formation of the soil, which may be deposited from water, wind or ice, residual formed by weathering or compacted by rolling, vibration or impact.

Worked examples

Example 5.1: Grading of soils   Table 5.3 gives the results of particle size tests on four different soils. The grading curves are shown in Fig. 5.6.

Soil A is predominantly medium sand; it is poorly (i.e. uniformly) graded with a relatively small range of sizes. It was probably deposited from a relatively fast flowing river.

Soil B is well graded with a very wide range of particle sizes from coarse gravel to fine silt with a little clay. It was probably deposited from a glacier and has not been sorted by wind or water.

Soil C is a silty clay. It could be deposited either in a shallow sea, in a lake or in an estuary.

Soil D is gap graded; there is a gap in the grading between coarse silt and medium sand. It possibly was originally a deposit with layers of fine to medium silt and coarse sand.

Table 5.3 Results of particle size tests – Example 5.1

BS sieve	Size from sedimentation (mm)	% smaller			
		Soil A	Soil B	Soil C	Soil D
63 mm			100		
20 mm			75		
6.3 mm		100	66		100
2 mm		96	60		74
600 μm		86	55		48
212 μm		10	45	100	45
63 μm		2	34	95	42
	0.020		22	84	40
	0.006		15	68	15
	0.002		8	42	4

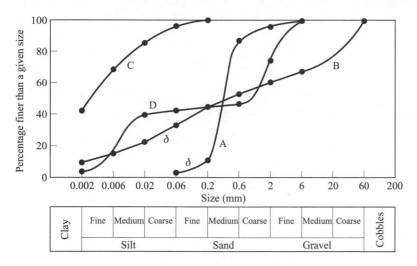

Figure 5.6 Grading curves for samples – Example 5.1.

Example 5.2: Calculations of the state of a soil sample A sample of saturated soil is 38 mm in diameter and 78 mm long and its mass is 142 g. After oven drying at 105°C its mass is 86 g.

Water content $w = \dfrac{W - W_d}{W_d} = \dfrac{142 - 86}{86} = 0.651 = 65.1\%$

Weight of saturated soil $W = 142 \times 9.81 \times 10^{-6}\,\text{kN}$

Volume of cylinder $V = \dfrac{\pi}{4} \times 38^2 \times 78 \times 10^{-9}\,\text{m}^3$

Unit weight $\gamma = \dfrac{W}{V} = 15.75\,\text{kN/m}^3$

From Eqs. (5.6) and (5.7)

$$\frac{\gamma}{\gamma_w} = \left(1 - \frac{1}{v}\right)\left(\frac{1}{w} + 1\right)$$

$$\frac{15.75}{9.81} = \left(1 - \frac{1}{v}\right)\left(\frac{1}{0.651} + 1\right)$$

$$v = 2.72$$

From Eq. (5.6),

$$G_s = \frac{v - 1}{w} = \frac{1.72}{0.651} = 2.65$$

Example 5.3: Atterberg limits and soil mineralogy   The Atterberg limits of a soil are $w_L = 70$ and $w_P = 35$ and it contains 80% by weight of clay. The water content of a sample is 45%.

Plasticity index $I_P = w_L - w_P = 70 - 35 = 35$

Liquidity index $I_L = \dfrac{w - w_P}{I_P} = \dfrac{45 - 35}{35} = 0.29$

Activity $A = \dfrac{I_P}{\% \text{ clay}} = \dfrac{35}{80} = 0.44$

The clay is likely to be predominantly kaolinite.

Example 5.4: Calculation of the state of a soil   A 1.5 kg sample of dry sand is poured into a Eureka can (see Fig. 5.7) and displaces 560 cm^3 of water. The volume of the soil grains is equal to the volume of the water displaced from the can and so

specific gravity $G_s = \dfrac{\text{weight of soil grains}}{\text{volume of soil grains}} = \dfrac{1.5 \times 10^3}{560} = 2.68$

A second 1.5 kg sample of the same dry sand is poured into an empty measuring cylinder 55 mm in diameter and occupies 950 cm^3 (see Fig. 5.8(a)). Therefore,

specific volume $v = \dfrac{\text{volume of soil}}{\text{volume of grains}} = \dfrac{950}{560} = 1.70$

unit weight of dry soil $\gamma_d = \dfrac{\text{weight of dry soil}}{\text{volume}} = \dfrac{1.5 \times 9.81 \times 10^{-3}}{950 \times 10^{-6}} = 15.5 \text{ kN/m}^3$

depth of dry sand $= \dfrac{\text{volume}}{\text{area}} = \dfrac{950 \times 10^{-6}}{\pi/4 \times 55^2 \times 10^{-6}} = 0.40 \text{ m}$

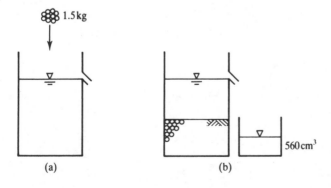

Figure 5.7 Eureka can experiment.

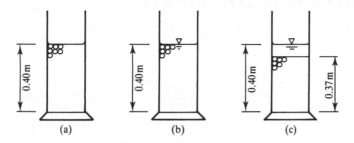

Figure 5.8 Sand in a measuring cylinder – Example 5.4.

When the cylinder is carefully filled with water up to the top level of the sand (see Fig. 5.8(b)),

$$\text{unit weight } \gamma = \left(\frac{G_s + v - 1}{v} \right) \gamma_w = \left(\frac{2.68 + 1.70 - 1}{1.70} \right) 9.81 = 19.5 \text{ kN/m}^3$$

The side of the measuring cylinder is tapped several times, causing the level of the sand to settle to a volume of 870 cm³. At the new denser state (see Fig. 5.8(c)),

$$\text{specific volume } v = \frac{\text{volume of soil}}{\text{volume of grains}} = \frac{870}{560} = 1.55$$

$$\text{unit weight } \gamma = \left(\frac{G_s + v - 1}{v} \right) \gamma_w = \left(\frac{2.68 + 1.55 - 1}{1.55} \right) 9.81 = 20.4 \text{ kN/m}^3$$

$$\text{depth of soil } z = \frac{\text{volume}}{\text{area}} = \frac{870 \times 10^{-6}}{\pi/4 \times 55^2 \times 10^{-6}} = 0.37 \text{ m}$$

References

BS 5930 (1999) *Code of Practice for Site Investigations,* British Standards Institution, London.

BS 1377 (1990) *Methods of Test for Soils for Civil Engineering Purposes,* British Standards Institution, London.

Kolbuszewski, J. J. (1948) 'An experimental study of the maximum and minimum porosities of sands', *Proceedings of 2nd International SMFE Conference, Rotterdam,* Vol. 1.

Further reading

Atkinson, J. H. and P. L. Bransby (1978) *The Mechanics of Soils,* McGraw-Hill, London.

Clayton, C. R. I., N. E. Simons and M. C. Matthews (1995) *Site Investigation,* Blackwell, London.

Head, K. H. (1980) *Manual of Soil Laboratory Testing,* Vol. 1, *Soil Classification and Compaction Tests,* Pentech Press, London.

Mitchell, J. K. and K. Soga (2005) *Fundamentals of Soil Behaviour,* 3rd Edition, Wiley, New York.

Chapter 6

Pore pressure, effective stress and drainage

6.1 Introduction

Soils consist of solid grains and water, and loads on foundations or on walls will arise from combinations of the stresses in the skeleton of soil grains and in the pore water. If there is no soil the normal stress on the hull of a ship is equal to the water pressure. If there is no water the stress on the bottom of a sugar basin arises from the weight of the dry sugar. The question then arises as to what combinations of the stresses in the skeleton of the grains and in the pore water determine the overall soil behaviour. To examine this we will look at the stresses and water pressures in the ground.

6.2 Stress in the ground

In the ground the vertical stress at a particular depth is due to the weight of everything above – soil grains, water, foundations – and so stresses generally increase with depth. In Fig. 6.1(a) the vertical stress σ_z is

$$\sigma_z = \gamma z \tag{6.1}$$

where γ is the unit weight of the soil (see Sec. 5.5). If the ground is below water level, in the bed of a lake or a sea, as in Fig. 6.1(b),

$$\sigma_z = \gamma z + \gamma_w z_w \tag{6.2}$$

and if there is a surcharge load q at the surface from a foundation or an embankment, as in Fig. 6.1(c),

$$\sigma_z = \gamma z + q \tag{6.3}$$

Remember that γ is the weight of everything (soil grains and water) in unit volume. Because σ_z arises from the total weight of the soil it is known as a total stress. Notice that the water in the lake in Fig. 6.1(b) applies a total stress at the ground surface in the same way that water in a glass applies total stresses to the bottom of the glass. The unit weight of soil does not vary very much and, typically, $\gamma \approx 20\,\text{kN/m}^3$ for saturated soil, $\gamma \approx 16\,\text{kN/m}^3$ for dry soil and for water $\gamma_w \approx 10\,\text{kN/m}^3$.

There are also total horizontal stresses σ_h, but there are no simple relationships between σ_z and σ_h. We will examine horizontal stresses in later chapters.

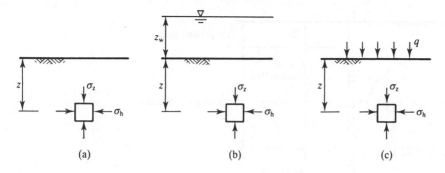

Figure 6.1 Total stresses in the ground.

6.3 Groundwater and pore pressure

The water in the pores of saturated soil has a pressure known as the pore pressure u. This is conveniently represented by the height of water h_w in a standpipe, as shown in Fig. 6.2. When everything is in equilibrium the pressures of water just inside and just outside the pipe are equal and so

$$u = \gamma_w h_w \qquad (6.4)$$

When the level of water in the pipe is below ground, as in Fig. 6.2(a), it is known as the water table or the phreatic surface. If the water in the soil is stationary the water table is horizontal like the surface of a lake. However, as we will see later, if the water table is not level there will be seepage as the groundwater moves through the pores of the soil. From Fig. 6.2(a) pore pressures at the water table are zero and positive below and a question is: what is the pore pressure above the water table?

  Figure 6.3 illustrates the variation of pore pressure in the region between the ground level and the water table. There may be a layer of dry soil at the surface where pore pressures are zero. This is actually relatively rare but can be found on beaches above the

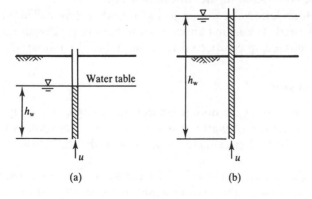

Figure 6.2 Pore water pressures in the ground.

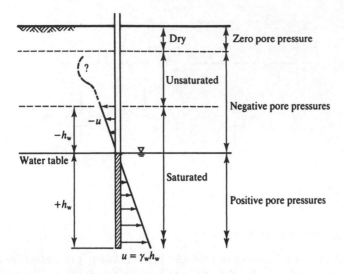

Figure 6.3 Pore pressures and suctions in the ground.

high-tide mark. Immediately above the water table the soil remains saturated because of capillary rise in the pore spaces. In this zone the pore pressures are negative and are given by

$$u = -\gamma_w h_w \qquad (6.5)$$

Between the dry and saturated zones there is a zone of unsaturated soil which contains soil grains, water and gas, usually air or water vapour. In this soil the pore water and gas exist at different pressures and the pore water pressures may increase or decrease as indicated in Fig. 6.3. There is, at present, no simple and satisfactory theory for unsaturated soils and in this book I will deal mostly with saturated or dry soils. Unsaturated soils will be considered further in Chapter 26.

In practice soils in slopes, foundations, retaining walls and other major civil engineering works are usually saturated, at least in temperate or wet climates. Unsaturated soils may occur very near the surface, in compacted soils and in hot dry climates.

6.4   Suctions in saturated soil

Saturated soils may very well have negative pore pressures or suctions. This implies that the water is in tension and the rise of water in soil above the water table is like the rise of water in a capillary tube with a diameter equivalent to the size of the pore spaces in the soil.

Figure 6.4(a) shows water rising to a height $-h_w$ in a capillary tube with diameter d_t. (Note that h_w is negative because the head of water in a standpipe is positive downwards in Fig. 6.3.) The suction just inside the meniscus is related to the tube

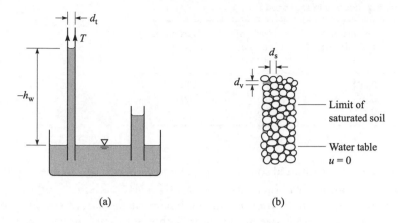

Figure 6.4 Suction in saturated soil.

diameter d_t. Resolving vertically on the column of water in the tube in Fig. 6.4(a)

$$T\pi d_t = -\frac{\pi d_t^2}{4}\gamma_w h_w \tag{6.6}$$

where T is the surface tension force between water and glass. Hence the height of water in the capillary tube is

$$-h_w = \frac{4T}{\gamma_w d_t} \tag{6.7}$$

and, since $u_w = \gamma_w h_w$ the pore water suction at the top of the water column is

$$-u_w = \frac{4T}{d_t} \tag{6.8}$$

(Since the pore pressure in Eq. (6.8) is negative the pore water is in suction.) In the soil in Fig. 6.4(b) the grain diameter is d_s, the specific volume is v and the mean pore space diameter d_v is

$$d_v = (v - 1)d_s \tag{6.9}$$

Hence, the height of saturated soil above the water table is

$$-h_w = \frac{4T}{\gamma_w(v - 1)d_s} \tag{6.10}$$

and the maximum pore water suction is

$$-u_w = \frac{4T}{(v - 1)d_s} \tag{6.11}$$

Table 6.1 Limiting suctions in saturated soil

Soil	Characteristic grain size (mm)	Height of saturated zone (m)	Suction (kPa)
Clay	0.001	60	600
Medium silt	0.01	6	60
Fine sand	0.1	0.6	6

From Eq. (6.11) suctions in saturated soil vary with the inverse of the grain size. Taking a value for T for water and quartz (i.e. glass) of about 7×10^{-5} kNm^{-1} and for soil with a specific volume of 1.5 the variations of the height of the saturated zone and the pore water suction with grain size are given in Table 6.1. This shows that even saturated soils can develop considerable suctions.

6.5   Effective stress

It is obvious that ground movements and instabilities can be caused by changes of total stress due to loading of foundations or excavation of slopes. What is perhaps not so obvious is that ground movements and instabilities can be caused by changes of pore pressure. For example, stable slopes can fail after rainstorms because the pore pressures rise due to infiltration of rainwater into the slope while lowering of groundwater due to water extraction causes ground settlements. (Some people will tell you that landslides occur after rainfall because water lubricates soil; if they do, ask them to explain why damp sand in a sandcastle is stronger than dry sand.)

If soil compression and strength can be changed by changes of total stress or by changes of pore pressure there is a possibility that soil behaviour is governed by some combination of σ and u. This combination should be called the *effective* stress because it is effective in determining soil behaviour.

The relationship between total stress, effective stress and pore pressure was first discovered by Terzaghi (1936). He defined the effective stress in this way:

> All measurable effects of a change of stress, such as compression, distortion and a change of shearing resistance, are due exclusively to changes of effective stress. The effective stress σ' is related to the total stress and pore pressure by $\sigma' = \sigma - u$.

Figure 6.5 shows Mohr circles of total stress and effective stress plotted on the same axes. Since $\sigma'_1 = \sigma_1 - u$ and $\sigma'_3 = \sigma_3 - u$ the diameters of the circles are the same. The points T and E represent the total and effective stresses on the same plane and clearly total and effective shear stresses are equal. Therefore, effective stresses are

$$\sigma' = \sigma - u \tag{6.12}$$

$$\tau' = \tau \tag{6.13}$$

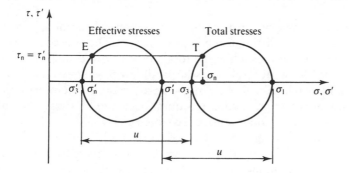

Figure 6.5 Mohr circles of total and effective stress.

From the definitions of the shear stress parameter q and the mean stress parameter p given in Chapter 2 and substituting $\sigma'_1 = \sigma_1 - u$, etc., it is easy to show that

$$p' = p - u \tag{6.14}$$

$$q' = q \tag{6.15}$$

From Eqs. (6.13) and (6.15) total and effective shear stresses are identical and most authors use shear stresses without primes all the time. In my work and teaching, and in this book, I use τ' and q' when I am considering analyses in terms of effective stress and τ and q for total stresses. I know that this is strictly unnecessary but I find that the distinction between total and effective shear stresses is helpful, particularly for teaching.

Notice that a sample of saturated soil sitting on the laboratory bench with zero total stress applied to it will have a negative pore pressure. From Eqs. (6.12) or (6.14) it will have a positive effective stress numerically equal to the negative pore pressure. We will see later that this effective stress, which arises purely from pore water suctions, accounts for the unconfined compressive strength of saturated soil. From Table 6.1 the suctions in fine grained soils are much larger than the suctions in coarse grained soils and this is why the unconfined compressive strength of a sandcastle is much smaller than the unconfined compressive strength of a cylinder of clay.

6.6   Importance of effective stress

The principle of effective stress is absolutely fundamental to soil mechanics and its importance cannot be overstated. This is the way in which soil behaviour due to loading is related to behaviour due to changes of groundwater pressure.

Although most texts on soil mechanics examine the validity of the principle and the meaning of effective stress by considering the interparticle forces and the intergranular contact areas, there really is no need to do this and the necessary assumptions are not supported by experimental evidence. No conclusive evidence has yet been found that invalidates Terzaghi's original postulate, at least for saturated soils at normal levels of engineering stress, and the principle of effective stress is accepted as a basic axiom of soil mechanics.

Because total and effective normal stresses are different (except when pore pressures are zero) it is absolutely essential to distinguish between the two. The effective stresses σ' and τ' are always denoted by primes while the total stresses σ and τ do not have primes. Any equation should have all total stresses, or all effective stresses, or total and effective stresses should be related correctly by the pore pressure. Engineers doing design calculations (or students doing examination questions) should always be able to say whether they are dealing with total or effective stresses.

From Figs. 6.1 and 6.2, and making use of Eqs. (6.1) to (6.4), you can calculate the vertical effective stress σ'_z at any depth in the ground for any position of the groundwater. If you try some examples you will discover that if the water table is below the ground level the effective stress depends on the position of the water table. If, on the other hand, the ground level is submerged, as in the bed of a river, lake or sea, the effective stress is independent of the depth of water; this means that the effective stresses in soil in the bed of a duck pond will be the same as those in the bed of the deep ocean where the water depth may exceed 5 km. In doing these calculations remember that free water which can slosh around (i.e. in a river, lake or sea) will apply a total stress to the soil (and to dams and submarines), but water in the pores of the soil has a pore pressure; these water pressures need not always be equal.

Submarines and fish illustrate effective stresses. Sea water applies total stresses to the skin of both. In a submarine the internal (pore) pressure is zero (atmospheric) so the skin of the submarine must be very strong, but in a fish the pressures in the blood and in the soft tissues are very nearly equal to the external water pressure so the skin and skeleton of the fish can be very weak and soft. In both cases the stresses on the skins are equivalent to effective stresses in soils.

6.7   Demonstrations of effective stress

The effective stress Eq. (6.12) can be written in terms of changes Δ so that

$$\Delta\sigma' = \Delta\sigma - \Delta u \tag{6.16}$$

This shows that effective stresses may be changed – causing measurable effects – by changing either the total stress with the pore pressure constant or by changing the pore pressure with the total stress constant. Note also that if the total stress and the pore pressure are changed equally the effective stress remains constant and the soil state does not change.

Figure 6.6(a) illustrates settlements $\Delta\rho$ caused by loading a foundation by $\Delta\sigma$ while the pore pressures in the ground remain constant so that $\Delta\sigma' = \Delta\sigma$. Figure 6.6(b) illustrates settlements $\Delta\rho$ caused by extraction of groundwater. Pumping lowers the water table by Δh_w so that pore pressures reduce by $\Delta u = \gamma_w \Delta h_w$. From Eq. (6.16), with $\Delta\sigma = 0$, the reduction of pore pressure causes an increase of effective stress $\Delta\sigma'$. The principle of effective stress states that if the change of foundation loading $\Delta\sigma$ is the same as the change of pore pressure Δu due to lowering of groundwater the settlements will be the same. In other words, it is simply the change of effective stress that affects the soil behaviour.

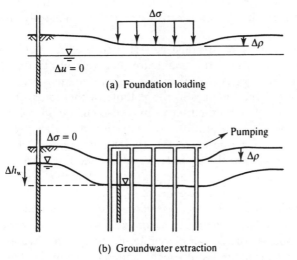

(a) Foundation loading

(b) Groundwater extraction

Figure 6.6 Settlements due to changing effective stresses.

A simple experiment which demonstrates the action of effective stresses is illustrated in Fig. 6.7. This shows the influence of pore pressure on the capacity of deep and shallow foundations. The soil should be fine to medium sand; if it is too coarse it will become unsaturated when the water table is lowered and if it is too fine pore pressures may not equalize in a reasonable time. Place the gravel and sand in water to ensure they are saturated and then open the valve to lower the water table to the gravel. Place a heavy foundation (a steel cylinder about 40 mm in diameter and 80 mm long works very well) and an eccentrically loaded pile as illustrated. Close the valve and raise the water table by pouring water into the standpipe; if the sand and gravel remained saturated it will only be necessary to fill the standpipe. As the water table and the pore pressures rise, effective stresses will fall and both foundations will fail.

If you then open the valve and drain the gravel pore pressures in the sand decrease and the sand will regain its strength.

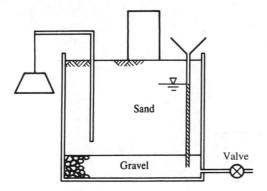

Figure 6.7 Rising groundwater experiment.

6.8   Volume change and drainage

As soil is loaded or unloaded due to changes of effective stress it will generally change in volume.

In silts and sands the grains themselves are relatively stiff so they do not change volume. In soils such as shelly sands where the grains are relatively weak, or where stresses are very large, the grains may break. In either case volume changes are largely due to rearrangement of the grains and changes in the volume of the voids. In clays which have high plasticity the clay particles themselves may also change in volume.

At small effective stress the spacing of the grains may be loose and at high stresses it will be dense, as shown in Fig. 6.8. If the pore pressure u_0 remains constant then the changes of total and effective stresses are the same ($\Delta\sigma' = \Delta\sigma$; see Eq. 6.16). If the volume of the soil grains remains constant then, in Fig. 6.8, the change of volume of the soil ΔV is the same as the volume of water expelled ΔV_w.

In saturated soil changes in volume must be due to seepage of water through the soil and so soil compression is rather like squeezing water from a sponge. In a laboratory, test water will seep to the boundaries of the sample while, in the ground, water will seep to the surface or to natural drainage layers in the soil. For example, Fig. 6.9 illustrates an embankment built on a bed of clay sandwiched between layers of sand which act as drains. As the embankment is constructed water will seep from the clay to the sand layers as indicated.

There must, of course, be sufficient time for the water to seep through the soil to permit the volume change to occur; otherwise the pore pressure will change. As a result

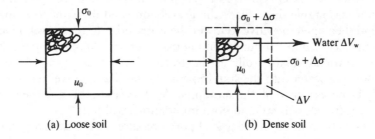

Figure 6.8 Volume changes in soil.

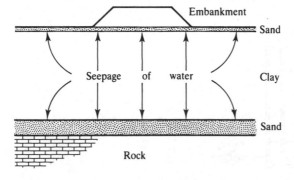

Figure 6.9 Drainage of clay beneath an embankment.

there must be some relationship between the rate at which the loads are applied, the rate of drainage and the behaviour of the soil and pore pressure.

6.9 Drained loading, undrained loading and consolidation

The relative rates at which total stresses are applied and at which the seepage takes place are of critical importance in determining soil behaviour. The limiting conditions are illustrated in Figs. 6.10 and 6.11.

Figure 6.10(a) illustrates an increment of total stress $\Delta\sigma$ applied slowly, over a long period of time. This could represent loading in a laboratory test or in the ground. If the loading is applied very slowly water will be able to seep from the soil as the total stresses increase. There will be no change of pore pressure, as shown in Fig. 6.10(c), and the volume changes will follow the change of loading, as shown in Fig. 6.10(b).

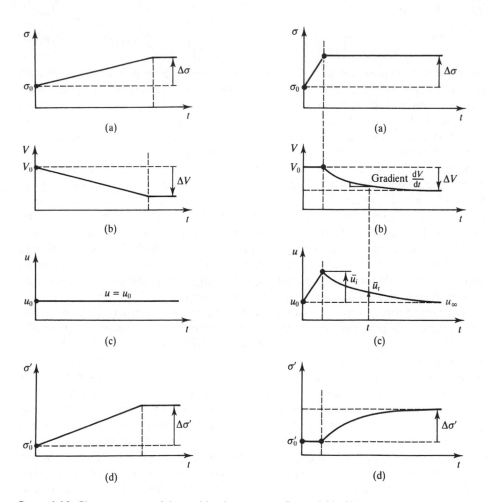

Figure 6.10 Characteristics of drained loading.

Figure 6.11 Characteristics of undrained loading and consolidation.

Because the pore pressures remain constant at u_0, the changes of effective stress follow the change of total stress, as shown in Fig. 6.10(d). When the stresses remain constant at $\sigma_0' + \Delta\sigma'$, the volume remains constant at $V_0 - \Delta V$. This kind of relatively slow loading is called drained because all the drainage of water takes place during the loading. The most important feature of drained loading is that the pore pressures remain constant at u_0, which is known as the steady state pore pressure.

Figure 6.11(a) illustrates the same increment of total stress $\Delta\sigma$ as in Fig. 6.10 but now applied so quickly that there was no time for any drainage at all and so the volume remains constant, as shown in Fig. 6.11(b). If the loading was isotropic with no shear distortion and undrained with no volume change then nothing has happened to the soil. From the principle of effective stress this means that the effective stress must remain constant, as shown in Fig. 6.11(d), and, from Eq. (6.16), the change in pore pressure is given by

$$\Delta\sigma' = \Delta\sigma - \Delta u = 0 \tag{6.17}$$

$$\Delta u = \Delta\sigma \tag{6.18}$$

This increase in pore pressure gives rises to an initial excess pore pressure $\overline{u}_i$, as shown in Fig. 6.11(c). Notice that the pore pressure u consists of the sum of the steady state pore pressure u_0 and the excess pore pressure $\overline{u}$; if the pore pressures are in equilibrium $u = u_0$ and $\overline{u} = 0$. Relatively quick loading is known as 'undrained loading' because there is no drainage of water during the loading. The most important feature of undrained loading is that there is no change of volume.

At the end of the undrained loading the pore pressure is $u = u_0 + \overline{u}_i$, where u_0 is the initial steady state, or equilibrium, pore pressure and $\overline{u}_i$ is an initial excess pore pressure. This excess pore pressure will cause seepage to occur and, as time passes, there will be volume changes as shown in Fig. 6.11(b). The volume changes must be associated with changes of effective stress, as shown in Fig. 6.11(d), and these occur as a result of decreasing pore pressures, as shown in Fig. 6.11(c). The pore pressures decay towards the long term steady state pore pressure u_∞. Fig. 6.11(c) shows $u_\infty = u_0$ but there are cases in which construction, especially of excavations changes the steady state groundwater and u_∞ can be greater or smaller than u_0.

At some time t the excess pore pressure is $\overline{u}_t$ and this is what drives the drainage and so, as the excess pore pressure decreases, the rate of volume change, given by the gradient dV/dt, also decreases, as shown in Fig. 6.11(b). Notice that while there are excess pore pressures in the soil, water pressures outside the surface of the soil will not be the same as the pore pressures; this means that the pore pressure in soil behind a new quay wall need not be the same as the pressure in the water in the dock.

This dissipation of excess pore pressure accompanied by drainage and volume changes is known as consolidation. The essential feature of consolidation is that there are excess pore pressures $\overline{u}$ that change with time. Usually, but not always, the total stresses remain constant. Consolidation is simply compression (i.e. change of volume due to change of effective stress) coupled with seepage. At the end of consolidation, when $\overline{u}_\infty = 0$, the total and effective stresses and the volume are all the same as those at the end of the drained loading shown in Fig. 6.10. Thus, the changes of effective stress for undrained loading plus consolidation are the same as those for drained loading.

The processes of undrained loading followed by consolidation can be represented by an experiment with a packet of crisps. Put the packet on the table and put a mass of a few kilos on it. The packet will inflate as the pressure in it increases. This represents undrained loading. Puncture the packet with a pin. The air will escape and as the pressure in the packet reduces the mass settles and you can hear crisps breaking as load is transferred from the air pressure to the crisps. This represents consolidation. The analogy is not exact but the experiment nicely illustrates the processes. If the crisp packet is punctured before the mass is applied the air escapes immediately and the load is taken by the crisps. This represents drained loading.

In the simple examples of drained and undrained loading illustrated in Figs. 6.10 and 6.11, the increment of loading was positive so that the soil compressed as water was squeezed out. Exactly the same principles apply to unloading where the increment is negative and the soil swells as water is sucked in by the negative excess pore pressure. You should sketch diagrams like Figs. 6.10 and 6.11 for an increment of unloading.

Remember the final steady state pore pressure at the end of consolidation u_∞ need not be same as the initial steady state pore pressure u_0 before the undrained loading. The excess pore pressure which causes consolidation is the difference between the current pore pressure and the final steady state pore pressure so after a long time $\bar{u}_\infty$ is always zero. Sometimes only the external water levels are changed as, for example, when a dam is filled or emptied. In this case there will be consolidation in the soil as the pore pressures adjust to the new external water levels.

Consolidation is any process in which effective stresses change as excess pore pressures dissipate towards their long term steady state values. If excess pore pressures are positive, effective stresses increase with consolidation and the soil compresses. On the other hand, if excess pore pressures are negative, effective stresses decrease with consolidation and the soil swells.

6.10   Rates of loading and drainage

When distinguishing between drained and undrained loading it is relative rates of loading and seepage that are important, not the absolute rate of loading. Seepage of water through soil, which will be covered in more detail in Chapter 14, is governed by the coefficient of permeability k. Figure 6.12 illustrates seepage with velocity V

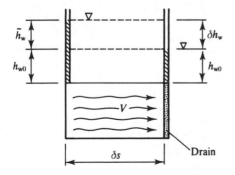

Figure 6.12 Seepage of water through soil.

through an element of soil δs long. At one end there is a drain where the pore pressure is $u_0 = \gamma_w h_{w0}$ and at the other end there is an excess pore pressure given by $\bar{u} = \gamma_w \bar{h}_w$. The difference in the levels of water in the standpipes is $\delta h_w = \bar{h}_w$ and the hydraulic gradient is given by

$$i = \frac{\delta h_w}{\delta s} \tag{6.19}$$

(Hydraulic gradient should really be defined in terms of the hydraulic potential P instead of the head h_w, but if the flow is horizontal these are the same; potential is introduced in Sec. 14.3.) The basic rule for seepage is Darcy's law, given by

$$V = ki \tag{6.20}$$

where the coefficient of permeability k has the units of velocity. The value of k is the seepage velocity of water through soil with unit hydraulic gradient.

Values for the coefficient of permeability for soils depend largely on the grain size (or more particularly on the size of void spaces through which the seepage takes place). Typical values for k for different grain sizes are given in Table 6.2 (For some natural clay soils the value of k may be considerably less than 10^{-8} m/s.). Notice the very large range (more than $\times 10^6$) of permeability for typical soils. Under a unit hydraulic gradient, water will travel 1 m through gravel in less than $10^2 = 100$ s and 1 m through clay in more than 10^8 s, which is about 3 years.

In civil engineering and related activities loads are applied to the ground at different rates and some typical examples are given in Table 6.3. Again, notice the very

Table 6.2 Values of coefficient of permeability of soils

Grain size	k (m/s)
Gravel	$>10^{-2}$
Sand	10^{-2}–10^{-5}
Silt	10^{-5}–10^{-8}
Clay	$<10^{-8}$

Table 6.3 Durations of typical engineering constructions

Event	Duration
Shock (earthquake, pile driving)	<1 s
Ocean wave	10 s
Excavate trench	10^4 s ≈ 3 h
Load small foundation	10^6 s ≈ 10 days
Large excavation	10^7 s ≈ 3 months
Embankment dam	10^8 s ≈ 3 years
Natural erosion	10^9 s ≈ 30 years

large range (more than $\times\ 10^9$) in the durations, or rates, of loading or unloading in these examples.

In any geotechnical calculation or analysis it is absolutely essential to state whether the calculation is for drained or undrained loading, and we will discover that different analyses are required for each in later chapters. What is important is the relative rates of loading and drainage – is there enough time during the loading to allow drainage to occur or is the loading so fast that there will be no drainage? Of course, in reality, neither condition will be satisfied absolutely and decisions must be made as to whether the construction is more nearly drained or undrained.

Many engineers will assume that loading or unloading of a coarse-grained soil will be drained and of a fine-grained soil will be undrained. These assumptions are adequate for loading rates which are not at the extremes of those in Table 6.3. Very rapid loading of coarse-grained soil is likely to be undrained. Earthquakes, pile driving and ocean waves may generate excess pore pressures in sands which can cause liquefaction failures and which explain the change of pile capacity after a delay in driving. Very slow loading of clay slopes due to natural erosion is likely to be drained and pore pressures and slope angles of many natural clay slopes correspond closely to the fully drained, steady state conditions.

6.11   Summary

1. In soils total stresses arise from the weight of the soil (including the soil grains and the pore water) and any other external loads from foundations, walls and free water. There are also pore pressures in the water in the voids.
2. The stresses that govern soil behaviour are effective stresses given by $\tau' = \tau$ and $\sigma' = \sigma - u$. As a result soils are affected equally by changes in total stress and pore pressure.
3. Pore pressures may be either positive below the water table or negative (suction) above it. Where pore pressures are negative effective stresses are larger than total stresses. The maximum suctions which can occur in saturated soil depend on the grain size of the soil.
4. Volume changes in saturated soil can only occur as water seeps through the pores and the rate of seepage is governed by the coefficient of permeability k. If soil is loaded slowly, compared with the rate of drainage, the pore pressures remain constant and volume changes occur during the loading, which is called drained.
5. If soil is loaded quickly, compared with the rate of drainage, the volume remains constant, excess pore pressures arise and the loading is called undrained. Subsequently, consolidation occurs as the excess pore pressures dissipate and water seeps from the soil, causing volume changes.

Worked examples

Example 6.1: Calculation of vertical stress   For the measuring cylinder of sand described in Example 5.4 (see Fig. 5.8), the total vertical stress, the pore pressure and the effective vertical stress at the base of the cylinder are:

(a)   When the sand is loose and dry:

$$z = 0.40 \text{ m}$$

$$\gamma_d = 15.5 \text{ kN/m}^3$$

$$u = 0$$

$$\sigma_z = \gamma_d z = 15.5 \times 0.40 = 6.2 \text{ kPa}$$

(b)   When the sand is loose and saturated:

$$z = 0.40 \text{ m}$$

$$\gamma = 19.5 \text{ kN/m}^3$$

$$\sigma_z = \gamma z = 19.5 \times 0.4 = 7.8 \text{ kPa}$$

$$u = \gamma_w h_w = 9.81 \times 0.4 = 3.9 \text{ kPa}$$

$$\sigma_z' = \sigma_z - u = 7.8 - 3.9 = 3.9 \text{ kPa}$$

(c)   When the sand is dense and saturated:

$$z = 0.37 \text{ m}$$

$$\gamma = 20.4 \text{ kN/m}^3$$

$$z_w = 0.03 \text{ m}$$

$$\sigma_z = \gamma z + \gamma_z z_w = (20.4 \times 0.37) + (9.81 \times 0.03) = 7.8 \text{ kPa}$$

$$u = \gamma_w h_w = 9.81 \times 0.40 = 3.9 \text{ kPa}$$

$$\sigma_z' = \sigma - u = 7.8 - 3.9 = 3.9 \text{ kPa}$$

Notice that densification of the soil by tapping the side of the cylinder did not change the total or effective stresses at the base of the cylinder. This is simply because the total weights of soil and water in the cylinder did not change.

Example 6.2: Calculation of stress in the ground   The deep clay deposit in Fig. 6.13 has unit weight $\gamma = 20 \text{ kN/m}^3$ and the soil remains saturated even if the pore pressures become negative. For the groundwater conditions, (a) water table 6 m below ground level and (b) with water to a depth of 3 m above ground level, the vertical effective stresses at a depth of 3 m are:

(a)   Water table at 6 m below ground level:

$$\sigma_z = \gamma z = 20 \times 3 = 60 \text{ kPa}$$

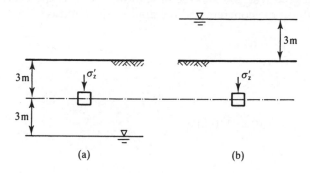

Figure 6.13 Stress in the ground – Example 6.2.

$$u = \gamma_w h_w = 10 \times -3 = -30 \text{ kPa}$$

$$\sigma'_z = \sigma_z - u = 60 + 30 = 90 \text{ kPa}$$

(b)   Water surface 3 m above ground level:

$$\sigma_z = \gamma z + \gamma_w z_w = (20 \times 3) + (10 \times 3) = 90 \text{ kPa}$$

$$u = \gamma_w h_w = 10 \times 6 = 60 \text{ kPa}$$

$$\sigma'_z = \sigma_z - u = 90 - 60 = 30 \text{ kPa}$$

Example 6.3: Calculation of stress in the ground below a foundation   The concrete bridge pier in Fig. 6.14 is 4 m tall, it has an area of 10 m² and carries a load of 1 MN. (The unit weight of concrete is $\gamma_c = 20$ kN/m³.) The pier is founded on the bed of a tidal river where there is at least 5 m of sand with a unit weight of 20 kN/m³. The river bed is at low tide level and at high tide there is 3 m depth of water.

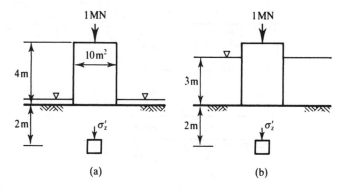

Figure 6.14 Stress in the ground – Example 6.3.

The total contact stress q between the soil and the base of the pier (i.e. the bearing pressure) arises from the weight of the concrete and the applied load and is

$$q = \gamma_c H_c + \frac{F}{A} = (4 \times 20) + \frac{1 \times 10^3}{10} = 180 \text{ kPa}$$

(a)   At low tide:

$$\sigma_z = \gamma z + q = (20 \times 2) + 180 = 220 \text{ kPa}$$

$$u = \gamma_w h_w = 2 \times 10 = 20 \text{ kPa}$$

$$\sigma_z' = \sigma_z - u = 220 - 20 = 200 \text{ kPa}$$

(b)   At high tide (note that q is reduced by uplift from the water pressure below the foundation):

$$\sigma_z = \gamma z + \gamma_w z_w + (q - \gamma_w z_w)$$

$$= (20 \times 2) + (10 \times 3) + (180 - (10 \times 3)) = 220 \text{ kPa}$$

$$u = \gamma_w h_w = 10 \times 5 = 50 \text{ kPa}$$

$$\sigma_z' = \sigma_z - u = 220 - 50 = 170 \text{ kPa}$$

Notice that in this case the increase of the water depth has reduced the effective stress in the ground; this is because of a reduction of the bearing pressure due to uplift.

Example 6.4: Calculation of stress below an embankment   The soil profile in Fig. 6.15 consists of 4 m clay over 2 m sand over rock: the unit weights of all the natural materials are 20 kN/m³and the steady state water table is at ground level. A wide embankment 4 m high is constructed from fill with a unit weight of 15 kN/m³. The total and effective vertical stresses at the centre of the clay and at the centre of the sand (a) before

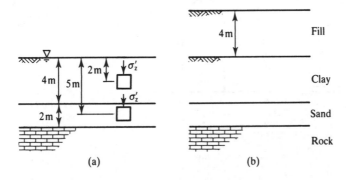

Figure 6.15 Stress below an embankment – Example 6.4.

the embankment is constructed, (b) immediately after it is completed and (c) after a very long time are:

(a) Before construction of the embankment:

- in the clay:

$$\sigma_z = \gamma z = 20 \times 2 = 40 \text{ kPa}$$

$$u = \gamma_w h_w = 10 \times 2 = 20 \text{ kPa}$$

$$\sigma_z' = \sigma_z - u = 40 - 20 = 20 \text{ kPa}$$

- in the sand:

$$\sigma_z = \gamma z = 20 \times 5 = 100 \text{ kPa}$$

$$u = \gamma_w h_w = 10 \times 5 = 50 \text{ kPa}$$

$$\sigma_z' = \sigma_z - u = 100 - 50 = 50 \text{ kPa}$$

(b) Immediately after construction of the embankment the sand is drained so the pore pressure remains constant. The embankment is wide so there are no horizontal strains, the clay is undrained and the effective stresses remain unchanged:

- in the clay:

$$\sigma_z = \sum \gamma z = (4 \times 15) + (2 \times 20) = 100 \text{ kPa}$$

$$\sigma_z' = 20 \text{ kPa, as in (a)}$$

$$u = \sigma_z - \sigma_z' = 100 - 20 = 80 \text{ kPa}$$

- in the sand:

$$\sigma_z = \sum \gamma z = (4 \times 15) + (20 \times 5) = 160 \text{ kPa}$$

$$u = 50 \text{ kPa}$$

$$\sigma_z' = \sigma_z - u = 160 - 50 = 110 \text{ kPa}$$

(c) After a very long time the excess pore pressures in the clay will have dissipated to the steady state conditions corresponding to the water table at original ground level:

- in the clay:

$$\sigma_z = 100 \text{ kPa, as in (b)}$$

$$u = 20 \text{ kPa, as in (a)}$$

$$\sigma'_z = \sigma_z - u = 100 - 20 = 80 \text{ kPa}$$

- in the sand there has been no change of total stress or pore pressure and the stresses are the same as those in (b).

Reference

Terzaghi, K. (1936) 'The shearing resistance of saturated soil and the angle between the planes of shear', *Proceedings of 1st International SMFE Conference, Harvard, Mass.*, Vol. 1, pp. 54–56.

Chapter 7

Laboratory testing of soils

7.1 Purposes of laboratory tests

Testing soil samples in the laboratory plays an important role in soil mechanics research and civil engineering practice. Almost all we know about soil behaviour has been learned from laboratory tests. Tests may be carried out on small samples of soil to examine the characteristics of the soil or on models of soil structures to examine how slopes, walls and foundations deform and collapse. In this chapter we will consider tests on soil samples. Laboratory tests are carried out for a number of purposes, the most important being:

1. For description and classification of a particular soil (see Chapter 5).
2. To investigate the basic mechanical behaviour of soils and to develop theories for soil behaviour (see Chapters 8 to 13).
3. To determine design parameters (i.e. numerical values for strength, stiffness and permeability) for geotechnical analyses (see Chapters 18 to 25).

Laboratory tests may be carried out on samples that are intact or have been completely reconstituted. In reconstituted samples the soil has been mixed at a relatively large water content and then recompressed. In this way any structure developed in the soil in the ground due to deposition or ageing is removed and the tests measure the fundamental behaviour and properties of the soil. Intact samples are recovered from the ground with minimum disturbance (see Chapter 17); they contain the *in situ* structure and retain the properties of the soil in the ground.

Most of the analyses of geotechnical structures described in Chapters 18 to 25 and used for routine design were developed for simple soils which behave more or less like the theories described in Chapters 8 to 13. These analyses may not be applicable to soils whose behaviour differs significantly from these simple theories, in which case special methods will be required which are outside the scope of this book. An important and often neglected purpose of soil testing is to examine soil for unexpected or strange behaviour. This is best done by comparing the behaviour of intact samples with the basic theories and with the behaviour of the same soil reconstituted and recompressed to the same state.

7.2   Standard tests and specifications

Many of the routine soil tests are very carefully and precisely specified in a number of national standards and codes of practice. In the United Kingdom the standard is BS 1377:1990 *Methods of Test for Soils for Civil Engineering Purposes*, and similar standards exist in other countries and regions. You should certainly look at a copy of the standards for soil testing relevant to your region to see exactly what they cover. Most of these standards follow what might be called a cookery book method: you do this, you do that and you serve up the result in this or in that way. There are, however, difficulties with the cookery book approach for soil testing which arise from the characteristics of soil strength and stiffness.

The values obtained from a particular test will obviously depend to a greater or lesser extent on details of the equipment and procedures used and for some tests, particularly those that measure the nature and state of a soil, it is essential that the tests follow standard procedures. This is because the parameters being measured (e.g. grading and Atterberg limits) are material parameters (they depend only on the nature of the grains) and different laboratories and different workers should obtain identical results for the same soil.

While it is possible and desirable to set standards for construction of equipment and for calibration of instruments to ensure that the accuracy of any observation is acceptable (or at least known), it is not so easy to specify tests that measure soil strength and stiffness because of the many important factors that affect these parameters. Instead, engineers should determine what parameters are required for a particular analysis, determine what factors will influence these within the theories described in Chapters 8 to 13 and then devise tests that take account of these. The engineer will need to specify not only the loading path applied in the test but also, equally importantly, the loads applied to the sample before the test starts.

I am not going to describe the standard equipment and soil tests in detail. Most of the standard apparatus and routine tests are described at length in a three-volume book by Head (1980, 1982 and 1986) and in various standards and codes of practice. All engineers concerned with groundworks should carry out simple classification, consolidation, shear and triaxial tests for themselves at least once in their career; they should also carry out simple foundation, slope stability and retaining wall experiments. The emphasis of this work should be on handling equipment and soil samples, good scientific practice and analysis and interpretation of test results within simple theories. They should also play around with soils and soil-like materials at home, in their garden and at the beach.

7.3   Basic classification tests

As discussed in Chapter 5, the nature of a soil is described principally by its grading (i.e. the distribution of particle sizes) and the mineralogy of the grains. The state is described by the current water content and unit weight together with the current stresses.

(a)   Measurement of grading

The distribution of particle sizes in a soil is found by sieving and sedimentation. Soil is first passed through a set of sieves with decreasing aperture size and the weight retained on each sieve recorded. The smallest practical sieve has an aperture size of about 0.07 mm, corresponding roughly to the division between silt and sand. Silt-sized particles can be separated by sedimentation making use of Stoke's law, which relates the settling velocity of a sphere to its diameter.

A rapid estimate of grading can be made by sedimentation in a jam jar or milk bottle. Take a sample about three quarters of the height of the container, fill the container with water and shake it up. Quickly stand the jar or bottle upright and leave it for several hours. You can see and estimate the grading of gravel, sand and silt; clay will remain in suspension for a long time and any material floating on the surface is likely to be organic (i.e. peat). This is a test frequently used by gardeners.

(b) Measurement of water content and unit weight

The water content of a soil is defined as

$$w = \frac{W_w}{W_s} \tag{7.1}$$

and the unit weight γ is defined as

$$\gamma = \frac{W}{V} \tag{7.2}$$

where W_w is the weight of water evaporated by heating soil to 105°C until the weight is constant, W_s is the weight of dry soil and $W = W_w + W_s$ is the weight of a sample with volume V. These weights can be measured by simple weighing and the volume of a cylindrical or cubic sample determined by direct measurement.

(c) Measurement of Atterberg limits

For coarse-grained soils the engineering properties are governed largely by the grading and, to a lesser extent, by the shape, texture and mineralogy of the grains, but the properties of fine-grained clay soils depend largely on the type of clay. The basic behaviour of clay soils can be assessed from the Atterberg limits (i.e. liquid limit, plastic limit and plasticity index) described in Sec. 5.6. The liquid limit determines the water content at which the soil has weakened so much that it starts to flow like a liquid. The plastic limit determines the water content at which the soil has become so brittle that it crumbles.

Liquid limit tests

The two alternative liquid limit tests are illustrated in Fig. 7.1. In the Casagrande test in Fig. 7.1(a) a small slope is failed by bumping a dish on to a rubber block. In the fall cone test a small cone-shaped foundation penetrates the soil. The precise details of the geometries, weights and so on are arranged so that the soil has a strength of approximately 1.7 kPa when it is at its liquid limit. In each case the sample has a high water content and is soft enough to be moulded into the container using a knife

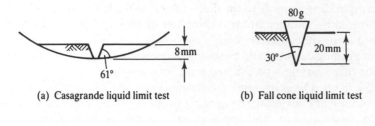

(a) Casagrande liquid limit test   (b) Fall cone liquid limit test

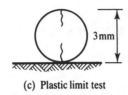

(c) Plastic limit test

Figure 7.1 The Atterberg limits tests.

or spatula. The tests are repeated with slightly different water contents until the precise requirements of the tests are met.

Plastic limit test

The test consists of rolling a 3 mm diameter thread of soil while the water evaporates and the water content decreases until the thread splits and crumbles. The failure of the thread corresponds to a strength of approximately 170 kPa. Notice that a strength of 170 kPa corresponds to the division between stiff and very stiff clay in Table 5.2. Remember the plasticity index I_P given by

$$I_P = w_L - w_P \qquad (7.3)$$

This is an important material parameter. Because the Atterberg limits determine the conditions of soil at certain well-specified strengths, the results can be used to estimate a number of other important soil properties, as discussed in Chapter 18. Further discussion of the Atterberg limits is given in Sec. 5.6.

7.4   Measurement of coefficient of permeability

Seepage of water through soil, discussed in Chapter 14, is governed by Darcy's law:

$$V = ki \qquad (7.4)$$

where k, the coefficient of permeability, is a soil parameter. The value of k depends principally on the grain size and specific volume (or more properly on the void size, which is related to the grain size and specific volume). Permeability can be measured

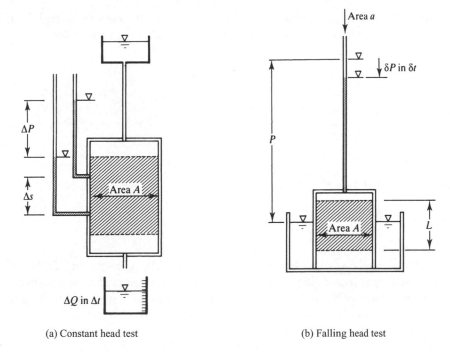

(a) Constant head test   (b) Falling head test

Figure 7.2 Permeameter tests.

in laboratory tests in a constant head permeameter, for soil with relatively large permeability, or in a falling head permeameter, for soils with relatively low permeability; these are illustrated in Fig. 7.2. In both cases water flows through a soil sample and the rates of flow and the hydraulic gradients are measured.

(a) Constant head permeability tests

In the constant head test illustrated in Fig. 7.2(a) water from a constant head tank flows through the sample in a cylinder and is collected in a measuring jar. Two standpipes measure the pore pressure and potential (see Sec. 14.3) at two points as shown. The flow is steady state and, from the observations,

$$V = \frac{\Delta Q}{A \Delta t} \tag{7.5}$$

$$i = \frac{\Delta P}{\Delta s} \tag{7.6}$$

and hence a value for k can be determined. In practice it is best to vary the rate of flow in stages and plot V against i; in this way you can verify Darcy's law and evaluate k.

(b) Falling head permeability tests

In the falling head test illustrated in Fig. 7.2(b) water flows through the sample as the level of water in the standpipe drops. Over a time interval δt the rate of flow is

$$q = -a\frac{\delta P}{\delta t} = Ak\frac{P}{L} \tag{7.7}$$

and hence, in the limit,

$$-\frac{\mathrm{d}P}{P} = \frac{Ak}{aL}\mathrm{d}t \tag{7.8}$$

Integrating with the limits $P = P_0$ at $t = 0$ we have

$$\ln\left(\frac{P_0}{P}\right) = \frac{Ak}{aL}t \tag{7.9}$$

and you can determine a value for k by plotting $\ln(P_0/P)$ against t and finding the gradient. Notice that in a falling head test the effective stresses change because the pore pressures change as the level of water in the standpipe falls. Any volume changes that occur as a result of these changes of effective stress have to be neglected.

Values of the coefficient of permeability measured in laboratory permeameter tests often do not represent the permeability in the ground, for a variety of reasons such as anisotropy (i.e. values of k different for horizontal and vertical flow) and small samples being unrepresentative of large volumes of soil in the ground, and in practice values of k measured from *in situ* tests are much better.

7.5   Principal features of soil loading tests

Soil strength and stiffness are investigated and measured in tests in which soil samples are loaded and unloaded and the resulting stresses and strains are measured. The requirements for testing soils are rather like those for testing metals, concrete and plastics, but the special features of soil strength and stiffness impose special requirements. The most important of these are:

1. Total stresses and pore pressures must be controlled and measured separately so effective stresses, which govern soil behaviour, can be determined.
2. Drainage of water into, or out of, the sample must be controlled so that tests may be either drained (i.e. constant pore pressure) or undrained (i.e. constant volume).
3. To investigate soil stiffness, measurements must be made of small strains (see Chapter 13), but to investigate soil strength it is necessary to apply large strains, sometimes greater than 20%.
4. Because soils are essentially frictional it is necessary to apply both normal and shear stresses. This can be done either by applying confining pressures to cylindrical or cubic samples or by applying normal stresses in direct shear tests (see Fig. 3.3); the relationships between the principal stresses on cylindrical samples and the normal and shear stresses on shear samples were discussed in Sec. 3.2.

During a test the total stresses could be changed, or held constant, and the resulting strains measured; such a test is called stress controlled. Alternatively, the strains could be changed, or held constant, and the resulting stresses measured; such a test is called strain controlled. In a particular test one set of stresses (i.e. axial or vertical) could be stress controlled and another set (i.e. radial or horizontal) could be strain controlled or vice versa.

Loads may be applied to soil samples by rigid plates or by fluid pressures acting on flexible membranes. In the first case the displacements and strains are uniform but the stresses may vary across the plate; in the second case the stresses will be uniform but the strains may vary. Rigid plates may be smooth, in which case shear stresses should be zero and so the faces of the sample are principle planes or they may be rough, in which case there will be both shear and normal stresses to be measured.

To control drainage and measure pore pressures the sample must be isolated within an impermeable membrane and the pore water connected through drainage leads to a pressure transducer and volume gauge, as shown in Fig. 7.3. (This shows details of drainage connections in a typical triaxial test apparatus but the general principles apply also to other soil testing apparatus.) There is a second drainage lead to the sample with a flushing valve. This is to allow water to be flushed through the drainage leads and the bottom drain for de-airing; this is an important requirement of soil testing. If both valves are closed the sample is undrained and if the drainage valve is open the sample is drained; the flushing valve is normally closed and it is only opened when the drainage leads are being flushed. The back pressure u_0 may be atmospheric or at some elevated pressure. Sometimes special tests are carried out in which the pore pressures are changed independently of the total stresses.

The general requirements of soil tests described above are often conflicting and a number of different soil tests have been developed for different specific purposes. The principal tests in routine use in practice are the oedometer test, the direct shear test and the triaxial test, which will now be described. If you read the literature of soil mechanics and become sufficiently interested to specialize in this area you will come across many other special tests; all you have to do is work out what are the boundary conditions and the abilities and limitations of the tests.

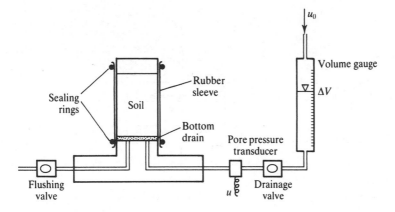

Figure 7.3 Control of drainage and measurement of pore pressure in soil tests.

7.6   One-dimensional compression and consolidation (oedometer) tests

One of the simplest forms of soil loading test is the one-dimensional oedometer test illustrated in Fig. 7.4. The soil sample is a disc contained in a stiff metal cylinder so that radial strains are zero. Porous discs at the top and bottom act as drains and so seepage of pore water is vertical and one-dimensional.

In the conventional apparatus illustrated in Fig. 7.4(a) the axial stress σ_a is applied by adding (or removing) weights so the loading is stress controlled and applied in stages. The axial strain ε_a is measured using a displacement transducer or a dial gauge. The pore pressures in the top drain u_t are zero. The pore pressures in the bottom drain u_b are usually zero but in some special oedometers the bottom drain may be closed and values of u_b measured.

In the Rowe cell illustrated in Fig. 7.4(b) the axial stress σ_a is applied by fluid pressure in a rubber diaphragm so the loading is stress controlled and may be either applied in stages or varied smoothly in continuous loading tests. The axial strain ε_a is measured using a displacement transducer mounted on the stiff top drainage lead. The top and bottom drains are connected to drainage apparatus like that illustrated in Fig. 7.3 so that either or both top and bottom faces of the sample may be drained (i.e. constant pore pressure) or undrained (i.e. the drainage valve is closed).

Oedometer tests may be used to investigate compression and swelling of soil (i.e. the relationship between effective stress and volumetric strain) or consolidation (i.e. the relationship between compression and seepage). Remember the distinctions between drained loading, undrained loading and consolidation discussed in Sec. 6.9. One-dimensional compression and swelling of soil is discussed in Sec. 8.5 and one-dimensional consolidation is discussed in Chapter 15.

7.7   Shear tests

The two forms of shear test used for soil testing are illustrated in Fig. 7.5. In the direct shear box test illustrated in Fig. 7.5(a) the sample is in a split box and is obliged to shear along the horizontal plane defined by the halves of the box. The normal stress σ_n is applied by weights and the shear stress τ_n is usually applied at a constant

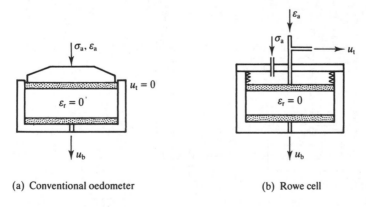

(a) Conventional oedometer   (b) Rowe cell

Figure 7.4 One-dimensional consolidation (oedometer) tests.

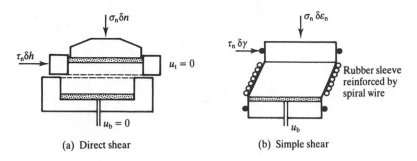

(a) Direct shear   (b) Simple shear

Figure 7.5 Shear tests.

rate of displacement. The vertical and horizontal displacements δn and δh are measured using displacement transducers or dial gauges. Drains are provided at the top and bottom and the pore pressures u_t and u_b are zero. Tests on clays could be undrained if they were carried out quickly, so there was negligible drainage during the test, but as the pore pressures in the sample are not measured effective stresses are unknown. It is fairly obvious looking at Fig. 7.5(a) that the states of stress and strain within the sample are likely to be highly non-uniform, particularly near the ends of the box.

The design of the simple shear apparatus avoids non-uniform strains by allowing the sides to rotate. The most common type, known as the NGI (Norwegian Geotechnical Institute) simple shear apparatus, is illustrated in Fig. 7.5(b). The sample is cylindrical and is sealed inside a rubber sleeve like a triaxial sample (see Sec. 7.8). The rubber sleeve has a spiral wire reinforcement which prevents radial strains but permits shear strains as shown. Applications of the normal and shear stresses and measurements of strains are generally similar to those used for direct shear tests. The drain at the bottom is connected to drainage apparatus like that shown in Fig. 7.3, so that tests may be drained or undrained with measurements of pore pressure.

Notice that if the shear stresses and horizontal displacements in the shear tests in Fig. 7.5 are zero, the conditions are just the same as those in the one-dimensional compression tests in Fig. 7.4.

A major problem with direct and simple shear tests arises with interpretation of the test results. In the apparatus illustrated in Fig. 7.5 only the shear stresses τ_n and σ_n on horizontal planes are measured and the stresses on the vertical planes τ_h and σ_h in Fig. 7.6(a) are unknown. This means that we can only plot one point T on the Mohr diagram shown in Fig. 7.6(b). There are many Mohr circles that pass through the point T; two possibilities are shown. In some special simple shear test apparatus the stresses τ_h and σ_h on the vertical planes are measured, and in this case the Mohr circle is properly defined, but for the conventional tests in Fig. 7.5 it is not certain that the stresses measured, τ_n and σ_n, are those on the most critical planes.

7.8   Conventional triaxial compression tests

The triaxial test is by far the most common and versatile test for soils. The conventional apparatus and the standard test procedures were described in detail by Bishop and

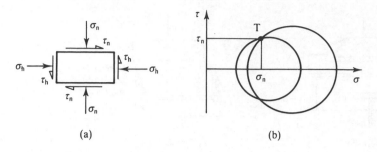

Figure 7.6 Interpretation of shear tests.

Henkel (1962) in their standard text. Most of these are still widely used today, although many of the instruments have been superseded by modern electronic devices.

The basic features of the conventional triaxial tests are shown in Fig. 7.7. The soil sample is a cylinder with height about twice the diameter; sizes commonly used in the United Kingdom are 38 and 100 mm diameters (originally $1\frac{1}{2}$ and 4 in). The sample is enclosed in a thin rubber sleeve sealed to the top platen and to the base pedestal by rubber O-rings. This is contained in a water-filled cell with a cell pressure σ_c. A frictionless ram passes through the top of the cell and applies a force F_a to the top platen; this is measured by a proving ring or by a load cell either inside or outside the cell, as shown. Axial displacements are measured by a displacement transducer attached to the loading ram. The cell and sample assembly are placed inside a loading

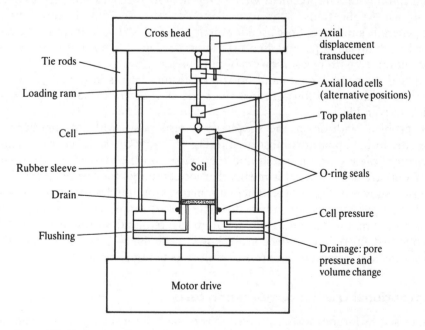

Figure 7.7 Conventional triaxial apparatus.

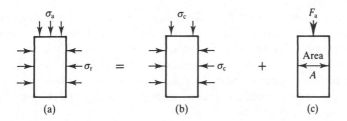

Figure 7.8 Stresses on a triaxial sample.

frame and a motor drive applies a constant rate of strain loading. There is a drain at the base of the sample connected to flushing and drainage apparatus like that shown in Fig. 7.3; if the drainage valve is open the sample is drained and if it is closed the sample is undrained. Radial strains are generally not measured directly but are calculated from measurements of the axial and volumetric strains.

The axial and radial total stresses on the sample, σ_a and σ_r, are shown in Fig. 7.8(a). The radial stress is

$$\sigma_r = \sigma_c \tag{7.10}$$

where σ_c is the cell pressure as shown in Fig. 7.8(b) but σ_c acts also on the top of the sample. From Fig. 7.8 the axial stress σ_a is given by

$$\sigma_a = \sigma_r + \frac{F_a}{A} \tag{7.11}$$

or

$$\frac{F_a}{A} = \sigma_a - \sigma_r = \sigma_a' - \sigma_r' \tag{7.12}$$

If you go back to Sec. 3.2 you will see that F_a/A is the same as the deviator stress q. A simple way to think of the stresses in a triaxial sample is to decompose σ_a and σ_r into an isotropic state $\sigma_a = \sigma_r = \sigma_c$ as in Fig. 7.8(b) plus a deviatoric state $q = F_a/A$ as in Fig. 7.8(c); thus the force in the ram F_a (divided by the area of the sample) applies a stress that deviates from an isotropic state. Note that A is the current area of the sample allowing for changes of axial and volumetric strain. If the loading ram is raised away from the top platen so that $F_a = 0$ the state of stress is isotropic, with $\sigma_a = \sigma_r$.

In a conventional triaxial test the sample would be isotropically compressed, either drained or undrained to the required initial state. The loading ram would then be lowered to touch the top platen, the axial strain set to zero and the sample sheared by increasing the deviator stress q, either drained or undrained, at a constant rate of strain. If the cell pressure σ_c is zero (in this case you need not fill the cell with water) the test is known as unconfined compression. There are a number of other special tests that can be carried out in the triaxial apparatus. These require special modifications to be made to the conventional apparatus, which are discussed in Sec. 7.9.

7.9   Hydraulic triaxial cells – stress path tests

Later we will discover that many features of soil strength and stiffness are governed by the initial state of the soil, its history of loading and unloading and the changes of axial and radial stress during loading or unloading. Consequently, in order to examine soil behaviour properly we will need to be able to control the axial and radial stresses, and perhaps the pore pressures, independently. In the conventional triaxial apparatus shown in Fig. 7.7 the axial stress is applied by strain-controlled loading and it is difficult to vary the axial stress in a controlled way.

Tests in which the paths of the effective stresses (i.e. the graph of σ_a' against σ_r' or the graph of q' against p') are varied, are called stress path tests and are carried out in hydraulic triaxial cells, illustrated in Fig. 7.9. Details of the sample, platens and drainage arrangements are the same as those for the conventional triaxial cell shown in Fig. 7.7, the principal difference being in the application of the axial stress. Another difference to notice is that the loading ram should be connected to the top platen so that extension tests can be carried out where $\sigma_a' < \sigma_r'$ and the force in the ram F_a is negative. (Note that σ_a' and σ_r' are always positive because uncemented soils cannot sustain tensile stresses and, in any case, the platens are not generally attached to the sample.)

A simple hydraulic triaxial cell can be made by adding a hydraulic cylinder to the loading ram, as illustrated in Fig. 7.9(a). Alternatively, special hydraulic triaxial cells are widely used in which a frictionless hydraulic ram is incorporated into the base of the cell, as illustrated in Fig. 7.9(b). In both cases the axial forces F_a should be measured independently using a load cell because it is inaccurate to calculate the value from measurements of the pressures in the hydraulic rams. Conventional strain-controlled triaxial tests can be carried out in both cells, in the first case by locking the hydraulic cylinder and using the motor drive in the loading frame as in a conventional test or, in the second case, by pumping fluid into the hydraulic ram at a constant rate from a screw ram.

In many modern hydraulic triaxial cells all the instruments are electronic and readings are made on a logger controlled by a PC and the pressures in the axial ram, in the cell and in the pore pressure leads are applied through electronic pressure converters.

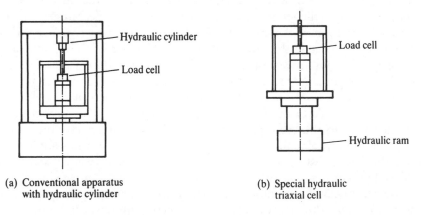

(a) Conventional apparatus
    with hydraulic cylinder

(b) Special hydraulic
    triaxial cell

Figure 7.9 Hydraulic triaxial apparatus.

In this case the PC can be used to control the test and to record the results. Details of this equipment are beyond the scope of this book.

With a hydraulic triaxial cell like those shown in Fig. 7.9 the axial and radial stresses or strains and the pore pressure or volumetric strains can be changed independently. You can illustrate the test path by plotting total and effective stress paths using the axes σ_a vs σ_r, and σ_a' vs σ_r'. However, because we are interested in shear and volumetric effects in soil behaviour it is more illustrative to plot stress paths using the axes q and p (or q' and p'). From Eqs. (3.3) and (3.4), changes of total stress are given by

$$\delta q = \delta\sigma_a - \delta\sigma_r \tag{7.13}$$

$$\delta p = \tfrac{1}{3}(\delta\sigma_a + 2\delta\sigma_r) \tag{7.14}$$

and, from Eqs. (6.14) and (6.15),

$$\delta q' = \delta q \tag{7.15}$$

$$\delta p' = \delta p - \delta u \tag{7.16}$$

Hence, if you know $\delta\sigma_a$, $\delta\sigma_r$ and δu, you can easily plot stress paths using the axes q vs p and q' vs p'.

Figure 7.10 illustrates four simple total stress paths and also defines terms like compression, extension, loading and unloading. Note that in a triaxial apparatus σ_a and σ_r must always be positive; however, we can have $\sigma_a < \sigma_r$ (provided that the loading ram is attached to the top platen) and so q and q' can be positive or negative.

In Fig. 7.10 the four total stress paths correspond to increasing or decreasing either σ_a or σ_r while the other is held constant. Using Eqs. (7.13) and (7.14) with either $\delta\sigma_r = 0$ or $\delta\sigma_a = 0$, you should show that the gradients dq/dp are 3 or $-\tfrac{3}{2}$. In Fig. 7.10

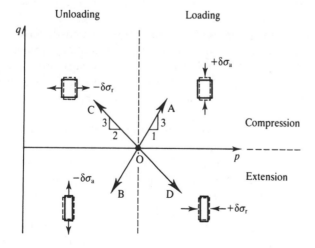

Figure 7.10 Typical stress paths available in hydraulic triaxial tests.

a distinction is made between loading or unloading (corresponding to increasing or decreasing p) and compression or extension (corresponding to positive or negative values of q). Notice that for compression the sample becomes shorter and fatter and for extension it becomes longer and thinner; the path OA corresponds to a conventional triaxial test with constant cell pressure, while path OD is like squeezing a toothpaste tube. During drained tests where the pore pressure u remains constant the total and effective stress paths are parallel, but during undrained tests in which the pore pressure generally changes, the total and effective stress paths are different.

7.10   Comments on soil testing

Although the routine soil tests described in this chapter are relatively simple there is a lot that can, and often does, go wrong with soil tests. Probably the most significant sources of error in measurements of soil parameters and behaviour in laboratory tests are:

1. Malfunctions and errors in the apparatus and in the instruments.
2. Incorrect detailed procedures in performing the tests.
3. Doing the wrong test or measuring the wrong parameter for a particular application.

The last of these is simply a matter of sound understanding of the basic theories involved, rather than blindly following a cookery book approach. The purpose of this book is to develop this sound understanding. The first two are largely a matter of care and attention and experience. In assessing the quality of a set of test results it is essential to distinguish very carefully and clearly between the accuracy and the resolution of the instruments. The resolution (or precision) of an observation is the smallest increment that can be discerned, while the accuracy is the limit within which you can be absolutely confident of the data. For a typical dial gauge measuring small displacements, the resolution and accuracy are both about 0.001 mm, but the resolution and accuracy of electronic instruments are often very different.

For a typical electronic load cell, pressure transducer or displacement transducer the resolution is linked to the electronics which converts an analogue signal (usually a small voltage) to a digital signal. For a 16-bit converter, using 1 bit for the sign, the resolution is 1 in $2^{15} (\approx 30\ 000)$ of the full-scale reading, so for a pore pressure transducer with a range of 0 to 1000 kPa the resolution is about 0.03 kPa. The accuracy depends on the linearity (or non-linearity) of the calibration constant between pressure and voltage and on the stability of the electronic signals. With most instruments commonly used in soil testing you will be doing well to achieve an accuracy better than ± 1 kPa, which is very different from the resolution.

The most difficult measurements to make are of small strains less than about 0.1% in triaxial and shear tests. With conventional instruments for measurement of axial and volumetric strain like those shown in Figs. 7.3 and 7.7 errors arise due to leakage and compliance (movements) in the apparatus and often these errors are greater than the measurements being made. Measurement of small and very small strains using local gauges and dynamic methods are described in Chapter 13.

Another factor is in detection of malfunctions in instruments. It is usually fairly easy to see whether a dial gauge or proving ring is not working properly, but it is much less easy to detect malfunctions in electronic instruments provided that they continue to produce reasonable output signals. The consequence of this is that use of electronic instrumentation in soil testing does not necessarily improve the accuracy of the results compared with old-fashioned instruments and may even reduce the accuracy considerably unless the instruments are frequently checked and recalibrated. The moral of all this is that you should always be suspicious of the accuracy of all laboratory tests.

7.11   Summary

1. Laboratory tests are carried out for description and classification of soils, to investigate their basic mechanical properties and to determine values for the stiffness and strength parameters.
2. The principal tests for description and classification are grading by sieving or sedimentation and the Atterberg limit tests which determine the liquid and plastic limits.
3. The principal loading tests are one-dimensional compression (oedometer) tests, shear tests and triaxial tests. These may be drained or undrained and they may be stress controlled or strain controlled.
4. Special loading or unloading stress path tests are carried out in hydraulic triaxial cells. In these tests the axial and radial stresses or strains and the pore pressure can be varied independently to follow the desired stress path.

Worked examples

Example 7.1: Interpretation of a constant head permeameter test   A constant head permeameter has a diameter of 100 mm and the standpipe tapping points are 150 mm apart. Results of a test on a relatively coarse-grained soil are given in Table 7.1.

Table 7.1 Results of constant head permeability test – Example 7.1

Volume of water collected in 1 min (cm^3)	Difference in standpipe levels (mm)
270	75
220	60
160	45
110	30

The seepage velocity V is given by Eq. (7.5) and the hydraulic gradient i is given by Eq. (7.6). For the first observation,

$$i = \frac{75}{150} = 0.5$$

$$V = \frac{\Delta Q}{A\Delta t} = \frac{270 \times (0.01)^3}{(\pi/4) \times 0.1^2 \times 60} = 5.7 \times 10^{-4}\,\text{m/s}$$

Figure 7.11 shows values of V plotted against i. These fall close to a straight line through the origin, which demonstrates that the basic form of Darcy's law (Eq. 7.4)

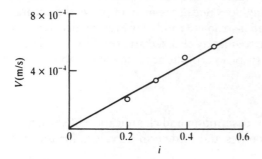

Figure 7.11 Results of constant head permeability test – Example 7.1.

is correct. The coefficient of permeability given by the gradient of the line is

$$k \approx 1 \times 10^{-3} \text{m/s}$$

Example 7.2: Interpretation of a falling head permeameter test   A falling head permeameter has a diameter of 100 mm, the sample is 100 mm long and the area of the
standpipe is 70 mm^2. Results of a test on a relatively fine-grained soil are given in
Table 7.2.

Table 7.2 Results of falling head permeability test – Example 7.2

Time (s)	Height of water in standpipe above overflow (m)	(P_0/P)	$\ln(P_0/P)$
0	1.60	1	0
60	1.51	1.06	0.06
120	1.42	1.13	0.12
240	1.26	1.27	0.24
480	0.99	1.62	0.48

At any instant the potential P is the height of water in the standpipe (above the
overflow) and $P_0 = 1.60$ m at $t = t_0$. Figure 7.12 shows the $\ln(P_0/P)$ plotted against
time. The data points fall close to a straight line. Hence, from Eq. (7.9) the coefficient
of permeability is given by

$$k = \frac{aL}{A} \frac{\ln(P_0/P)}{t} = \frac{70 \times (0.001)^2 \times 0.1}{(\pi/4) \times (0.1)^2} \times \frac{0.1}{100} \approx 1 \times 10^{-6} \text{m/s}$$

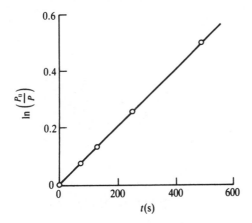

Figure 7.12 Results of falling head permeability test – Example 7.2.

Example 7.3: Interpretation of a drained triaxial test   The first three columns of Table 7.3 give data from a drained triaxial compression test in which the cell pressure was held constant at $\sigma_c = 300$ kPa and the pore pressure was held constant at $u = 100$ kPa. At the start of the test the sample was 38 mm in diameter and 76 mm long and its specific volume was $v = 2.19$.

Table 7.3 Results of drained triaxial test – Example 7.3

Axial force F_a (N)	Change of length ΔL (mm)	Change of volume ΔV (cm³)	ε_s	ε_v	v	q' (kPa)	p' (kPa)
0	0	0	0	0	2.19	0	200
115	−1.95	−0.88	0.022	0.010	2.17	100	233
235	−5.85	−3.72	0.063	0.042	2.10	200	267
325	−11.70	−7.07	0.127	0.080	2.01	264	288
394	−19.11	−8.40	0.220	0.095	1.98	287	296
458	−27.30	−8.40	0.328	0.095	1.98	286	296

The initial dimensions of the sample were

$$A_0 = \frac{\pi}{4}D_0^2 = 1.134 \times 10^{-3}\,\mathrm{m}^2$$

$$V_0 = A_0 L_0 = 88.46 \times 10^{-6}\,\mathrm{m}^3$$

At any stage of the test

$$\varepsilon_a = -\frac{\Delta L}{L_0}$$

$$\varepsilon_v = -\frac{\Delta V}{V_0}$$

$$v = v_0(1 - \varepsilon_v)$$

and

$$\sigma_r = \sigma_c = 300\,\text{kPa} \quad \sigma_a = \sigma_r + \frac{F_a}{A}$$

where the current area is $A = A_0(1 - \varepsilon_v)/(1 - \varepsilon_a)$. From Eqs. (3.5) to (3.8),

$$\varepsilon_s = \varepsilon_a - \frac{1}{3}\varepsilon_v$$

and

$$q' = (\sigma_a' - \sigma_r') = q \qquad p' = \tfrac{1}{3}(\sigma_a' + 2\sigma_r') = p - u$$

or

$$q' = \frac{F_a}{A} \qquad p' = p_0 + \tfrac{1}{3}q' - u$$

where $p_0 = 300$ kPa. The test results are given in the right-hand side of Table 7.3 and are plotted in Fig. 7.13 as O → A.

Example 7.4: Interpretation of an undrained triaxial test   The first three columns in Table 7.4 give data from an undrained triaxial compression test in which the cell pressure was held constant at $\sigma_c = 300$ kPa. At the start of the test the sample was 38 mm diameter and 76 mm long, the pore pressure was $u_0 = 100$ kPa and the specific volume was $v = 2.19$.

For an undrained test $\varepsilon_v = 0$ (by definition), but otherwise the calculations are the same as those given in Example 7.3. The test results are given in the right-hand side of Table 7.4 and are plotted in Fig. 7.13 as O → B.

Example 7.5: Stress paths   The left-hand side of Table 7.5 gives the initial states and increments of axial and radial total stresses for a set of drained and undrained tri-axial stress path tests. In the drained tests the pore pressure was $u = 0$. The soil can be assumed to be isotropic and elastic so that shearing and volumetric effects are decoupled.

The stress paths corresponding to tests lasting for 10 hours are shown in Fig. 7.14. The right-hand side of Table 7.5 gives the states at the start and at the end of each path. For the undrained test $\delta p' = 0$ (because $\delta\varepsilon_v = 0$ and shear and volumetric effects are decoupled). For the drained tests the changes of q' and p' are found from Eqs. (7.13) and (7.14).

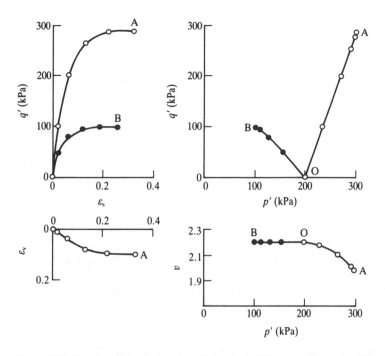

Figure 7.13 Results of drained and undrained triaxial tests – Examples 7.3 and 7.4.

Table 7.4 Results undrained triaxial test – Example 7.4

Axial force F_a (N)	Change of length ΔL (mm)	Pore pressure (kPa)	ε_s	q' (kPa)	p' (kPa)
0	0	100	0	0	200
58	−1.95	165	0.026	50	152
96	−4.29	200	0.056	80	127
124	−9.36	224	0.123	96	108
136	−14.04	232	0.185	98	101
148	−19.50	232	0.257	97	100

Table 7.5 Loading in stress path tests – Example 7.5

Sample	σ_a (kPa)	σ_r (kPa)	$d\sigma_a/dt$ (kPa/h)	$d\sigma_r/dt$ (kPa/h)	Drainage	σ_{ae} (kPa)	σ_{re} (kPa)	q'_0 (kPa)	p'_0 (kPa)	q'_e (kPa)	p'_e (kPa)
A	200	200	10	0	Drained	300	200	0	200	100	233
B	200	200	−10	0	Undrained	100	200	0	200	−100	200
C	250	175	−10	−10	Drained	150	75	75	200	75	100
D	250	175	0	−10	Drained	250	75	75	200	175	133

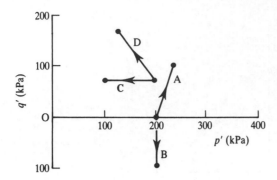

Figure 7.14 Stress paths – Example 7.5.

Further reading

Atkinson, J. H. and G. Sallfors (1991) 'Experimental determination of soil properties. General Report to Session 1', *Proceedings of 10th ECSMFE, Florence*, Vol. 3, pp. 915–956.

Baldi, G., D. W. Hight and G. E. Thomas (1988) 'A re-evaluation of conventional triaxial test methods', in *Advanced Triaxial Testing of Soil and Rock*, R. T. Donaghe, R. C. Chaney and M. L. Silver (eds), ASTM, STP 977, pp. 219–263.

Bishop, A. W. and D. J. Henkel (1962) *The Triaxial Test*, Edward Arnold, London.

BS 1377 (1990) *Methods of Test for Soils for Civil Engineering Purposes*, British Standards Institution, London.

Head, K. H. (1980) *Manual of Soil Laboratory Testing*, Vol. 1, *Soil Classification and Compaction Tests*, Pentech Press, London.

Head, K. H. (1982) *Manual of Soil Laboratory Testing*, Vol. 2, *Permeability, Shear Strength and Compressibility Tests*, Pentech Press, London.

Head, K. H. (1986) *Manual of Soil Laboratory Testing*, Vol. 3, *Effective Stress Tests*, Pentech Press, London.

Chapter 8

Compression and swelling

8.1 Introduction

As soils are loaded or unloaded isotropically (i.e. with equal all-round stresses) or anisotropically they will compress and swell. As we saw in Chapter 6, volume changes in saturated soils involve rearrangement of grains together with breakage of weak sand grains and swelling or compression of clay grains accompanied by seepage of water as shown in Fig. 8.1. Volume changes occur in unsaturated soils as air is expelled. Compression of unsaturated soil is important mostly during compaction of excavated soil in to embankment fill. This is described in Chapter 26.

To account for seepage flow it is necessary to consider the relative rates of loading and drainage as discussed in Sec. 6.10; this is equally true for laboratory tests and for loadings of structures in the ground. In laboratory tests the sample may be loaded undrained and then allowed to consolidate under constant total stress; this is the basis of the conventional incremental loading oedometer test described in Chapter 7. In this case measurements of effective stress can only be made at the end of consolidation when all the excess pore pressures have dissipated (unless the excess pore pressures are measured separately). Alternatively, the loading could be applied at a continuous rate and the excess pore pressures measured. It is simplest, however, to load samples fully drained at a rate that is slow enough to ensure that any excess pore pressures are negligible so that effective stresses can be determined. I will consider the behaviour of soil during incremental and continuous loading consolidation tests in Chapter 15; for the present I will consider only fully drained states where excess pore pressures are zero. The idealized behaviour described in this chapter is based on experimental data given by Atkinson and Bransby (1978) and by Muir Wood (1991).

8.2 Isotropic compression and swelling

The general behaviour of soil during isotropic compression and swelling is illustrated in Fig. 8.2. This shows soil in which the grains are loosely packed, initially at p_0' at O compressed to A, unloaded to B and reloaded through C to D where the grains are more densely packed. This behaviour is similar to that illustrated in Fig. 3.12 and C is a yield point.

Soil compression is primarily caused by rearrangement of the grains and so the stiffness will increase from loose states (where there are plenty of voids for grains

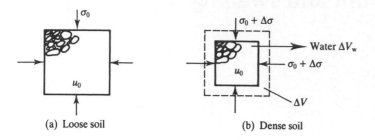

Figure 8.1 Volume changes in soil.

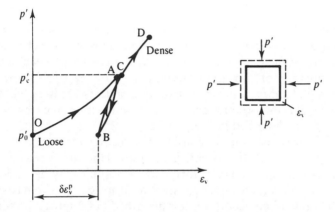

Figure 8.2 Isotropic compression and swelling of soil.

to move into) to dense states (where there is much less opportunity for grains to rearrange). As shown in Fig. 8.2, the stress–strain line is curved. Thus the mechanisms of volume change in soils due to rearrangement of the grains largely accounts for the non-linear bulk stiffness behaviour. For the unloading–reloading loop ABC the soil is very much stiffer (i.e. the volume changes are less) than for first loading because the grains will obviously not 'un-rearrange' themselves on unloading. Behaviour similar to that shown in Fig. 8.2 is also found for soils which have weak grains (such as carbonate or shelly sands) that fracture on loading. In this case most of the compression during first loading is associated with grain fracture but obviously the grains do not 'unfracture' on unloading. Soils which contain a high proportion of plastic clay may swell significantly on unloading due to volume changes in the clay grains themselves. From Eq. (3.12) the instantaneous bulk modulus at any point is the gradient of the curve for first loading or for unloading or reloading, given by

$$K' = \frac{\mathrm{d}p'}{\mathrm{d}\varepsilon_\mathrm{v}} \qquad\qquad (8.1)$$

and the value of K' is not a soil constant.

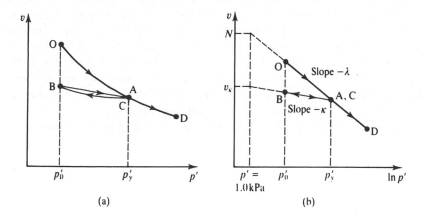

Figure 8.3 Isotropic compression and swelling.

The behaviour shown in Fig. 8.2 is repeated in Fig. 8.3(a) but plotted as specific volume instead of volumetric strain and with p' plotted horizontally; this is the conventional representation of soil compression and swelling. Figure 8.3(b) shows the same behaviour but now with the stress on a logarithmic scale. In Fig. 8.3(b) the compression curve from Fig. 8.3(a) is now linear, which is a very good approximation for the behaviour of many soils over a wide range of loadings. This idealization is good for most clays and for sands. For coarse-grained soils volume changes during the first loading are often accompanied by breakage of the soil grains and it is usually necessary to apply large stresses (greater than 1000 kPa) to identify the full range of behaviour. The unloading–reloading loop A → B → C in Fig. 8.3(a) is approximated by the straight line AB in Fig. 8.3(b). This approximation is less good than the one for the line OACD. In many soils, especially fine-grained ones, the differences between unloading and reloading are substantial and both unloading and reloading lines are non-linear even when p' is plotted to a logarithmic scale. Stiffness of soil will be discussed further in Chapter 13. The idealization for isotropic compression and swelling shown in Fig. 8.3(b) is, however, widely used in basic soil mechanics theories and in practice.

The line OACD corresponding to first loading is known as the normal compression line (NCL) and is given by

$$v = N - \lambda \ln p' \tag{8.2}$$

where λ is the gradient and N is the value of v at $p' = 1.0$ kPa where $\ln p' = 0$. The line ABC is known as a swelling line and is given by

$$v = v_\kappa - \kappa \ln p' \tag{8.3}$$

where κ is the gradient and v_κ is the value of v at $p' = 1.0$ kPa. The swelling line ABC meets the normal compression line at C which is a yield point and the yield stress is p'_y.

The parameters λ, κ and N are material parameters and their values depend only on the nature of the grains. Relationships between these and other material parameters and soil classifications are discussed in Chapter 18. Soil could be unloaded from any point on the normal compression line and there are any number of swelling lines. For each line there is a particular value of v_κ and a particular value for the yield stress p'_y. Using Eqs. (8.2) and (8.3) it is possible to calculate the current specific volume of any isotropically compressed sample given the history of loading and unloading and to calculate the recoverable and irrecoverable volume changes.

From Eq. (8.2), differentiating with respect to p' and dividing by v we have

$$-\frac{\mathrm{d}v}{v} = \frac{\lambda}{vp'}\mathrm{d}p' = \mathrm{d}\varepsilon_v \tag{8.4}$$

and, comparing with Eq. (8.1),

$$K' = \frac{vp'}{\lambda} \tag{8.5}$$

which is appropriate for first loading. Similarly, for unloading and reloading, we have $K' = vp'/\kappa$. Notice that the bulk modulus K' contains λ or κ, which are material parameters and so are constants for a particular soil, and vp' which changes during loading and unloading. As a result K' is not a constant and isotropic compression and swelling lines are non-linear, as shown in Figs. 8.2 and 8.3(a).

8.3   Overconsolidation and yield stress ratio

In Fig. 8.4 the state of a soil during first loading, after deposition, travels down the normal compression line OACD and soil that has been unloaded or reloaded travels on a swelling and recompression line such as ABC characterized by v_κ or p'_y. The state of the soil can reach any point below and to the left of the normal compression line by unloading, but the state cannot reach the region above and to the right. Hence the

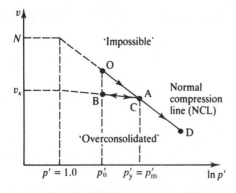

Figure 8.4 Overconsolidation.

normal compression line is a boundary to all possible states for isotropic compression: later we will see that this state boundary line forms part of a state boundary surface. This is also a yield surface like that illustrated in Fig. 3.16 and the NCL is a yield curve like that shown in Fig. 3.12.

At any state such as B inside the boundary surface the soil is overconsolidated and its overconsolidation ratio is

$$R_p = \frac{p'_m}{p'_0} \tag{8.6}$$

where p'_0 is the current stress and p'_m is the stress at the point C which is the maximum stress which the sample at B has experienced in the past. Notice that any isotropic state can be described by only two of the parameters p', v and R_p.

For a normally consolidated soil the state lies on the normal compression line and $R_p = 1.0$. Figure 8.5 shows two states, R_1 and R_2, that have the same overconsolidation ratio. From the geometry of the figure, or from Eq. (8.6),

$$\ln R_p = \left(\ln p'_{y1} - \ln p'_{01} \right) = \left(\ln p'_{y2} - \ln p'_{02} \right) \tag{8.7}$$

so that the line through R_1 and R_2, where the overconsolidation ratio is the same, is parallel to the normal compression line.

Soils at points N_1 and R_2 have the same current stress, and so would be at the same depth in the ground, but they have very different stiffnesses related to λ and κ respectively. Similarly, soils at points R_2 and N_2 have nearly the same specific volume and water content but, again, they have very different stiffnesses. Soils at points N_1 and N_2 are both normally consolidated; they will have different stiffnesses for loading and for unloading. This means that soil stiffness is not directly related either to the water content or to the current stress (or depth in the ground) and the overconsolidation ratio is an important factor in determining soil behaviour.

In Fig. 8.5, the state of the sample at R_1 where the stress is p'_{01} can move to R_2 only by loading to the NCL at N_1 where it yields at the yield stress p'_{y1}, further compression

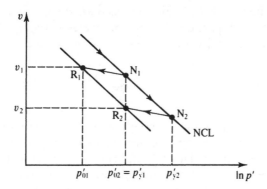

Figure 8.5 Overconsolidation ratio.

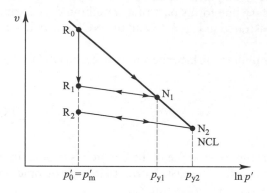

Figure 8.6 Changes of state due to creep or vibration.

along the NCL to N_2 where the yield stress is p'_{y2} and unloading to R_2 where the stress is p'_{02}. The state of a soil can however move directly from R_1 to R_2 by creep in fine-grained soils and vibration in coarse-grained soils. Moreover the position of the NCL can shift as a result of soil structure. These mechanisms will be described further in Chapter 16.

Figure 8.6 shows the state of a sample of soil initially normally consolidated at R_0 where the stress $p'_0 = p'_m$ moving directly to R_1 where the stress is same, by creep or vibration. From the definition of overconsolidation ratio in Eq. (8.6) the overconsolidation ratios at R_0 and R_1 are both the same and are equal to 1.0 since the stresses have not changed. This means that the overconsolidation ratio defined in Eq. (8.6) does not properly describe the current state of a soil.

The state of a soil can be better described by the yield stress ratio

$$Y_p = \frac{p'_y}{p'_0} \tag{8.8}$$

where p'_0 is the current stress and p'_y is the yield stress which is the stress at the intersection of the swelling line through R_1 with the NCL. Notice that as the state moves from R_1 to R_2 in Fig. 8.6, either by loading, yielding and unloading or by creep or vibration the yield stress ratio increases because the yield stress increases from p'_{y1} to p'_{y2}.

8.4   State of soils on the wet side and on the dry side of critical

Clays may be normally consolidated or, depending on how far the state is from the normal consolidation line, lightly or heavily overconsolidated. There is a critical overconsolidation ratio, shown as a broken line in Fig. 8.7(a), which separates lightly and heavily overconsolidated soils. (We will see later that this line is below the critical state line (CSL) which corresponds to states at which soil fails during shearing.) The precise value for the critical overconsolidation ratio depends principally on the

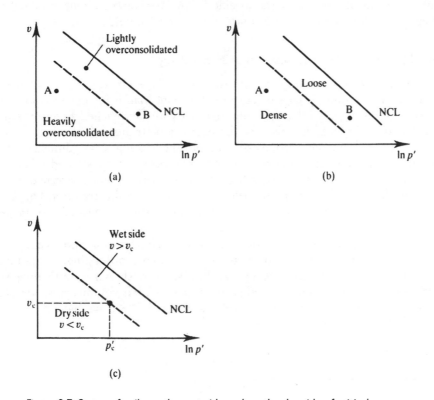

Figure 8.7 States of soils on the wet side and on the dry side of critical.

nature of the soil; most soils will be lightly overconsolidated at $R_p < 2$ and heavily overconsolidated at $R_p > 3$.

Sands and gravels may be loose or dense depending on the position of the state with respect to the critical overconsolidation line, as shown in Fig. 8.7(b). Notice that the state is defined by a combination of specific volume and pressure. In Fig. 8.7(b) the state at A is dense while the state at B is loose although the specific volume at B is smaller than at A: this is because the stress at B is considerably greater than at A. Similarly, in Fig. 8.7(a) the state at A is heavily overconsolidated while the state at B is only lightly overconsolidated although the specific volume at B is smaller than that at A. The regions in which clays are normally consolidated or lightly overconsolidated and sands that are loose are said to be on the wet side of the critical line, as shown in Fig. 8.7(c), and the regions where clays are heavily overconsolidated and sands are dense are said to be on the dry side. We will find later that there are fundamental differences in the behaviour of soils when they are sheared from states initially on the wet side or initially on the dry side of the critical line.

Do not misunderstand the terms wet side and dry side. The soil is always either saturated or dry and it is simply that at a given stress, such as p'_c in Fig. 8.7(c), the specific volume (or water content) on the wet side is higher than v_c (i.e. the soil is wetter than at the critical state) while the specific volume (or water content) on the dry side is lower than v_c (i.e. the soil is drier than at the critical state).

8.5 One-dimensional compression and swelling

In the ground the stresses are not generally isotropic as the horizontal and vertical stresses are different. A common case where a relative wide load from an embankment or spread foundation is on a relatively thin layer of clay sandwiched between stiff sand is illustrated in Fig. 8.8. In this case the horizontal strains below most of the embankment are approximately zero, as shown, and the loading is one-dimensional. In the laboratory one-dimensional conditions occur in oedometer tests and in shear box tests before the shear stresses are applied. Although here we are concerned with one-dimensional loading, the conditions below the foundation illustrated in Fig. 8.8 and in the one-dimensional laboratory tests correspond to one-dimensional drainage as well.

The general behaviour of soil during one-dimensional compression and swelling is illustrated in Fig. 8.9. This corresponds to the same sequence of loading, unloading and reloading illustrated in Fig. 8.2, except that the results are shown as vertical stress σ_z' rather than mean stress p' and vertical strain ε_z rather than volumetric strain ε_v; note, however, that for one-dimensional straining where $\varepsilon_h = 0$ we have $\varepsilon_z = \varepsilon_v$.

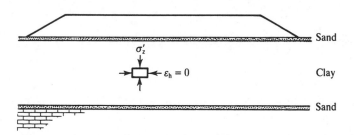

Figure 8.8 One-dimensional states beneath a wide embankment.

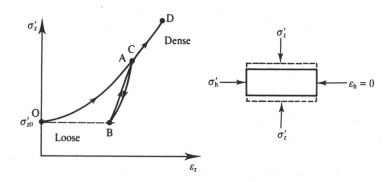

Figure 8.9 One-dimensional compression and swelling.

The one-dimensional compression modulus M' is given by

$$M' = \frac{d\sigma'_z}{d\varepsilon_z} = \frac{1}{m_v} \tag{8.9}$$

and, as before, C is a yield point. A parameter often quoted in practice is the one-dimensional coefficient of compressibility m_v.

Note that the compression and swelling lines in Fig. 8.9 are non-linear and so M and m_v (like K in Eq. 8.5) are not constants but depend on the current stress and are different for loading and unloading.

Figure 8.10(a) shows the same behaviour as that in Fig. 8.9 and is equivalent to Fig. 8.3(a) for isotropic compression. Figure 8.10(b) shows the same behaviour with σ'_z plotted to a $\log_{10}$ scale and specific volume replaced by voids ratio. (The axes e and $\log \sigma'_z$ are commonly used in practice for plotting the results of one-dimensional tests.) All the essential features for isotropic compression and swelling described in Sec. 8.2 are repeated for one-dimensional compression and swelling. The principal differences are that the parameter N for isotropic compression is replaced by e_0 and the parameters λ and κ are replaced by C_c and C_s. The normal compression line OACD is given by

$$e = e_0 - C_c \log \sigma'_z \tag{8.10}$$

and the swelling and recompression line ABC is given by

$$e = e_\kappa - C_s \log \sigma'_z \tag{8.11}$$

Since $\delta v = \delta e$ and $\log_{10} x = 0.43 \ln x$ we have $C_c = 2.3\lambda$ and $C_s = 2.3\kappa$.

For overconsolidated soil at a point such as B in Fig. 8.10(a) the yield stress ratio Y_0 is given by

$$Y_0 = \frac{\sigma'_y}{\sigma'_0} \tag{8.12}$$

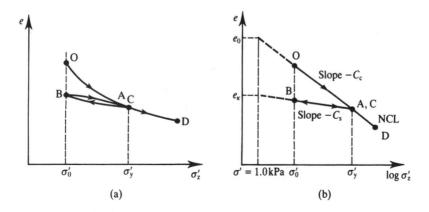

(a)   (b)

Figure 8.10 One-dimensional compression and swelling.

where σ_0' is the current stress and σ_y' is the yield point which lies at the intersection of the swelling line through B with the normal compression line. Compare the definition of Y_0 for one-dimensional overconsolidation with the definition of Y_p in Eq. (8.8) for isotropic over-consolidation.

8.6 Horizontal stress in one-dimensional loading

During the increase and decrease of σ_z' in one-dimensional loading and unloading the horizontal stresses σ_h' changes since ε_h is held constant. The variations of σ_z' and σ_h' are illustrated in Fig. 8.11(a). The ratio

$$K_0 = \frac{\sigma_h'}{\sigma_z'} \tag{8.13}$$

is known as the coefficient of earth pressure at rest (i.e. corresponding to zero horizontal strain) and the variation of K_0 with yield stress ratio Y_0 is illustrated in Fig. 8.11(b). For states OACD on the normal compression line $Y_0 = 1$ and the value of K_0 is K_{0nc} for normally consolidated soil: for many soils this can be approximated by

$$K_{0nc} = 1 - \sin \phi_c' \tag{8.14}$$

where ϕ_c' is the critical friction angle (see Chapter 9). For overconsolidated states ABC the value of K_0 increases with overconsolidation and K_0 may well exceed 1.0 as the horizontal stress exceeds the vertical stress at large values of Y_0. Figures 8.11(a) and (b) illustrate substantial hysteresis in K_0 during unloading and reloading, but if this is neglected then K_0 is found to vary with Y_0 and an approximate empirical relationship is

$$K_0 = K_{0nc}\sqrt{Y_0} \tag{8.15}$$

During one-dimensional loading and unloading σ_z' and σ_h' are generally unequal and so there are shear stresses in the soil and any comparison between isotropic and

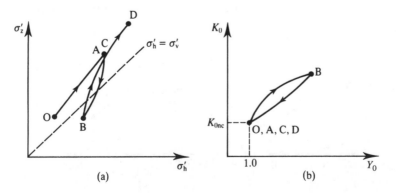

Figure 8.11 Horizontal stresses during one-dimensional loading and unloading.

one-dimensional compression and swelling will have to take account of the shear stresses. The link between these can be developed by going back to Sec. 3.2. From Eqs. (3.3) and (3.4) with $\sigma'_a = \sigma'_z$ and $\sigma'_r = \sigma'_h$, and making use of Eq. (8.13), we have

$$q' = \sigma'_z(1 - K_0) \tag{8.16}$$

$$p' = \tfrac{1}{3}\sigma'_z(1 + 2K_0) \tag{8.17}$$

Figure 8.12 shows the behaviour of soil in isotropic and one-dimensional compression and swelling together; the subscripts 1 refer to one-dimensional behaviour. These show normal compression lines OACD and $O_1A_1C_1D_1$ with the same gradients λ and values of v at $p' = 1\,\text{kPa}$ of N and N_0. The swelling and recompression lines ABC and $A_1B_1C_1$ have approximately the same gradients, κ, and the same yield stresses, p'_y, but different values of v_κ. (The gradients κ are actually slightly different because the value of K_0 changes during one-dimensional swelling and recompression.)

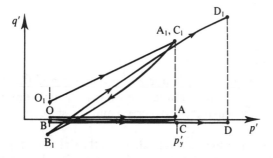

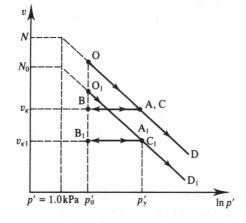

Figure 8.12 Soil behaviour during isotropic and one-dimensional compression and swelling.

8.7 Summary

1. The basic mechanism of compression in soils is by rearrangement of the grains. In coarse-grained soils this may be accompanied by fracturing of the soil grains and in fine-grained soils by compression or swelling of clay particles.
2. The behaviour of soil during isotropic compression and swelling is given by

$$v = N - \lambda \ln p' \tag{8.2}$$

$$v = v_\kappa - \kappa \ln p' \tag{8.3}$$

The parameters λ, κ and N are material parameters; their values depend only on the nature of the grains. Equation (8.2) is for normally consolidated soil and Eq. (8.3) is for overconsolidated soil.

3. Equations (8.2) and (8.3) demonstrate that the stiffness of soil is non-linear (i.e. the bulk modulus is not a constant) when it is both normally consolidated and overconsolidated.
4. Equation (8.2) represents the normal compression line. The state of a soil cannot usually lie outside this line and moves below the line on unloading when the soil becomes overconsolidated. The yield stress ratio Y_p is given by

$$Y_p = \frac{p'_y}{p'_0} \tag{8.8}$$

where p'_y is the current yield stress.

5. Normally the state of soil is changed only by loading and unloading and the state moves on the current swelling and recompression line or on the normal compression line. The state of a clay may also change due to creep and the state of a sand may change due to vibration.
6. There is a critical state line which separates the wet side from the dry side. Lightly overconsolidated clays and loose sands are on the wet side of the critical line while heavily overconsolidated clays and dense sands are on the dry side.
7. The behaviour of soil during one-dimensional compression and swelling is similar to that for isotropic loading and is given by

$$e = e_0 - C_c \log \sigma'_z \tag{8.10}$$

$$e = e_\kappa - C_s \log \sigma'_z \tag{8.11}$$

The parameters e_0, C_c and C_s are material parameters.

Worked examples

Example 8.1: Analysis of an isotropic compression test   Table 8.1 gives results obtained from an isotropic test. The data are shown plotted in Fig. 8.13. Scaling from the diagram, $\lambda = 0.20$ and $\kappa = 0.05$. Projecting the lines back to $p' = 1.0 \, \text{kPa}$ (i.e. $\ln p' = 0$),

Table 8.1 Results of isotropic compression tests – Example 8.1

Mean effective stress p′ (kPa)	ln p′ (kPa)	Specific volume v
20	3.00	2.65
60	4.09	2.43
200	5.30	2.19
1000	6.91	1.87
200	5.30	1.95
60	4.09	2.01

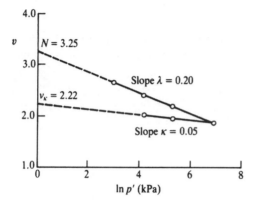

Figure 8.13 Isotropic compression and swelling – Example 8.1.

$N = 3.25$ and $v_\kappa = 2.22$. The bulk modulus K' is not a constant: from Eq. (8.1), for the second and last increments between $p' = 60\,\text{kPa}$ and $p' = 200\,\text{kPa}$.

$$K' = \frac{\Delta p'}{\Delta \varepsilon_v} = \frac{200 - 60}{-(2.19 - 2.43)/2.43} = 1.42\,\text{MPa}$$

$$K' = \frac{\Delta p'}{\Delta \varepsilon_v} = \frac{60 - 200}{-(2.01 - 1.95)/1.95} = 4.55\,\text{MPa}$$

Example 8.2: Determination of soil behaviour during isotropic compression   A soil has the parameters $\lambda = 0.20$, $\kappa = 0.05$ and $N = 3.25$. A sample is subjected to the sequence of isotropic loading and unloading given in the second column in Table 8.2. At each stage the overconsolidation ratio R_p is given by Eq. (8.6). For the normally consolidated state ($R_p = 1$) the specific volume is given by Eq. (8.2). At the point C, the specific volume is given by both Eqs. (8.2) and (8.3) and hence

$$v_\kappa = N - (\lambda - \kappa) \ln p' = 3.25 - (0.20 - 0.05) \ln 600 = 2.29$$

For the overconsolidated states the specific volume is given by Eq. (8.3).

Table 8.2 Isotropic compression and swelling – Example 8.2

Point	Mean effective stress p′ (kPa)	Overconsolidation ratio R_p	Specific volume v
A	60	1	2.43
B	200	1	2.19
C	600	1	1.97
D	300	2	2.01
E	150	4	2.04

Further reading

Atkinson, J. H. and P. L. Bransby (1978) *The Mechanics of Soils,* McGraw-Hill, London.

Muir Wood, D. M. (1991) *Soil Behaviour and Critical State Soil Mechanics,* Cambridge University Press, Cambridge.

Chapter 9

Critical state strength of soil

9.1 Behaviour of soil in shear tests

In simple terms the strength of a material is the maximum shear stress that it can sustain; materials loaded just beyond the maximum stress will fail. Failure may be sudden and catastrophic leading to a complete loss of strength (which is what happens when you break a biscuit, which is brittle) or it may lead to a very large plastic straining (which is what happens if you mould plasticine, which is ductile). For most soils, failure of slopes and foundations involves large straining without complete loss of strength and failing soil structures can usually be stabilized by unloading them.

The essential features of soil strength can most easily be seen in ideal shearing tests, as illustrated in Fig. 9.1. The shear and normal effective stresses are τ' and σ' and, at a particular stage of the test, there are increments of strain $\delta\gamma$ and $\delta\varepsilon_v$. These are similar to the conditions in the direct shear box test and the simple shear test described in Chapter 7 and in soil in thin slip surfaces that occur during failure of slopes as described in Chapter 21. The conventional direct and simple shear tests are, however, not ideal because the stresses and deformations are likely to be non-uniform and the states of stress and strain are not completely defined by the measurements on only one plane. Although a shear test is not ideal for measuring soil properties it is, however, convenient for demonstrating the basic characteristics of soil strength.

Typical stress–strain curves for soils on the wet side of critical (i.e. normally consolidated or lightly overconsolidated clays or loose sands) marked W and for soils on the dry side (i.e. heavily overconsolidated clays or dense sands) marked D, tested drained with constant σ', are shown in Fig. 9.1(b) and the corresponding volumetric strains are shown in Fig. 9.1(c). (Remember the distinctions between the wet side of the critical line and the dry side, discussed in Sec. 8.4.) The behaviour shown in Fig. 9.1 is typical for normally consolidated or overconsolidated clays as well as for loose or dense sands. Soils on the wet side compress as the shear stresses increase while soils on the dry side dilate (expand) after a small compression. Both ultimately reach critical states at which the shear stress is constant and there are no more volumetric strains. Soils on the dry side reach peak shear stresses before reaching the critical state. Remember that strength is the maximum shear stress which a material can sustain so for soil there is a peak strength and a critical state strength. There is also a residual strength which will be discussed in Sec. 9.2. At any stage of shearing the angle of dilation ψ (see Sec. 2.7) is

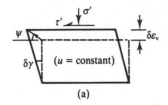

(a)

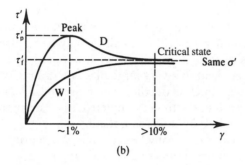

(b)

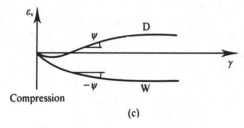

(c)

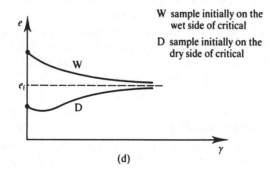

W sample initially on the
   wet side of critical

D sample initially on the
   dry side of critical

(d)

Figure 9.1 Typical behaviour of soils in drained shear tests.

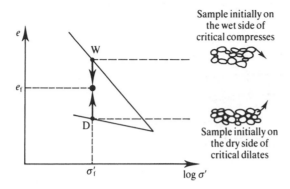

Figure 9.2 Compression and dilation during shearing.

defined by

$$\tan \psi = -\frac{d\varepsilon_v}{d\gamma} \qquad (9.1)$$

This is the gradient of the volume change curve as shown in Fig. 9.1(c) and it also gives the direction of movement of the top of the sample as shown in Fig. 9.1(a). The negative sign is introduced into Eq. (9.1) so that dilation (negative volumetric straining) is associated with positive angles of dilation.

Figure 9.1(d) shows the change of voids ratio e rather than the volumetric strains shown in Fig. 9.1(c), although, of course, they are related. Both samples have the same effective normal stresses but the initial voids ratio of the sample on the wet side is higher than that of the sample on the dry side. Notice, however, that both samples reach their critical states at the same voids ratio e_f.

As volume changes in soils are principally due to rearrangement of particles it is easy to see why soils on the wet side compress while soils on the dry side dilate. In Fig. 9.2 the grains of the loose or normally consolidated soil at W are spaced well apart and, on shearing, they can move into the neighbouring void spaces, while the grains of the dense or overconsolidated soil at D must move apart during shear. This is an example of the coupling between shear and volumetric effects in soils.

9.2   Peak, critical state and residual strengths

As shown in Fig. 9.1, soils initially on the dry side of the critical line reach peak states before the critical state. The peak state will normally be reached at strains of the order of 1% while the critical state will be reached after strains greater than 10% (in some soils the critical states are not reached until the strains have exceeded 50% or so). Notice that the peak state coincides with the point of maximum rate of dilation (i.e. at maximum ψ). Soils on the wet side compress throughout, shearing up to the ultimate state, and there is no peak.

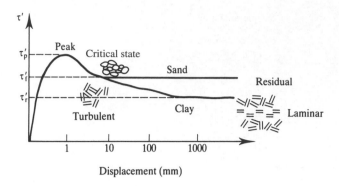

Figure 9.3 Residual strength of clay at very large displacements.

For soils that have a peak state it is not easy to decide whether the strength of the soil – the maximum shear stress it can sustain – should be the peak state that can be sustained only for relatively small strains or the critical state. I will leave this question for the time being and, for the present, I will discuss the conditions at the peak state and the conditions at the critical state separately.

There is another aspect of soil shearing that must be considered here and that is the development of residual strength at very large displacements on slip planes (Skempton, 1964). Figure 9.3 illustrates the behaviour of a sand and a plastic clay soil over large displacements; note the logarithmic scale, which allows the diagram to represent displacements exceeding 1 m. (Tests of this kind can be carried out in a direct shear box by moving the box backwards and forwards or in a special ring shear apparatus in which an annulus of soil can be sheared continuously.) The behaviour illustrated is for tests in which the effective stresses and the initial states were chosen so that the peak and ultimate states of the clay and the sand soil happened to be the same. At the critical state, at displacements of about 10 mm corresponding to shear strains of about 10% as shown in Fig. 9.1, the movements of grains are essentially turbulent, involving relative movements and rotations of both clay and sand grains. At larger displacements, however, the strains become localized into distinct zones of intense shearing and the shear stresses applied to the clay soil decrease.

The lowest shear stress reached after very large displacements is called the residual strength. It is associated with laminar flow of flat clay grains which have become oriented parallel to a very thin shearing zone, as illustrated in Fig. 9.3. In sands and other soils with rotund (i.e. not flat) grains there is no opportunity for laminar flow and the residual strength is the same as the critical state strength. In clays the residual strength may be as little as 50% of the critical state strength and it is important for design of works on old landslides. Choices of soil strength for design of slopes, foundations and retaining walls are discussed in Chapter 18.

9.3   Critical states

We now come to the essence of soil mechanics, which is the critical state. The idealized behaviour described in this chapter is based on experimental data given by Atkinson

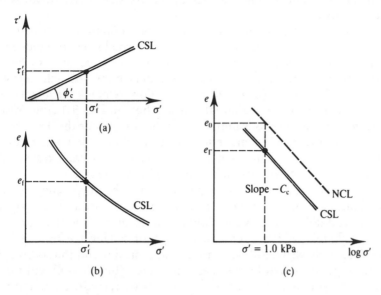

Figure 9.4 Critical states of soils.

and Bransby (1978) and by Muir Wood (1991). From Figs. 9.1 and 9.3 the critical state is the state reached after strains of at least 10% and is associated with turbulent flow. The relationships between the shear stress, the normal effective stress and the voids ratio of soils at the critical states are illustrated in Fig. 9.4.

Figure 9.4(a) and (b) shows the critical state line (CSL). This shows that, at the critical state, there is a unique relationship between the shear stress, the normal effective stress and the voids ratio. Figure 9.4(c) is the same as Fig. 9.4(b) but with the normal stress on a logarithmic scale. Also shown on Fig. 9.4(c) is the one-dimensional normal compression line from Fig. 8.10(b).

The critical state line is given by

$$\tau_f' = \sigma_f' \tan \phi_c' \tag{9.2}$$

$$e_f = e_\Gamma - C_c \log \sigma_f' \tag{9.3}$$

where the subscripts f denote that the stresses and the voids ratio are those at failure at the critical state. It is essential to recognise that Eqs. 9.2 and 9.3 contain effective, not total stresses.

In Fig. 9.4(c) the normal compression and critical state lines are parallel and both have the same gradient, C_c. The parameter e_Γ defines the position of the critical state line in the same way that e_0 defines the position of the normal compression line. Equation (9.2) is the friction failure criterion discussed in Sec. 3.3 and ϕ_c' is the critical friction angle. The critical state line shown in Fig. 9.4(c) is directly above the critical overconsolidation line shown in Fig. 8.7. (The height of the critical state line above the critical overconsolidation line is τ_f' given by Eq. (9.2).) Later, in Chapter 11, we will see how the state of a soil initially on the wet side or the dry side moves towards the critical state line during shearing.

It is essential to emphasize that at the critical state soil continues to distort (i.e. suffer shear strains) without any change of shear stress or normal stress or voids ratio (i.e. it is distorting at constant state) and the strains are associated with turbulent flow. The essential features of the critical states are that, during shearing, all soils will ultimately reach their critical states (provided that the flow remains turbulent) and the critical states are independent of the initial states. Thus, in Fig. 9.1, the critical shear stresses τ'_f are the same for the soils initially on either the wet or the dry sides of critical, because they have the same normal effective stress σ'_f and the voids ratios e_f at the critical states will also be the same. Later we will see how we can explain fully the behaviour of soils from knowledge of their initial and ultimate states.

The existence of unique critical states for soils is, at first sight, surprising, but it is quite logical. Firstly, during continuous shear straining any soil must ultimately reach a constant state because, if it did not, it would continue to compress and strengthen or dilate and weaken indefinitely, which is, of course, impossible. During shearing from the initial to the critical states there will be relatively large strains and the soil will be essentially reworked or reconstituted by the shear straining. Thus the soil will forget its initial state and it is reasonable to suppose that the new, reconstituted, soil will achieve unique states independent of the initial states.

Since the critical state is independent of the initial state, the parameters ϕ'_c, e_Γ and C_c in Eqs. (9.2) and (9.3) depend only on the nature of the grains of the soil: they are material parameters. Relationships between these, and other, material parameters and soil classifications are discussed in Chapter 18.

The critical state lines illustrated in Fig. 9.4 are a very good idealization for the critical states of most clays and sands. For coarse-grained soils volume changes during first loading and during shearing are often accompanied by fracture of the soil grains, and it is often necessary to apply large stresses (greater than 1000 kPa) to identify the full range of behaviour.

9.4 Undrained strength

The critical state strength of soil given by Eq. (9.2) relates the ultimate shearing resistance to the corresponding normal effective stress. This can be used to determine soil strength provided that the pore pressure is known so that $\sigma'(= \sigma - u)$ can be calculated. Pore pressures in the ground will generally only be determinable for cases of drained loading and the strength for undrained loading – the undrained strength – must be calculated differently.

Figures 9.5(a) and (b) show the critical state line for soil and are the same as Figs. 9.4(a) and (b). Figure 9.5(c) combines these and shows the corresponding relationship between the critical state shear stress and the voids ratio: this shows the strength decreasing with increasing voids ratio. In saturated soil the voids ratio is simply related to the water content by Eq. (5.6) so Fig. 9.5(c) shows that there is a unique relationship between critical state strength and water content. It is common experience that the strength of soil decreases as the water content increases.

If a sample of saturated soil is taken from the ground and tested, or if it is tested in the ground, without any change in water content the strength measured will represent

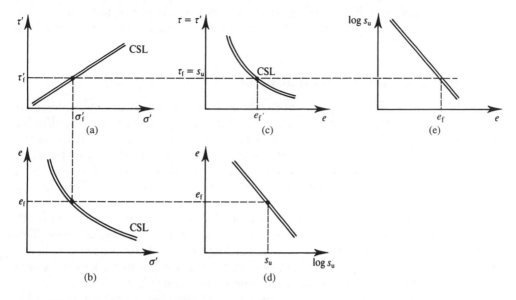

Figure 9.5 Undrained strength of soil.

the strength of the soil in the ground at its current water content. It will also represent the strength of soil near a foundation or excavated slope provided there is no change of water content during construction. This is the essence of the undrained strength: it is the strength of soil tested or loaded without any change in water content.

The undrained strength s_u is given by

$$\tau_f = s_u \tag{9.4}$$

which is the same as the cohesion failure criterion given by Eq. (3.9). From Eqs. (9.2) and (9.3) and noting that $\tau_f' = \tau_f = s_u$ we have

$$\log\left(\frac{s_u}{\tan \phi_c'}\right) = \frac{e_\Gamma - e}{C_c} \tag{9.5}$$

Equation 9.5 gives a linear relationship between the logarithm of undrained strength and voids ratio (or water content) as illustrated on Figs. 9.5(d) and (e).

Since undrained strength changes with changing voids ratio it is not a material parameter but, instead, it is a state-dependent parameter. Later we will discover some more state dependent parameters whose value depends on the current state of the soil. Although undrained strength s_u is a state dependent parameter it is related to voids ratio through Eq. (9.5) which contains the material parameters ϕ_c', e_Γ and C_c.

In many books you will find the undrained strength given by the Mohr–Coulomb criterion in terms of total stress as

$$\tau_f = c_u + \sigma \tan \phi_u \tag{9.6}$$

Comparing Eqs. (9.4) and (9.6) we have $c_u = s_u$ and $\phi_u = 0$. In saturated soil ϕ_u must always equal zero. If you find triaxial test results which give ϕ_u not equal to zero then either the sample or the drainage leads were not saturated or there was something wrong with the test.

9.5   Total and effective stress analyses

There are now two different criteria of the strength of soils which determine the shear stress at the critical state. The first, given by Eq. (9.2), relates the strength to the effective normal stress through a friction angle. In order to use this equation it is necessary to be able to calculate the effective stress which requires knowledge of the pore pressure. In general the pore pressure will be known only if the soil is drained. Analyses using Eq. (9.2) to determine strength are known as effective stress analyses and they are used when the soil is fully drained.

The second, given by Eq. (9.4), gives the strength directly as the undrained strength s_u and, for a given water content, this is independent of the total normal stress. This equation can be used when the soil is undrained and the voids ratio does not change during construction. Analyses using Eq. (9.4) to determine strength are known as total stress analyses and they are used when saturated soil is undrained.

It is important to get this right. You can do an effective stress analysis if the soil is fully drained and you know the pore pressure. You can do a total stress analysis if the soil is saturated and undrained. You must not mix these. If you are uncertain whether the soil is drained or not you should do both analyses and consider the worst case.

We will meet examples of total and effective stress analyses for foundations, slopes and retaining walls in later chapters of this book.

9.6   Normalizing

Representation of the critical state line, as in Figs. 9.4 and 9.5, is relatively straightforward because, at the critical state τ_f', σ_f' and e_f are uniquely related and there is only one critical state line. When we come to deal with peak states and other states before the critical, the situation is a little more complex and it will be convenient to have a method of normalizing stresses and voids ratios or specific volumes.

In Fig. 9.6 there is a point A where the state is σ_a' and e_a and there may also be some shear stresses (not necessarily at the critical state) τ_a'. In Sec. 8.3 we found that the overconsolidation ratio or the current state was an important factor in determining soil behaviour and so all the states with the same overconsolidation ratio should ideally have the same equivalent state after normalization. This can be achieved in a variety of ways and the two most common are illustrated in Fig. 9.6.

We have already seen that the positions of the normal compression and critical state lines are defined by the parameters e_0 and e_Γ and so the line of constant overconsolidation ratio containing A and A' is given by

$$e_\lambda = e_a + C_c \log \sigma_a' \tag{9.7}$$

Notice that e_λ contains both e_a and σ_a' and e_λ decreases with increasing overconsolidation ratio.

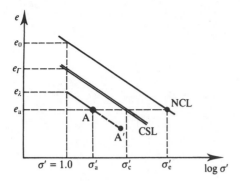

Figure 9.6 Parameters for normalizing shear test results.

Figure 9.7(b) shows the one-dimensional normal compression and critical state lines plotted with axes e_λ and τ' normalized with respect to the current stress σ'. Both lines appear as single points; at the normal compression point $\tau'/\sigma' = 0$ and $e_\lambda = e_0$ while at the critical state point $\tau'/\sigma' = \tan \phi_c'$ and $e_\lambda = e_\Gamma$. It seems fairly obvious that there will be important states between these, represented by the broken line, and we will explore these later.

A second method of normalizing is to make use of a critical stress. In Fig. 9.6 the critical stress σ_c' is on the critical state line at the same voids ratio as A and

$$\log \sigma_c' = \frac{e_\Gamma - e_a}{C_c} \tag{9.8}$$

(There is another equivalent pressure σ_e' on the normal compression line which is often used as a normalizing parameter. In this book I want to use σ_c' because the critical state line is unique for a given soil, while there are different normal compression lines for isotropic and one-dimensional compression and the position of the normal compression lines of natural soils can be influenced by structure and other effects.) Figure 9.7(a) shows the normal compression and critical state lines plotted with normalized stresses τ'/σ_c' and σ'/σ_c'. Again both lines appear as single points and the broken line corresponds to the broken line in Fig. 9.7(b). The position of the critical state line is determined by $\tau'/\sigma' = \tan \phi_c'$ and $\sigma'/\sigma_c' = 1.0$. From the geometry of Fig. 9.6 the position of the normal compression line is given by

$$\log \left(\frac{\sigma_e'}{\sigma_c'} \right) = \frac{e_0 - e_\Gamma}{C_c} \tag{9.9}$$

9.7 Critical state strength of soils measured in triaxial tests

So far we have considered strength of soils in ideal shear tests. As it is impossible to control and measure pore pressures in the conventional shear box apparatus the

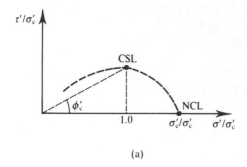

(a)

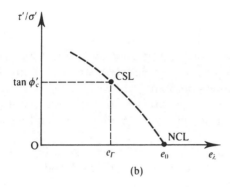

(b)

Figure 9.7 Normalized critical state and normal consolidation lines.

tests were drained so the pore pressure was zero and total and effective stresses were equal. We also considered undrained tests in which pore pressures were not measured and the undrained strength was related to the constant voids ratio. A more common and more useful test to examine soil behaviour is the triaxial test described in Chapter 7. In the triaxial test a cylindrical sample is subjected to total axial and radial stresses while the pore pressures and the sample volume can be controlled and measured independently so that it is possible to determine the effective stresses and the strains.

Relationships between stresses and strains in shear and triaxial tests were discussed in Chapter 3. For shear tests the shear and normal stresses and strains are τ', σ', γ and ε_v and for triaxial tests the equivalent parameters are q', p', ε_s and ε_v; these can be related through Mohr circle constructions, as described in Chapter 3.

All the features of soil behaviour in shear tests shown in Fig. 9.1 are seen in the results of triaxial tests plotted as q' against ε_s and ε_v against ε_s. In triaxial tests soils reach critical states where they continue to distort at a constant state (i.e. with constant effective stresses and constant volume) and soils initially on the dry side of the critical state line have peak states before the critical state is reached.

The critical state lines obtained from drained and undrained triaxial tests are shown in Fig. 9.8, which may be compared with Fig. 9.4 which shows the critical state line

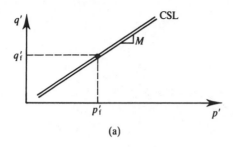

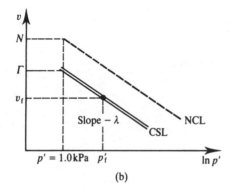

Figure 9.8 Critical state line for triaxial tests.

for shear tests. The critical state line in Fig. 9.8 is given by

$$q'_f = Mp'_f \tag{9.10}$$

$$v_f = \Gamma - \lambda \ln p'_f \tag{9.11}$$

where, as before, the subscripts f denote failure at the critical states. Comparing Eq. (9.10) with Eq. (9.2), the critical stress ratio M is equivalent to the critical friction angle ϕ'_c. In Fig. 9.8(b) the gradients of the critical state line and the isotropic normal compression line are both λ and the lines are parallel.

The parameters M, λ and Γ which describe the critical states in triaxial tests are equivalent to the parameters ϕ'_c, C_c and e_Γ which describe the critical states in shear tests. Both sets of parameters are material parameters. They depend only on the nature of the grains of the soil. (However, values of M and Γ measured in triaxial compression $(\sigma'_a > \sigma'_r)$ are a little different to the values measured in triaxial extension tests $(\sigma'_a < \sigma'_r)$.

Results of triaxial tests may be normalized like the results of shear tests. The normalizing parameters, shown in Fig. 9.9, are the critical pressure p'_c and the equivalent specific volume v_λ; these are comparable to σ'_c and e_λ in Fig. 9.6. (The equivalent pressure on the isotropic normal compression line p'_e is often used as a normalizing parameter for triaxial tests but, again, I want to use p'_c because the critical state line is unique.)

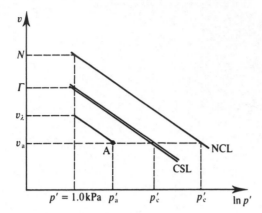

Figure 9.9 Parameters for normalizing triaxial test results.

From the geometry of Fig. 9.9,

$$v_\lambda = v_a + \lambda \ln p'_a \qquad (9.12)$$

$$\ln p'_c = \frac{\Gamma - v_a}{\lambda} \qquad (9.13)$$

Figure 9.10 shows critical state and normal compression lines normalized with respect to p'_c and v_λ: these correspond to Fig. 9.7 for shear tests. Again a broken line has been drawn representing important states between the normal compression and critical state lines; we will consider these states in later chapters. Note that, for triaxial tests, there will be two critical state lines, one for compression and one, with negative values of q', for extension.

The undrained strength s_u is uniquely related to the voids ratio, and hence to the specific volume. From Eqs. (9.10) and (9.11), noting that $s_u = \frac{1}{2}(\sigma'_a - \sigma'_r)_f = \frac{1}{2}q'_f$ we have

$$\ln\left(\frac{2s_u}{M}\right) = \frac{\Gamma - v}{\lambda} \qquad (9.14)$$

which is comparable to Eq. (9.5). Undrained strength may be measured in unconfined compression tests (i.e. tests with $\sigma_r = 0$) or in triaxial tests with any confining pressure provided that the voids ratio does not change. Figure 9.11 shows Mohr circles of total and effective stress for confined and unconfined compression tests on samples with the same voids ratio. The Mohr circles of effective stress are identical; they both touch the lines given by $\tau_f = s_u$ and $\tau'_f = \sigma'_f \tan \phi'_c$. The Mohr circles of total stress have the same diameter (because the voids ratios of the samples are the same) but they are in different positions, so the pore pressure in the unconfined compression test sample is negative. It is this negative pore pressure that produces positive effective stresses and gives rise to the unconfined compressive strength; this accounts for the strength of a sandcastle and the stability of a trench with steep sides.

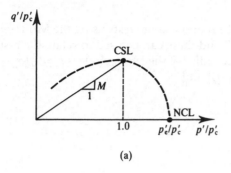

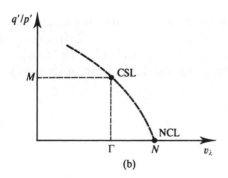

Figure 9.10 Normalized critical state and normal consolidation lines.

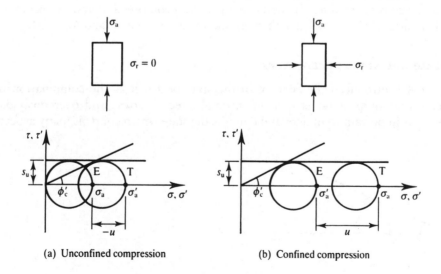

Figure 9.11 Undrained strength in compression tests.

9.8   Relationships between strength measured in shear and triaxial tests

The relationships between stress ratios in shear and triaxial tests using the Mohr circle constructions were introduced in Chapter 2 and these can be used to relate the results of triaxial and shear tests. From Fig. 9.12 the radius of the Mohr circle is $t' = \frac{1}{2}(\sigma_a' - \sigma_r')$ and the position of its centre is $s' = \frac{1}{2}(\sigma_a' + \sigma_r')$ and

$$\frac{t'}{s'} = \sin \phi' = \frac{(\sigma_a' - \sigma_r')}{(\sigma_a' + \sigma_r')} \tag{9.15}$$

$$\frac{\sigma_a'}{\sigma_r'} = \frac{(1 + \sin \phi')}{(1 - \sin \phi')} = \tan^2(45 + \tfrac{1}{2}\phi') \tag{9.16}$$

and, at the critical state $\phi' = \phi_c'$. Relationships between ϕ_c' and M can be obtained from Eqs. (9.10) and (9.16) with $q' = \sigma_a' - \sigma_r'$ and $p' = \frac{1}{3}(\sigma_a' + 2\sigma_r')$, noting that for compression $\sigma_a' > \sigma_r'$ while for extension $\sigma_a' < \sigma_r'$, so that in Eq. (9.16) σ_a'/σ_r' for compression must be replaced with σ_r'/σ_a' for extension. Readers are invited to work through the algebra and demonstrate that

$$M_c = \frac{6 \sin \phi_c'}{3 - \sin \phi_c'} \tag{9.17}$$

$$M_e = \frac{6 \sin \phi_c'}{3 + \sin \phi_c'} \tag{9.18}$$

where M_c is for triaxial compression and M_e is for triaxial extension. The critical friction angle ϕ_c' is approximately the same for triaxial compression and extension, so Eqs. (9.17) and (9.18) demonstrate that M_c and M_e are not equal and $M_c > M_e$.

9.9   State and state parameters

In Chapter 8 I introduced the concept of the state of a soil as the combination of its current voids ratio or water content, normal effective stress and overconsolidation ratio. It is important to understand that it is the state which controls many aspects

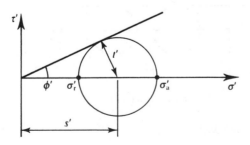

Figure 9.12 Stress ratios in triaxial tests.

of the strength and stiffness of soil and it is necessary to determine both voids ratio or water content and normal effective stress; water content or stress alone are not sufficient.

In Sec. 8.4 I considered the relationship between the current state and a critical line, which is below the critical state line, and introduced the ideas of states on the wet side of critical and on the dry side of critical, as shown in Fig. 8.7. While this qualitative description is important it is necessary to quantify state as the distance of current state from the critical state line.

Earlier in this Chapter I described procedures for normalizing test results for shear and triaxial tests and these are illustrated in Figs. 9.6 and 9.9. What I am going to do now is combine the ideas of state and normalizing to define state parameters.

Figure 9.13(a), which is like Fig. 9.6, shows a state at A with voids ratio e_a and normal effective stress σ'_a together with a critical state line. The vertical and horizontal distances of the point A from the critical state line are

$$S_v = e_\Gamma - e_\lambda \tag{9.19}$$

and

$$\log S_s = \log \sigma'_c - \log \sigma'_a \tag{9.20}$$

or

$$S_s = \frac{\sigma'_c}{\sigma'_a} \tag{9.21}$$

where S_v and S_s are alternative state parameters. From the geometry of Fig. 9.13(a) these state parameters are related by

$$S_v = C_c \log S_s \tag{9.22}$$

If the state A is on the critical state line $S_v = 0$ and $S_s = 1$. For states on the dry side of critical S_v is positive and $S_s > 1$: for soils on the wet side of critical S_v is negative and $S_s < 1$.

From Fig. 9.13(b) the state parameters for triaxial tests are

$$S_v = \Gamma - v_\lambda \tag{9.23}$$

and

$$S_s = \frac{p'_c}{p'_a} \tag{9.24}$$

From the geometry of Fig. 9.13(b) these state parameters are related by

$$S_v = \lambda \ln S_s \tag{9.25}$$

The state parameter S_v is similar to the state parameter defined by Been and Jefferies (1985). The state parameter S_s is similar to the overconsolidation ratio given by

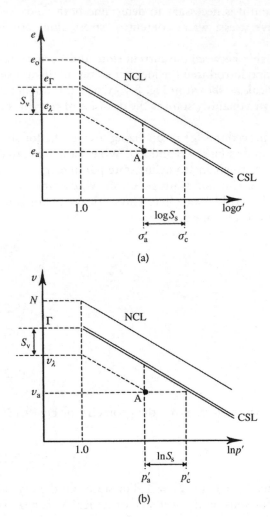

Figure 9.13 (a) State parameters for shear tests. (b) State parameters for triaxial tests.

Eq. (8.6) or (8.12) except the state parameter is related to the critical state line while the overconsolidation ratio is related to the normal compression line.

9.10 Simple experimental investigations of critical states

In any theoretical or experimental study of soil, and in many design studies, it is essential to determine the position of the critical state line as accurately as possible. This is needed to determine the ultimate strength for many of the design analyses described in Chapters 19 to 25 and it is also required to determine the ultimate states of soil samples. It is important to be able to distinguish between states on the wet side of critical from states on the dry side of critical, and the critical state parameters λ and Γ (or C_c and e_r) are required for normalizing soil test data.

However, if you try to measure critical states of soils in the conventional shear or triaxial tests described in Chapter 7, or if you use test results obtained by other people, you must be very careful to ensure that the samples really have reached their critical states, as defined m Sec. 9.3. Remember that for soils to reach their critical states they must be straining with no change of state (i.e. at constant shear stress, constant effective normal stress and constant volume) and with turbulent flow. This means that, if the stresses or pore pressures change at all, or if there are any volume changes, the states measured in the tests will not be the critical states. Very often soil tests are terminated when the apparatus runs out of travel, usually at strains of 20% or so. In many cases these strains are not large enough to reach the critical states in soils initially on the wet side of critical and are sufficient to cause slip planes to form in soils initially on the dry side. As discussed later, if there are any distinct slip surfaces in a test sample stresses and strains become non-uniform and cannot be measured reliably. We will discover later (in Chapter 11) that it is possible to find the critical states of soils from tests on normally consolidated and overconsolidated samples by considering the stresses and volume changes at strains before the critical state is reached.

Tests to determine the critical states of soils should be carried out on lightly over-consolidated samples for which the initial specific volume or voids ratio is close to the critical state value. Some people will tell you that soils do not reach unique critical states or that the critical state lines are curved, but usually their test data are suspect because the samples were not at their rigorously defined critical states.

There are some simple experiments that can be done to illustrate the critical states of soils and to obtain reasonable values of soil parameters. Because these simple experiments do not control pore pressures or drainage, tests on sands will be drained and will examine the critical friction angle ϕ'_c, while tests on clays will be undrained and will examine the undrained strength s_u.

We will see later that if there is no seepage the critical angle i_c of a failing slope is equal to the critical friction angle ϕ'_c, and so observation of slopes is a convenient method of determining the friction angle of soil. One test is to put dry sand in a horizontal rotating cylinder; as the cylinder rotates the angle of the continuously failing slope is the critical angle, as shown in Fig. 9.14. Another test is to pour dry sand into a cone and measure the cone angle.

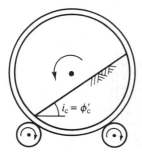

Figure 9.14 Rotating cylinder tests for ϕ'_c.

9.11   True cohesion in soils

In Figs. 9.4(a) and 9.8(a) the critical state lines were drawn passing through the origin, so at the critical state, soil has no strength when the effective normal stress is zero. If soils are cemented so the grains are glued together they will have a strength at zero effective stress but the strains required to reach the critical state are enough to break these cemented bonds.

Critical state lines obtained from the results of laboratory tests on soils always pass through the origin, at least within the accuracy of the results, which is typically about ± 2 to 3 kPa. It is very difficult to measure directly soil strength at zero effective stress. Some materials, such as dry sand, sugar and grain, are obviously cohesionless and you can pour them like water (although they will form cones), but it is not so obvious that fine-grained materials, such as clays, dry cement and flour, are cohesionless. The problem is that any moisture present will give rise to pore suctions which will raise the effective stresses, and hence the strength.

You can only really examine true cohesion in soils if the pore pressures are zero, which is clearly the case in dry materials. Dry flour has no cohesion if it is loose, because you can blow it away, but if you compact it by squeezing it in your hand it has a small strength. This is a result of the relatively large specific surface of finely ground flour.

The pore pressures in saturated fine-grained soils become zero if a sample is submerged in water. Figure 9.15 illustrates the behaviour of initially cylindrical samples of soil with different cohesions after they have been submerged in water. (The samples should be completely reconstituted so that any cementing is destroyed.) If the cohesion is zero as in Fig. 9.15(b) the sample forms a cone. If the cohesion is positive as a result of small interparticle attractions the sample will remain as a cylinder, as shown in Fig. 9.15(a). If, however, the water becomes dirty, this must mean that there were small interparticle repulsive forces and the true cohesion was negative. Each of the three characteristic types of behaviour shown in Fig. 9.15 are observed in tests on soils (Atkinson, Charles and Mhach, 1990). Even though the true cohesion in soils may be positive or negative the values are usually very small, only a few kiloPascals, which is too small to measure reliably in conventional laboratory tests.

9.12   Summary

1.   During shearing soils ultimately reach a critical state where they continue to distort with no further change of state (i.e. at constant shear stress, constant normal effective stress and constant volume).

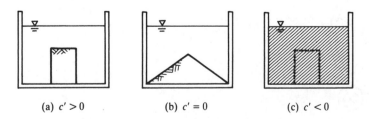

(a) $c' > 0$ (b) $c' = 0$ (c) $c' < 0$

Figure 9.15 Assessment of true cohesion in soils.

2. Before the critical state there may be a peak state and after large strains clay soils reach a residual state. The peak state is associated with dilation and the residual state is associated with laminar flow of flat clay particles.

3. The critical states of soils in shear tests are given by

$$\tau_f' = \sigma_f' \tan \phi_c' \tag{9.2}$$

$$e_f = e_\Gamma - C_c \log \sigma_f' \tag{9.3}$$

4. The critical state strength of soil is uniquely related to its voids ratio or water content so for undrained loading of saturated soil (i.e. at constant water content) the undrained strength s_u remains unchanged.

5. If the soil is drained and effective stresses can be determined you can use effective stress analyses and the critical state strength is given by ϕ_c'. If the soil is saturated and undrained you can use total stress analyses and the critical state strength is given by the undrained strength s_u.

6. To take account of different normal effective stresses and different voids ratios, stresses, should be normalized with respect to the critical stress σ_c' or the critical voids ratio e_λ given by

$$e_\lambda = e + C_c \log \sigma' \tag{9.7}$$

$$\log \sigma_c' = \frac{e_\Gamma - e}{C_c} \tag{9.8}$$

7. The critical states in shear tests are found also in triaxial tests and the critical state line is given by

$$q_f' = M p_f' \tag{9.10}$$

$$v_f = \Gamma - \lambda \ln p_f' \tag{9.11}$$

8. The critical state parameters ϕ_c', e_Γ and C_c (or M, Γ and λ) are material parameters: they depend only on the nature of the soil grains.

9. The state of a soil is described by the distance of the voids ratio − effective stress point from the critical state line and it is given by the either of the state parameters S_v or S_s.

Worked examples

Example 9.1: Determination of critical state soil parameters   A number of drained and undrained triaxial tests were carried out on normally consolidated and overconsolidated samples of the same soil. Table 9.1 gives values for the stress parameters q_f' and p_f' and the specific volume v_f when the samples had reached failure at their critical states.

The data are shown plotted in Fig. 9.16. Scaling from the diagram, $M = 0.98$, $\lambda = 0.20$. Substituting (say) $v = 1.82$ and $p' = 600\,\text{kPa}$ with $\lambda = 0.20$ into Eq. (9.11) we have $\Gamma = 3.10$.

Table 9.1 Critical states of a soil – Example 9.1

Test	p'_f (kPa)	q'_f (kPa)	v_f
A	600	588	1.82
B	285	280	1.97
C	400	390	1.90
D	256	250	1.99
E	150	146	2.10
F	200	195	2.04

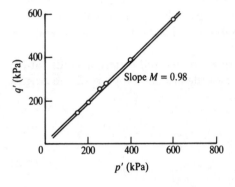
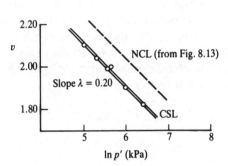

Figure 9.16 Critical states of a soil – Example 9.1.

Example 9.2: Determination of critical states of soils   A soil has the parameters $M = 0.98$, $\lambda = 0.20$ and $\Gamma = 3.10$. Four samples were isotropically compressed and swelled to the initial states shown in the first four columns of Table 9.2. In each case the pore pressure was $u_0 = 100\,\text{kPa}$. Each sample was tested by increasing q with the total mean stress p held constant: samples A and C were tested drained and samples B and D were tested undrained.

For the drained tests $p'_f = p'_0$ and for the undrained tests $v_f = v_0$. From Eq. (9.11) the values of v_f in drained tests and p'_f in undrained tests are given by

$$v_f = \Gamma - \lambda \ln p'_f \quad \text{or} \quad p'_f = \exp\left(\frac{\Gamma - v_f}{\lambda}\right)$$

Table 9.2 Results from tests on a soil – Example 9.2

Sample	p'_0 (kPa)	R_0	v_0	p'_f (kPa)	v_f	q'_f (kPa)	u_f (kPa)
A	600	1	1.97	600	1.82	588	100
B	600	1	1.97	284	1.97	278	416
C	150	4	2.04	150	2.09	147	100
D	150	4	2.04	200	2.04	196	50

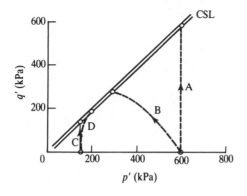
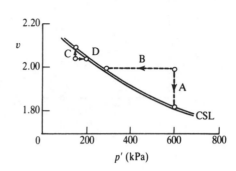

Figure 9.17 Results from tests on a soil – Example 9.2.

Notice that the tests were carried out with p constant so that $p_f = p_0$, the pore pressures at failure u_f in the undrained tests are given by

$$u_f = p_f - p'_f = p_0 - p'_f = p'_0 + u_0 - p'_f$$

From Eq. (9.10),

$$q_f = Mp'_f$$

The points corresponding to isotropic compression and to failure at the critical state are shown in Fig. 9.17; these are linked by lines that represent approximately the state paths for the tests.

Example 9.3: Normalized critical states   The initial states and the critical states given in Table 9.1 can be normalized with respect to the critical pressure p'_c given by Eq. (9.13) or with respect to the equivalent specific volume v_λ given by Eq. (9.12). The values for the normally consolidated samples A and C are given in Table 9.3 and the points representing the critical state and normal compression lines are given in Fig. 9.18.

The initial and final state points may be joined together as shown by a line that represents approximately the state paths followed by the drained and undrained tests. Notice that in both tests the value of p'/p'_c decreases from 2.11 to 1.00. In the drained test this is because p'_c increases from 284 to 600 kPa as the specific volume decreases

Table 9.3 Normalized initial and critical states of a soil – Example 9.3

Sample	Initial state				Critical state						
	p'_0 (kPa)	p'_c (kPa)	p'_0/p'_c	v_λ	q'_f	p'_f (kPa)	p'_c (kPa)	q'_f/p'_c	p'_f/p'_c	q'_f/p_f	v_λ
A	600	284	2.11	3.25	588	600	600	0.98	1.00	0.98	3.10
C	600	284	2.11	3.25	278	284	284	0.98	1.00	0.98	3.10

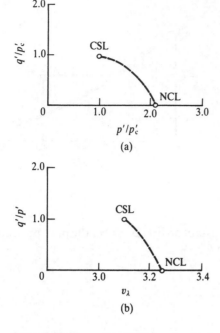

Figure 9.18 Normalized initial and critical states of a soil – Example 9.3.

while $p' = 600\,\text{kPa}$ remains constant, but in the undrained test p' decreases from 600 to 284 kPa as the pore pressure increases while $p'_c = 284\,\text{kPa}$ remains constant because the specific volume does not change.

Example 9.4: Critical state Mohr circles and friction angle   For the four tests given in Table 9.2 the principal stresses at the critical state can be calculated from Eqs. (3.3) and (3.4). Rearranging:

$$\sigma'_a = p' + \tfrac{2}{3}q' \quad \sigma'_r = p' - \tfrac{1}{3}q'$$

Values for σ'_a and σ'_r are given in Table 9.4 and the Mohr circles are given in Fig. 9.19. Scaling from the diagram, the critical friction angle is $\phi'_c = 25°$.

Table 9.4 Critical states of a soil – Example 9.4

Sample	q'_f (kPa)	p'_f (kPa)	σ'_{af} (kPa)	σ'_{rf} (kPa)
A	588	600	992	404
B	278	284	469	191
C	147	150	248	101
D	196	200	330	134

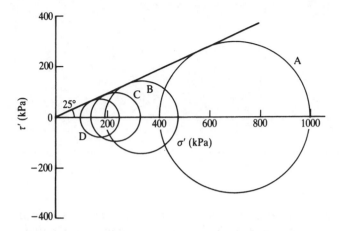

Figure 9.19 Mohr circles for critical states of a soil – Example 9.4.

References

Atkinson, J. H., J. A. Charles and H. K. Mhach (1990) 'Examination of erosion resistance of clays in embankment dams', *Quarterly Journal of Engineering Geology*, **23**, 103–108.
Been, K. and M. G. Jefferies (1985) A state parameter for sands, *Geotechnique*, **35**, 2.
Skempton, A. W. (1964) 'Long term stability of clay slopes', *Geotechnique*, **14**, 77–101.

Further reading

Atkinson, J. H. and P. L. Bransby (1978) *The Mechanics of Soils*, McGraw-Hill, London.
Muir Wood, D. M. (1991) *Soil Behaviour and Critical State Soil Mechanics*, Cambridge University Press, Cambridge.
Schofield, A. N. and C. P. Wroth (1968) *Critical State Soil Mechanics*, McGraw-Hill, London.

Chapter 10

Peak states

10.1   Introduction

Figure 10.1 shows the states of soil samples at the same effective stress σ' but at different voids ratios and overconsolidation ratios: at N the soil is normally consolidated, at W it is lightly overconsolidated or loose and the state is on the wet side of the critical state, and D_1 and D_2 are two states on the dry side where the soil is heavily overconsolidated or dense. For samples W and N on the wet side of critical the state parameter S_v (see Sec. 9.9) is positive and for samples D_1 and D_2 the state parameter is negative. Figure 10.2 shows the behaviour of these samples during drained shear tests and is similar to Fig. 9.1. At the critical states at C the samples have the same shear stress τ_f', the same normal stress σ_f' and the same voids ratio e_f, but at the peak states the shear stresses and voids ratios are different. The idealized behaviour described in this chapter is based on experimental data given by Atkinson and Bransby (1978) and by Muir Wood (1991).

Peak states from shear tests on samples with different values of normal effective stress, overconsolidation ratio and voids ratio generally fall within the region OAB in Fig. 10.3 which is above the critical state line, and at first sight there is no clear relationship for the peak states as there was for the critical states. There are three ways of examining the peak states: the first is to make use of the Mohr–Coulomb equation, the second is to fit a curved line to the peak state points and the third is to include a contribution to strength from dilation. I am going to consider each of these three methods. They are simply different ways to describe the same peak strengths; although the equations are different the soil behaviour remains the same.

10.2   Mohr–Coulomb line for peak strength in shear tests

Figure 10.4 shows peak states of two sets of samples which reached their peak states at voids ratios e_1 and e_2. These can be represented by the Mohr–Coulomb equation written with effective stresses

$$\tau_p' = c_{pe}' + \sigma_p' \tan \phi_p' \tag{10.1}$$

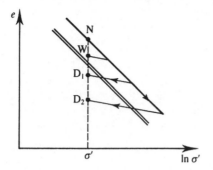

Figure 10.1 Initial states of samples at the same stress but different voids ratios.

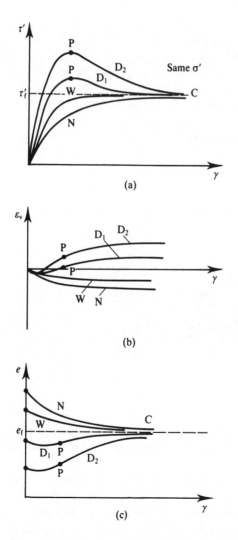

Figure 10.2 Behaviour of samples in drained shear tests.

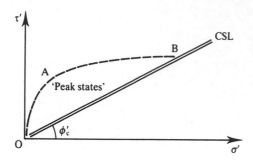

Figure 10.3 Region of peak states.

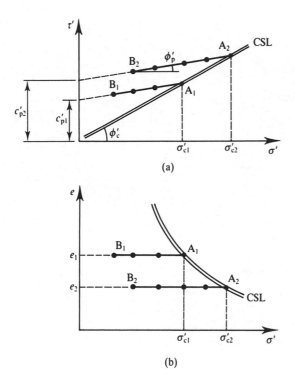

Figure 10.4 Mohr–Coulomb lines for peak states.

where the subscripts p are there to make it clear that Eq. (10.1) relates to the peak state and the subscript e in c'_{pe} is there because the cohesion intercept depends on the voids ratio.

There are a number of important things to notice about the peak states shown in Fig. 10.4. The peak friction angle ϕ'_p is less than the critical friction angle ϕ'_c and the peak state lines meet the critical state line at points such as A_1 and A_2. For any states to the right of the critical state line in Fig. 10.4(b) the soil is on the wet side of critical and, on shearing, it compresses and reaches its critical state without a peak, as in Fig. 10.2. As a consequence peak states are associated with dense or overconsolidated soils on the dry side which dilate on shearing.

In Fig. 10.4 the peak state lines have been terminated at low stresses at points such as B_1 and B_2 and peak states at low stresses are not given by Eq. (10.1). This means that the cohesion intercept c'_{pe} is not the shear stress which the soil can sustain at zero stress and it is merely a parameter required to define the Mohr–Coulomb equation. Since, in this case, these peak states apply equally for clean sand and reconstituted clays this cohesion intercept should not be associated with cementing or interparticle attraction in clays.

In order to take account of the different voids ratios e_1 and e_2 in Fig. 10.4, we can make use of the normalizing parameter σ'_c described in Chapter 9. Figure 10.5 shows the peak state lines from Fig. 10.4 normalized with respect to σ'_c. Now all the peak state lines for different voids ratios reduce to the single line BA and A is the critical state point.

The equation of the line BA is

$$\frac{\tau'_p}{\sigma'_c} = c'_p + \frac{\sigma'_p}{\sigma'_c} \tan \phi'_p \tag{10.2}$$

where

$$c'_p = \frac{c'_{pe}}{\sigma'_c} \tag{10.3}$$

From Eqs. (9.8) and (10.3) the peak cohesion intercept c'_{pe} in Eq. (10.1) is

$$\log\left(\frac{c'_{pe}}{c'_p}\right) = \frac{e_\Gamma - e}{C_c} \tag{10.4}$$

and so c'_{pe} decreases with increasing voids ratio and, for a given normal effective stress, the peak strength decreases with increasing water content. From the geometry of Fig. 10.5

$$c'_p = \tan \phi'_c - \tan \phi'_p \tag{10.5}$$

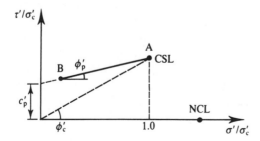

Figure 10.5 Normalized peak and critical states for shear tests.

and Eq. (10.2) becomes

$$\frac{\tau_p'}{\sigma_c'} = \left(\tan \phi_c' - \tan \phi_p'\right) + \left(\frac{\sigma_p'}{\sigma_c'}\right) \tan \phi_p' \tag{10.6}$$

In Eqs. (10.2) and (10.6) the parameters c_p', ϕ_p' and ϕ_c' do not depend on the voids ratio, or water content, which is contained in the critical stress σ_c' through Eq. (9.8), so they are material parameters but they are only material parameters if the Mohr–Coulomb equation is a good representation for the peak strength of soil.

10.3 Mohr–Coulomb line for peak strength in triaxial tests

In triaxial tests the peak states depend on the specific volume in the same way as for shear tests. Figure 10.6(a) shows the peak state line for the particular specific volume v in Fig. 10.6(b). In the region AB this is given by

$$q_p' = G_{pv} + H_p p_p' \tag{10.7}$$

where H_p is the gradient and G_{pv} is the intercept on the q' axis. The broken line OT at a gradient $dq'/dp' = 3$ represents the condition $\sigma_r' = 0$. Since uncemented soils cannot

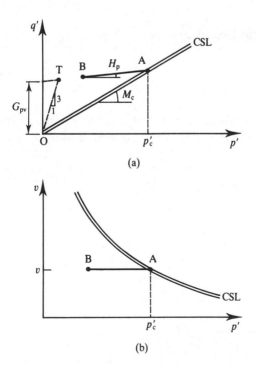

(a)

(b)

Figure 10.6 Peak states in triaxial tests.

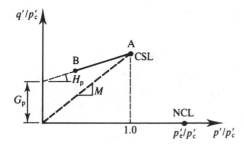

Figure 10.7 Normalized peak and critical states for triaxial tests.

sustain tensile (negative) effective stresses, this represents a limit to possible states; the line OT is known as the tension cut-off and it is equivalent to the τ' axis for shear tests in Fig. 10.4. The parameter G_{pv} is simply a parameter that defines the position of the peak state line and is not necessarily the peak strength at low effective stress.

After normalization with respect to p'_c the results of triaxial tests appear as shown in Fig. 10.7, which is similar to Fig. 10.5 for shear tests. The critical state and normal compression lines reduce to single points and the peak states fall on a single line given by

$$\frac{q'_p}{p'_c} = G_p + H_p \left(\frac{p'_p}{p'_c} \right) \tag{10.8}$$

where the gradient is H_p and the intercept is G_p.

From the geometry of Fig. 10.7

$$G_p = M - H_p \tag{10.9}$$

and Eq. (10.8) becomes

$$\frac{q'_p}{p'_c} = (M - H_p) - \left(\frac{p'_p}{p'_c} \right) H_p \tag{10.10}$$

Equation (10.10) for triaxial tests is equivalent to Eq. (10.6) for shear tests. In both there are two independent parameters ϕ'_p and ϕ'_c or M and H_p. The voids ratio, or water content, is contained in the critical stress σ'_c through Eq. (9.8) or in the critical pressure p'_c through Eq. (9.13).

10.4   A power law equation for peak strength

Uncemented soil can have no strength when the normal effective stress is zero. This is why you can pour dry sand from a jug. This means that the peak strength envelope must pass through the origin where $\tau'_p = \sigma'_p = 0$ and it must meet the critical state line

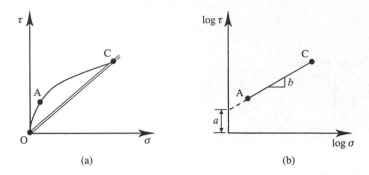

Figure 10.8 Power law peak strength envelope for shear tests.

where $\tau_c' = \sigma_c' \tan \phi_c'$. The only smooth envelope which can meet these requirements is the curved line OAC shown in Fig. 10.8(a).

A convenient equation for this line is a power law

$$\tau_p' = a\sigma_p'^{b} \tag{10.11}$$

or

$$\log \tau_p' = \log a + b \log \sigma_p' \tag{10.12}$$

so the curved peak strength envelope is linear if the stresses are plotted with logarithmic axes, as shown in Fig. 10.8(b). If $b = 1$ then Eq. (10.11) is linear and the parameter a is equivalent to $\tan \phi_c'$ in Eq. (9.2). The parameter b describes how curved the peak envelope is: the smaller the value of b the greater the curvature. Both parameters a and b are state dependent parameters and they depend on the voids ratio or water content.

To take account of voids ratio Eq. (10.11) can be normalized with respect to the critical stress σ_c'. The power law for the peak failure envelope becomes

$$\frac{\tau_p'}{\sigma_c'} = A\left(\frac{\sigma_p'}{\sigma_c'}\right)^B \tag{10.13}$$

Since the peak envelope ends at the critical state point, $A = \tan \phi_c'$ and Eq. (10.13) becomes

$$\frac{\tau_p'}{\sigma_c'} = \tan \phi_c'\left(\frac{\sigma_p'}{\sigma_c'}\right)^B \tag{10.14}$$

or

$$\log\left(\frac{\tau_p'}{\sigma_p'}\right) = \log(\tan \phi_c') + B \log\left(\frac{\sigma_p'}{\sigma_c'}\right) \tag{10.15}$$

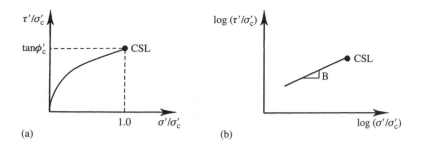

Figure 10.9 Power law peak strength envelope for shear tests.

and these are shown in Fig. 10.9. Equation (10.14) describes the peak strength of soil in shear tests. The voids ratio or water content is contained within the critical stress σ'_c through Eq. (9.8). Equation (10.14) contains two parameters ϕ'_c and B which are material parameters and they depend only on the nature of the soil grains.

For triaxial tests the power law peak strength envelope is

$$\frac{q'_p}{p'_c} = M \left(\frac{p'_p}{p'_c} \right)^{\beta} \tag{10.16}$$

Equation (10.16) shown in Fig. 10.10 describes the peak strength of soil in triaxial tests. The voids ratio or water content is contained with in the critical stress p'_c through Eq. (9.13). Equation (10.16) contains two parameters M and β which are material parameters and they depend only on the nature of the soil grains.

Figure 10.11(a) shows points *a*, *b* and *c* which are the peak strengths in three shear tests on the same soil. The test results have been normalized with respect to the critical stress σ'_c to take account of different water contents. The curved envelope which passes close to the three points, through the origin and through the critical state point is a power law curve. A linear Mohr–Coulomb envelope has been drawn as a best fit to

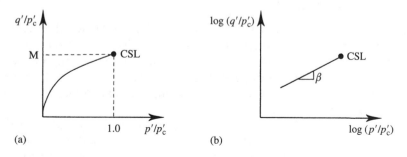

Figure 10.10 Power law peak strength envelope for triaxial tests.

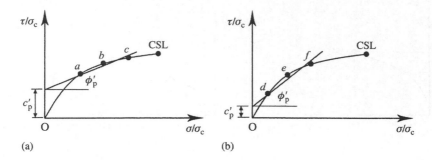

Figure 10.11 Comparison between linear Mohr–Coulomb and curved power law failure criteria for peak strengths.

the test points. Figure 10.11(b) shows different points d, e and f which are close to the same power law curve as that in Fig. 10.11(a) with a linear Mohr–Coulomb envelope drawn as the best fit to the test points.

Neither of the linear Mohr–Coulomb envelopes passes through the origin or the critical state point. Although both Mohr–Coulomb envelopes are close to the test points which they represent they have different parameters c'_p and ϕ'_p. These are not, after all, material parameters because they depend on the range of effective stress over which the peak strengths were measured.

The linear Mohr–Coulomb failure criterion is the one which is nearly always used in current geotechnical engineering practice, mainly because it has been the basis of soil mechanics for a very long time. However, the curved power law envelope passes through the origin and through the critical state point which is required for uncemented soils. Peak strengths measured in triaxial tests over a wide range of effective stresses are close to curved power law envelopes. On the present evidence the curved power law envelope is to be preferred to the linear Mohr–Coulomb envelope. Notice that the Mohr–Coulomb envelope is above the curved power law envelope, and so it is unconservative, for stresses outside the range of the tests.

10.5   Peak states and dilation

If you examine Fig. 10.2 you will notice that the samples D_1 and D_2 which have peak strengths also dilate during shearing; their volume increases. You will also notice that the rate of dilation given by the gradient $(\delta\varepsilon_v/\delta\gamma)$ of the volumetric strain curve in Fig. 10.2(b) is maximum at the point of peak shear stress. When the soil is at its peak strength the shear stresses are both overcoming friction and expanding the sample against the normal effective stresses.

Figure 10.12(a) shows a sample of soil which was originally on the dry side of critical at its peak state in a shear test. The effective stresses are τ' and σ' and there are small increments of displacement δh and δv. As given in Chapter 2 the angle of dilation ψ is

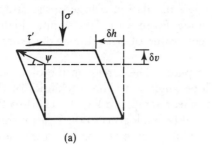
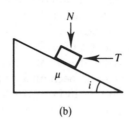

(a)   (b)

Figure 10.12 A model for shearing and dilation.

given by

$$\tan \psi = -\frac{\delta \varepsilon_v}{\delta \gamma} = \frac{\delta v}{\delta h} \tag{10.17}$$

and so the angle ψ describes the direction of movement of the top of the sample as shown in Fig. 10.12(a).

   Figure 10.13 shows how the stress ratio τ'/σ' and height change with horizontal displacement. This is similar to the behaviour of samples D_1 and D_2 in Fig. 10.2. Firstly the sample compresses from O to A. At the point A the stress ratio is equal to $\tan \phi_c'$ and the angle of dilation ψ is zero; the sample is neither compressing nor

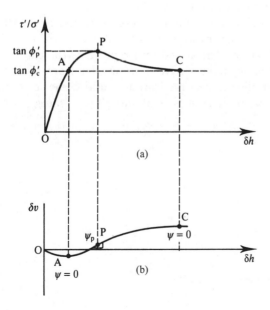

Figure 10.13 Shearing and dilation in shear tests.

dilating but, for an instant it is straining at constant volume. At C at the critical state the stress ratio is again equal to $\tan \phi'_c$ and the angle of dilation ψ is zero; the sample is straining at constant stress and at constant volume. From A to C the sample dilates. The maximum rate of dilation given by the largest value of ψ occurs at the point P where the stress ratio is a maximum.

Figure 10.12(b) shows a frictional block on a plane. The forces on the block are T and N, the angle of friction is μ and the slope angle is i. The mechanics of the sliding block are similar to the mechanics of the shear sample in Fig. 10.12(a) so the relationships between τ' and σ' in Fig. 10.12(a) will be analogous to the relationships between T and N in Fig. 10.12(b), the angle of dilation ψ is analogous to the slope angle i and μ is analogous to ϕ'.

From Fig. 10.12(b), resolving vertically and horizontally and after some algebra

$$\frac{T}{N} = \tan(\mu + i) \tag{10.18}$$

Following the analogy between the shear test sample and the sliding block, the soil behaviour can be represented by

$$\frac{\tau'}{\sigma'} = \tan(\phi'_c + \psi) \tag{10.19}$$

Equation (10.19) corresponds with the observations from Fig. 10.13. At the points A and C $\psi = 0$ and the stress ratio is $\tau'/\sigma' = \tan \phi'_c$ and at the point P both τ'/σ' and ψ have their maximum values. In fact Eq. (10.19) relates stress ratio τ'/σ' to angle of dilation ψ throughout the whole test from O to the ultimate critical state at C. This is the essence of the stress–dilatancy theory (Taylor, 1948).

Figure 10.14 shows a stress path O–A–P–C for the shear test illustrated in Fig. 10.13. The peak stress ratio, and indeed any stress ratio is the sum of a component due to friction and a component due to dilation. The stress path represents the changes of stress throughout the test in which the normal effective stress was constant. The stress ratio τ'/σ' is given by Eq. (10.19) at all stages of the test: at A and C the stress ratio $\tau'/\sigma' = \tan \phi'_c$ because $\psi = 0$; for the path O–A $\psi < 0$ and $\tau'/\sigma' < \tan \phi'_c$; for the

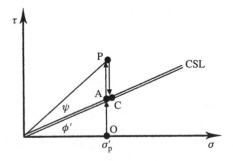

Figure 10.14 Peak strength of dilating soil.

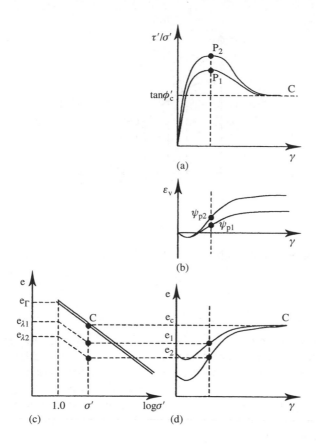

Figure 10.15 Dilation and peak strength related to state.

path A–D C $\psi > 0$ and $\tau'/\sigma' > \tan\phi'_c$; at P both ψ and τ'/σ' have their maximum values.

The maximum angle of dilation and the peak strength are related to the voids ratio at the peak states. Figures 10.15(a) and 10.15(b) show the behaviour of two samples of the same soil in shear tests and they are the same as that shown in Fig. 10.13. The peak states at P are a consequence of the rate of dilation ψ through Eq. (10.19). In Fig. 10.15(b) the volumetric strains start from zero and end at different strains but the voids ratios start from different initial values and end at the same critical voids ratio as shown in Fig. 10.15(d). The specific volumes at the peak state are e_1 and e_2 and these have different values of e_λ, as shown in Fig. 10.15(c). Sample 2, which has the larger peak strength, has the smaller voids ratio: it is more dense and its overconsolidation ratio is larger.

Figure 10.16(a) shows the peak states P_1 and P_2 with axes τ'/σ' and e_λ. The critical state line is at $\tau'/\sigma' = \tan\phi'_c$ and $e_\lambda = e_\Gamma$ and the normal compression line is at $\tau'/\sigma' = 0$ and $e_\lambda = e_0$. The peak strength points and the critical state line are on a smooth curve and this can be extended to the normal compression line. Figure 10.16(a) is similar to Fig. 9.7(b).

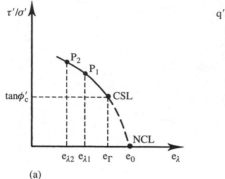
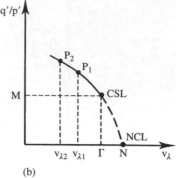

(a)   (b)

Figure 10.16 Normalized peak states.

For triaxial tests the equation which is analogous to Eq. (10.19) is

$$\frac{q'}{p'} = M - \frac{d\varepsilon_v}{d\varepsilon_s} \tag{10.20}$$

(The negative sign is required because $d\varepsilon_v$ is negative for dilation.) Again the stress ratio q'/p' is the sum of a friction component M and a component due to dilation. The relationship between the peak strength and the specific volume for triaxial tests is qualitatively similar to that for shear tests. Figure 10.16(b) shows peak strengths measured in two tests plotted with axes q'/p' and v_λ. The critical state line is at $q'/p' = M$ and $v_\lambda = \Gamma$ and the normal compression line is at $q'/p' = 0$ and $v_\lambda = N$. Figure 10.16(b) is similar to Fig. 9.10(b).

10.6   Comparison between the Mohr–Coulomb and the stress–dilatancy theories

Remember that the Mohr–Coulomb and stress–dialatancy theories are two different ways of describing the same soil behviour. The relationships between the two are illustrated in Fig. 10.17.

This shows four peak state points. The points B_1 and C_1 have the same voids ratio e_1 and they lie on the same Mohr–Coulomb line given by c'_{p1} and ϕ'_p. B_2 and C_2 are similar points at the same voids ratio e_2 and they lie on the Mohr–Coulomb line given by c'_{p2} and ϕ'_p. Points B_1 and B_2 have the same normal stress σ'_b, but B_2 is more heavily overconsolidated than B_1 and has a lower voids ratio. Since B_2 and B_1 will reach the same critical states at B_c, sample B_2 must dilate more (i.e. have a larger value of ψ_p) than sample B_1. C_1 and C_2 are similar points. Points B_1 and C_2 have the same overconsolidation ratio but different voids ratios and normal stresses; their peak states are given by Eq. (10.19) with the same value of ψ_p.

These simple analyses demonstrate the importance of considering voids ratio, or water content and dilation, as well as shear and normal effective stresses when analysing test results and soil behaviour.

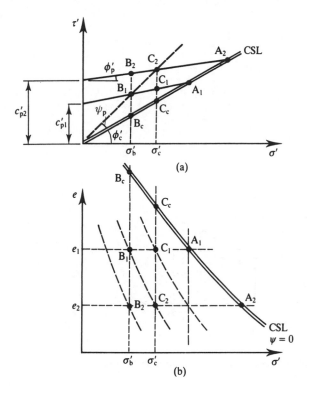

Figure 10.17 Peak states of soils with different states.

10.7 Relationship between peak strength and state parameter

Figure 10.16(a) shows the relationship between the peak strength and the normalizing parameters e_λ or v_λ. Instead of using these normalizing parameters we could use the volume state parameter S_v defined in Fig. 9.13 and by Eqs. (9.19) and (9.23).

Figure 10.18(b) shows stress–strain curves for two samples of the same soil in shear tests. The samples had different initial voids ratios, as shown in Fig. 10.15. They have the same critical state strength but different peak strengths. Figure 10.18(a) shows the voids ratios at the peak states and the corresponding state parameters S_v. Sample 2 has the larger peak strength and the larger state parameter because its state is furthest from the critical state line.

Figure 10.18(c) shows the relationship between the peak stress ratio and the state parameter at the peak. When $S_v = 0$ the peak strength and the critical state strength are the same and the peak strength increases steadily with increasing values of the state parameter. There will be a similar increase in peak strength with state parameter for triaxial tests.

Figure 10.18(c) shows a linear relationship between peak stress ratio and state parameter but the relationship may not be linear for all soils. This will be discussed further in Sec. 11.6.

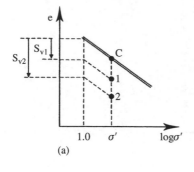

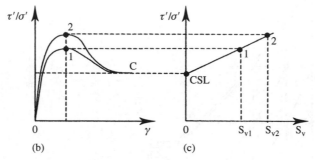

Figure 10.18 Relationship between peak stress ratio and state parameter.

10.8 Summary

1. During shearing overconsolidated soils with states on the dry side of critical reach peak states where the shear stress and stress ratio are larger than those at the critical state.
2. After the peak stresses have been normalized with respect to the critical stress or the critical voids ratio to take account of different voids ratios or water contents the peak states of a particular soil all fall on a single envelope.
3. There are three alternative criteria which may be used to describe the peak states of soils; these are the Mohr–Coulomb, power law and stress–dilatancy criteria.
4. The Mohr–Coulomb criterion is given by

$$\frac{\tau'_p}{\sigma'_c} = \left(\tan \phi'_c - \tan \phi'_p\right) + \left(\frac{\sigma'_p}{\sigma'_c}\right)\tan \phi'_p \tag{10.6}$$

The power law criterion is given by

$$\frac{\tau'_p}{\sigma'_c} = \tan \phi'_c \left(\frac{\sigma'_p}{\sigma'_c}\right)^B \tag{10.14}$$

The stress–dilatancy criterion equation is given by

$$\frac{\tau'}{\sigma'} = \tan(\phi'_c + \psi) \tag{10.19}$$

5. In Eqs. (10.6) and (10.14) the parameters ϕ'_c, ϕ'_p, and B are material parameters and they depend only on the nature of the soil grains. The voids ratio or water content is included in the critical stress σ'_c. In Eq. (10.19) the angle of dilation ψ depends on the initial state.
6. The peak state of a soil is related to the state parameter.

Worked examples

Example 10.1: Determination of peak state parameters   Table 10.1 shows data obtained at the peak state from a series of shear tests on the same soil as that in the examples in Chapter 9 (Note that the samples reached their peak state at one of only two different voids ratios.) The peak states are plotted in Fig. 10.19. The points fall close to two straight lines given by $c'_{pe} = 60$ kPa and $c'_{pe} = 130$ kPa with $\phi'_p = 15°$ in both cases.

The test results can be normalized with respect to the equivalent stress σ'_c given by Eq. (9.8). The soil parameters are $C_c = 0.46$ and $e_\Gamma = 2.17$. The normalized stresses are given in Table 10.1 and these are plotted in Fig. 10.20. The data now all fall close to a single straight line given by $c'_p = 0.2$ and $\phi'_p = 15°$.

Table 10.1 Peak states of a soil – Example 10.1

Sample	τ'_p (kPa)	σ'_p (kPa)	e_p	σ'_c (kPa)	τ'_p/σ'_c	σ'_p/σ'_c
A	138	300	1.03	300	0.46	1.00
C	123	240	1.03	300	0.41	0.80
E	108	180	1.03	300	0.36	0.60
G	93	120	1.03	300	0.31	0.40
B	264	540	0.89	606	0.44	0.90
D	228	420	0.89	606	0.38	0.70
F	198	300	0.89	606	0.33	0.50

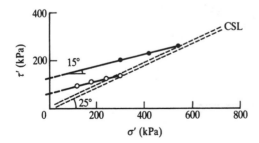

Figure 10.19 Peak states of a soil – Example 10.1.

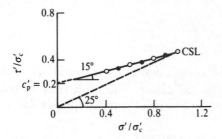

Figure 10.20 Normalized peak states of a soil – Example 10.1.

Example 10.2: Curved peak state envelope   A further set of four shear tests was car-ried out in addition to those described in Example 10.1 and the results are given in Table 10.2. The peak states are plotted in Fig. 10.21 together with the data from Table 10.1. The points fall close to two curved lines, although at large, stresses, for which the data are those given in Table 10.1, the lines are very nearly straight. As before, the stresses can be normalized with respect to the equivalent stress σ'_c. The normalized stresses are given in Table 10.2 and plotted in Fig. 10.22(a). The data

Table 10.2 Peak states of a soil – Example 10.2

Sample	τ'_p (kPa)	σ'_p (kPa)	e_p	σ'_c (kPa)	τ'_p/σ'_c	σ'_p/σ'_c
J	63	60	1.03	300	0.21	0.20
L	30	15	1.03	300	0.10	0.05
H	156	180	0.89	606	0.26	0.30
K	84	60	0.89	606	0.14	0.10

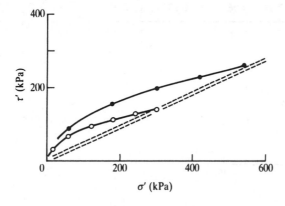

Figure 10.21 Peak states of a soil – Example 10.2.

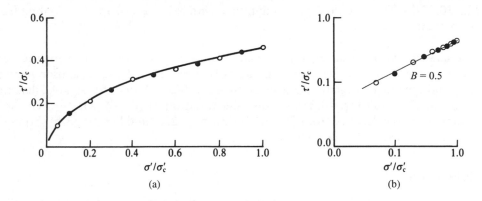

Figure 10.22 Normalized peak states of a soil – Example 10.2.

Table 10.3 Peak states and dilation of a soil – Example 10.3

Sample	τ_p' (kPa)	σ_p' (kPa)	e_p	t'/s'	$\phi + \psi$	ψ	e_λ	S_v
A	138	300	1.03	0.46	24.7	−0.3	2.17	0.00
B	264	540	0.89	0.49	26.1	1.1	2.15	0.02
C	123	240	1.03	0.51	27.1	2.1	2.12	0.05
D	228	420	0.89	0.54	28.5	3.5	2.10	0.07
E	108	180	1.03	0.60	31.0	6.0	2.07	0.10
F	198	300	0.89	0.66	33.4	8.4	2.03	0.14
G	93	120	1.03	0.78	37.8	12.8	1.99	0.18
H	156	180	0.89	0.87	40.9	15.9	1.93	0.24
J	63	60	1.03	1.05	46.4	21.4	1.85	0.32
K	84	60	0.89	1.40	54.5	29.5	1.71	0.46
L	30	15	1.03	2.00	63.5	38.5	1.57	0.60

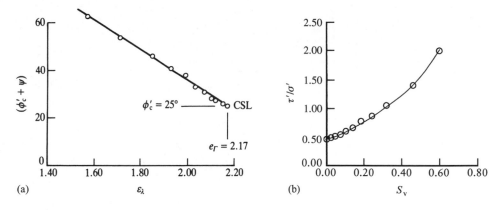

Figure 10.23 Peak states of a soil related to state parameter – Example 10.3.

now all fall close to a single curved line. The data are plotted to logarithmic scales in Fig. 10.22(b). They are close to a straight line with B in Eq. (10.14) and, scaling from the diagram, $B = 0.5$.

Example 10.3: Dilation and equivalent state   Table 10.3 shows the data for the peak states for the set of shear tests given in Tables 10.1 and 10.2. The critical friction angle is ϕ'_c and the angle of dilation ψ is given by Eq. (10.19). Values for e_λ are calculated from Eq. (9.7) and Fig. 10.23(a) shows the variation of $\phi' + \psi$ with e_λ. The state parameter S_v is calculated from Eq. (9.19) with $e_\Gamma = 2.17$ and Fig. 10.23(b) shows the variation of peak stress ratio with S_v.

Reference

Schofield, A. N. (1980) 'Cambridge geotechnical centrifuge operations', *Geotechnique*, 30, 3, 227–268.

Further reading

Atkinson, J. H. and P. L. Bransby (1978) *The Mechanics of Soils*, McGraw-Hill, London.
Muir Wood, D. M. (1991) *Soil Behaviour and Critical State Soil Mechanics*, Cambridge University Press, Cambridge.
Schofield, A. N. and C. P. Wroth (1968) *Critical State Soil Mechanics*, McGraw-Hill, London.
Taylor, D. W. (1948) *Fundamentals of Soil Mechanics*, Wiley, New York.

Chapter 11

Behaviour of soil before failure

11.1 Introduction

In laboratory triaxial or shear tests, and in the ground, soil is loaded from some initial state and it will ultimately reach a critical state. (If the soil is clay it may go on to a residual state.) In Chapter 8 I described the behaviour of soils during isotropic and one-dimensional compression and swelling and I showed how these were related. If the history of isotropic or one-dimensional loading and unloading is known the initial state described by the current effective stress, specific volume and overconsolidation ratio are fixed. Notice that the conditions in a shear test when the shear stress is zero are the same as those in a one-dimensional compression test and the conditions in a triaxial test when $q = 0$ (i.e. when $\sigma_a = \sigma_r$) are the same as those in an isotropic compression test.

In Chapter 9, I described the behaviour of soils in drained shear tests in which the changes of shear stress τ' and volumetric strain ε_v with shear strain γ are shown in Fig. 9.1 and I described the critical states in shear tests. I then showed that there were qualitatively similar peak and critical states in triaxial tests. If either the effective normal stress or the specific volume at the critical state is known the critical state strength is fixed.

We can now consider how the state moves from the initial to the critical for different loadings. In Chapter 9 I suggested that there might be unique states on the wet side of critical between the normal compression line and the critical state line, as shown in Figs. 9.7 and 9.10. In Chapter 10 I showed that there were unique peak states on the dry side of critical.

In this chapter I will consider these intermediate states in more detail. I will mostly consider the behaviour of soils in triaxial tests because the analyses will lead to development of Cam Clay in the next chapter, but of course the behaviour of soils in shear tests is qualitatively similar to that in triaxial tests.

11.2 Wet side and dry side of critical

During drained shearing soil may either compress or dilate, as illustrated in Figs. 9.1 and 10.2, and during undrained shearing pore pressures may either increase or decrease. What actually happens depends on the position of the initial state with respect to the critical state line. We can now see the significance of the distinction made in

Sec. 8.4 between states on the wet side of the critical state (i.e. normally consolidated or lightly overconsolidated clays or loose sands) and states on the dry side (i.e. heavily overconsolidated clays or dense sands).

Figures 11.1 and 11.2 illustrate the idealized behaviour of soils initially on the wet side or on the dry side during undrained or drained triaxial tests. In Fig. 11.1 the initial total stress is at the point I with $q = 0$ and a total mean stress p. Sample W is normally consolidated and its initial effective stress state is on the wet side of critical: it has an initial pore pressure u_{0w}. Sample D is overconsolidated and its initial effective stress state is on the dry side of critical: it has an initial pore pressure u_{0d}. Samples W and D both have the same initial specific volume v_0.

The samples are loaded undrained in a test in which the mean total stress p is constant. (A test in which p is constant can be carried out in a hydraulic triaxial cell (see Chapter 7) by simultaneously reducing the cell pressure and increasing the axial stress in the proportion $\delta\sigma_a = 2\delta\sigma_r$.) The total stress path is I → F in Fig. 11.1(a). During any shearing test the states must move towards, and ultimately reach, the critical state line. For undrained loading the states must remain at constant volume and both samples reach the critical state line at F_u, where they have the same undrained strength because they have the same specific volume. You can see from Fig. 11.1 that for the soil initially on the wet side the pore pressure increases on shearing, while for the soil initially on the dry side the pore pressure reduces. Notice that the overconsolidated soil reached a peak stress ratio at P but the deviator stress at F_u is greater than that at P.

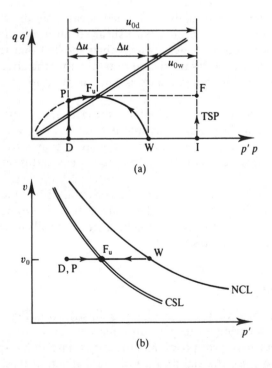

Figure 11.1 Behaviour of soil during undrained shearing.

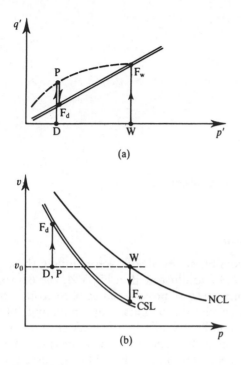

Figure 11.2 Behaviour of soil during drained constant p' shearing.

Notice also that the loading path $D \rightarrow P$ for the overconsolidated soil before the peak state is linear and vertical (i.e. $\delta p' = 0$).

Figure 11.2 shows the same two initial states but with the paths for drained shearing with constant p'. Again both paths must move towards, and ultimately reach, the critical state line. The soil initially on the wet side compresses on shearing and ultimately fails at F_w. The soil initially on the dry side first shears at constant volume to the peak state at P but then it dilates and the shear stress reduces as the specific volume increases. The shear stresses at the failure points, F_w and F_d, are different because the effective stresses and specific volumes are different.

The principal distinction between soils that compress on drained shearing or where pore pressures increase on undrained shearing and soils that dilate or where pore pressures decrease, is in whether the initial state lies to the right (i.e. on the wet side) or to the left (i.e. on the dry side) of the critical state line as illustrated in Fig. 11.3. Soils initially on the wet side compress during drained shearing or the pore pressures increase during undrained shearing, while soils initially on the dry side dilate or pore pressures decrease.

The distinction between the dry side and the wet side of critical is very important in determining the basic characteristics of soil behaviour. Soils must be heavily overconsolidated ($R_p > 3$) to be on the dry side while soils that are normally or lightly overconsolidated ($R_p < 2$) will be on the wet side.

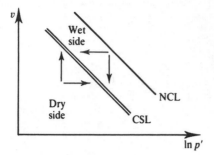

Figure 11.3 States on the wet side and the dry side of critical.

11.3   State boundary surface for soil

We have already found cases where the possible states of soils were limited; these are shown in Fig. 11.4. As discussed in Sec. 8.3 and illustrated in Fig. 8.4, the isotropic normal compression line represents a boundary to all possible states of isotropic compression; the state can move below (i.e. inside) the boundary by unloading, but it cannot move outside the normal compression line. Similarly, the peak envelope must represent a boundary to all possible states since, by definition, this represents the limiting or peak states. Remember that the peak state line in Fig. 11.4(a) corresponds to one specific volume. There will be other peak state lines corresponding to other

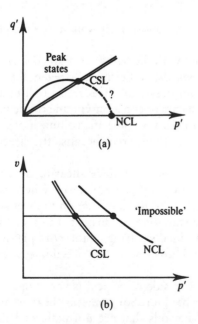

Figure 11.4 Part of a state boundary surface on the dry side.

specific volumes and together these will form a peak state surface. The surface can be reduced to a line by normalization as described in Sec. 9.6. The normalized peak state boundary surface can be represented by either the Mohr–Coulomb equation or a power law equation and these are shown in Figs. 10.7 and 10.10 respectively. The peak state boundary surface on the dry side of critical shown in Fig. 11.4(a) is close to, but not the same as, the power law peak state envelope in Fig. 10.10(a).

The peak state surface is a boundary on the dry side of critical and it is now necessary to examine whether there is a well-defined state boundary on the wet side. If there is it will join the isotropic normal compression and critical state lines and it might look like the broken line in Fig. 11.4.

Figure 11.5 shows paths for three different initial states all on the wet side of critical. P and V are on the isotropic normal compression line; P is sheared drained with p' constant and V is sheared undrained and the paths cross at S. R is initially anisotropically compressed and it is compressed further at a constant stress ratio $q'/p' = \eta'$ so that the state passes through the point S. (Notice that the normal compression line is like this path but with $\eta' = 0$ and so is the critical state line but with $\eta' = M$.) We can easily arrange for all the stress paths in Fig. 11.5(a) to pass through the same point S, but the question is whether they all have the same specific volume at S in Fig. 11.5(b).

The best way to examine this is to normalize the states with respect to the critical pressure p'_c or with respect to the equivalent volume v_λ. The resulting normalized state boundary surface is shown in Fig. 11.6. As before, the critical state and isotropic normal compression lines reduce to single points and the anisotropic compression line RS reduces to a single point S. Also shown in Fig. 11.6 are the parts of the state

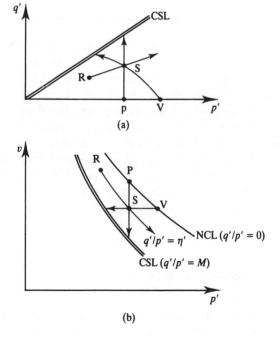

Figure 11.5 State paths for normally consolidated soil.

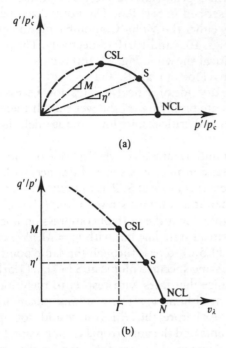

Figure 11.6 Part of a state boundary surface on the wet side.

boundary surface on the dry side of critical, corresponding to the peak states, from Figs. 10.10 and 10.16.

The state boundary surface in Fig. 11.6 has been drawn as a smooth curve linking the wet side and the dry side. Later, in Chapter 12, this will be represented by a simple mathematical expression. Do not forget that the line shown in Fig. 11.6, which has normalized axes, is really a three-dimensional surface in the set of axes $q':p':v$. This surface is rather difficult to draw, which is why it is easier to normalize the results first. Figure 11.7 illustrates the three-dimensional surface; this shows constant specific volume sections as full lines and constant stress ratio (i.e. constant η') sections as broken lines. Data from soil tests that demonstrate the existence of unique state boundary surfaces were given by Atkinson and Bransby (1978) and by Muir Wood (1991). The part of the state boundary surface on the wet side of critical (i.e. between the normal compression line and the critical state line) is sometimes known as the Roscoe surface and the part on the dry side corresponding to peak states is sometimes known as the Hvorslev surface (Atkinson and Bransby, 1978).

11.4   Elastic behaviour at states inside the state boundary surface

The state boundary surface is a boundary to all possible states of a reconstituted soil. The state cannot exist outside the surface – by definition – although later we will find cases of structured soils where unstable states outside the boundary surface for

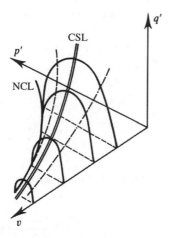

Figure 11.7 A state boundary surface for soil.

reconstituted soil can occur. If soil with a state on the surface is unloaded the state moves inside the surface and, on reloading the state will move back to, but not outside, the surface. Thus, the state boundary surface can also be a yield surface like that shown in Fig. 3.16. If the boundary surface is a yield surface then while the state is on the surface there are simultaneous elastic and plastic strains, but if the state is brought inside the boundary surface, by unloading, the strains are assumed to be purely elastic. This is a highly idealized model for soil behaviour and we now know that there are inelastic strains when the state is inside the boundary surface. These aspects of soil stress–strain behaviour will be considered in Chapter 13.

The idealized behaviour of soil during isotropic compression and swelling was considered in Secs. 8.2 and 8.3 (see Figs. 8.2 to 8.6) and is illustrated in Fig. 11.8. This shows a sequence of loading and unloading from A to D where the overconsolidation ratios are the same but the specific volumes are different. Between B and C the state was on the normal compression line (i.e. on the state boundary surface) and

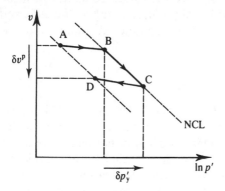

Figure 11.8 Elastic and plastic compression.

the soil yielded and hardened as the yield stress increased by $\delta p_y'$ with an irrecoverable plastic volume change δv^p. Along AB and CD the state was inside the boundary surfaces and the behaviour is taken to be elastic.

The stress–strain behaviour of an isotropic elastic material is decoupled (i.e. the shearing and volumetric effects are separated) and from Eq. (3.27),

$$\delta \varepsilon_s = \frac{1}{3G'} \delta q' \tag{11.1}$$

$$\delta \varepsilon_v = \frac{1}{K'} \delta p' \tag{11.2}$$

Another expression for the elastic volumetric strains can be obtained from the equation for the swelling and recompression lines (see Sec. 8.2) as

$$\delta \varepsilon_v = \frac{\kappa}{vp'} \delta q' \tag{11.3}$$

where κ is the slope of the lines AB and CD in Fig. 11.8. A similar expression for shearing can be written as

$$\delta \varepsilon_s = \frac{g}{3vp'} \delta q' \tag{11.4}$$

where g is a soil parameter which describes shear stiffness in the same way that κ describes volumetric stiffness. (The basic assumption made here is that $G'/K' = \kappa/g =$ constant, which implies that Poisson's ratio is a constant.)

With the simple idealization that soil is isotropic and elastic, shear and volumetric effects are decoupled and volume changes are related only to changes of p' and are independent of any change of q'. This means that, inside the state boundary surface, the state must remain on a vertical plane above a particular swelling and recompression line. This vertical plane is sometimes known as an elastic wall (see Fig. 11.9). Notice that an elastic wall is different from a constant volume section (except for the case of a soil with $\kappa = 0$). Since the soil yields when the state reaches the boundary surface a yield curve is the intersection of an elastic wall with the state boundary surface, as shown in Fig. 11.9. Remember that there will be an infinite number of elastic walls, each above a particular swelling and recompression line, and an infinite number of yield curves.

11.5   Soil behaviour during undrained loading

Figure 11.10 shows the behaviour during undrained loading of sample W initially on the wet side of the critical state and sample D initially on the dry side, both with the same initial specific volume. The broken line in Fig. 11.10(a) is the yield curve which is the intersection of the elastic wall with the state boundary surface so the samples yield at Y_W and Y_D where the stress paths meet the yield curve. Thereafter the stress paths follow the yield curve and reach the critical state line at the same point at F_u because their specific volumes remain constant, as shown in Fig. 11.10(d). Figure 11.10(a)

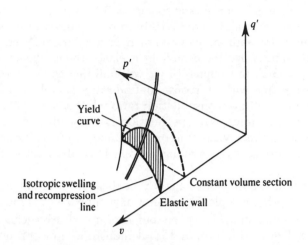

Figure 11.9 Elastic wall and yield curve.

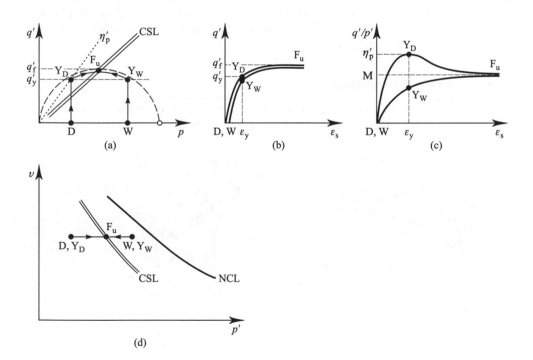

Figure 11.10 Behaviour of soils during undrained loading.

has been drawn so the deviator stresses q'_y at the yield points are the same for both samples but this would not generally be the case.

For undrained loading $\delta\varepsilon_v = 0$, for isotropic elastic soil $\delta p' = 0$ from Eq. 11.3 and the paths $W \rightarrow Y_W$ and $D \rightarrow Y_D$ are vertical. Figure 11.10(b) shows the deviator stress q' plotted against the shear strain ε_s for both samples and there is very little difference between them. Before the yield points when the behaviour is elastic the gradients of the stress–strain curves are $3G_u = 3G'$. In Chapter 13 we will find that soil stiffness increases with mean effective stress and with overconsolidation ratio. Sample W has a large mean effective stress and a small overconsolidation ratio while sample D has a smaller mean effective stress and a larger overconsolidation ratio, so their elastic stiffnesses can be nearly the same and they yield at about the same strain ε_y. Notice that neither sample has a peak deviator stress even though sample D yields on the peak failure envelope.

Figure 11.10(c) shows the stress ratio q'/p' plotted against the shear strain. Sample D reaches a peak stress ratio η'_p at Y_D but sample W does not have a peak stress ratio. The yield point is at Y_W at the same strain ε_y as Y_D but it is not clearly defined on the stress ratio–strain curve. Figures 11.10(b) and (c) demonstrate how soil test data can give very different looking curves when they are plotted in different ways. It is always a good idea to plot test data in different ways to explore soil behaviour fully. As most soil test data are now captured electronically and stored in spreadsheets plotting them in different ways is relatively straightforward. Test data should not, however, be plotted randomly; the axes of graphs should be chosen to investigate behaviour within a particular theory.

11.6   Stress ratio and dilation

Figure 11.11(a) and (b) shows the variations of stress ratio and volumetric strain with shear strain for an ideal soil sheared from states initially either on the wet side or on

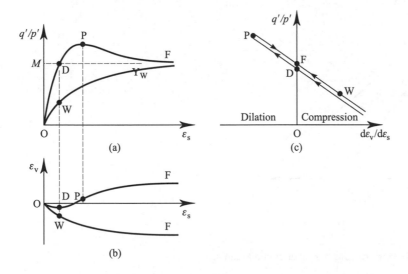

Figure 11.11 Stress ratio and dilation of soil.

the dry side and are similar to Figs 9.1 and 10.2. In Sec. 10.5 I showed that at the peak state the stress ratio was related to the rate of dilation by

$$\frac{q'}{p'} = M - \frac{d\varepsilon_v}{d\varepsilon_s} \tag{11.5}$$

Since elastic strains are relatively small compared to the plastic strains. Eq. (11.5) also applies to states before and after the peak and to soils on the wet side and on the dry side (except at states close to the start of the shearing where the behaviour is essentially elastic). Figure 11.11(c) shows Eq. (11.5) as q'/p' against $d\varepsilon_v/d\varepsilon_s$ for the normally consolidated soil and for the overconsolidated soil. There are two points, D and F where the rates of volume change are zero and $q'/p' = M$. Consequently, by plotting soil test data as q'/p' against $d\varepsilon_v/d\varepsilon_s$ the position of the critical state point F can be found even if the loading is terminated before the samples have reached their critical states. It is best to conduct tests on both normally consolidated and overconsolidated samples of clay or on loose and dense samples of sand to obtain data on both sides of the critical state.

11.7 Slip planes and apparent errors in test results

During shearing at and beyond the peak state overconsolidated clays and dense sands on the dry side of critical often have non-uniform strains and develop strong discontinuities like those shown in Fig. 2.10. These are usually called slip planes although they have finite thickness which may be only a few grains thick. Strains in slip planes were described in Sec. 2.8. Soil in a slip plane has volumetric strains which are very different to the mean volumetric strain in the whole sample. In a nominally undrained test in which no water enters of leaves the sample water may move into a slip plane from nearby soil so there is local drainage. Once slip planes appear you cannot rely on measurements of volumetric strains made with the instruments described in Chapter 7.

If a soil is on the dry side of critical it will dilate on shear and if a slip plane starts to form the water content of the soil in it will increase, the soil will weaken and the slip plane will grow. However, if the soil is on the wet side of critical it will compress during shear and if a slip plane starts to form the soil will strengthen and the slip plane will stop growing. Slip planes are seen most often in soils whose states are on the dry side of critical. As there must be some drainage of water into the slip plane from the surrounding soil, volumetric strains and water contents measured in the usual way at the boundaries of the sample are different from those in the soil in the slip plane. The quantity of this local drainage depends on the permeability of the soil and on the rate of shearing in the test.

Figure 11.12 illustrates the behaviour of a sample of soil in a nominally undrained triaxial compression test. The drainage tap was closed so no water could enter or leave the sample but water could move by local drainage into a slip plane from nearby soil. If the soil is fully undrained it would follow the path $O \rightarrow Y \rightarrow F_u$ in Fig. 11.12 and if it is fully drained it would follow the path $O \rightarrow Y \rightarrow F_d$. (These are the same as those shown in Figs. 11.1 and 11.2.) If there is partial local drainage the soil follows an

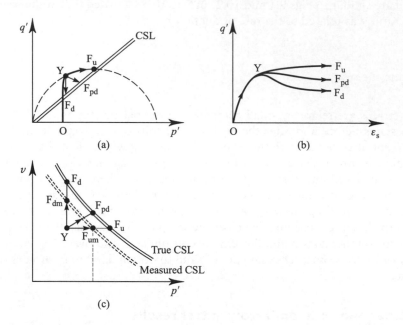

Figure 11.12 Influence of slip planes on critical state lines measured in tests.

intermediate path such as $O \to Y \to F_{pd}$ in Fig. 11.12, which is somewhere between the fully drained and fully undrained cases.

The influence of partial local drainage starts at the yield point at Y when the soil starts to dilate. There is a peak deviator stress which is close to the yield stress: in a fully undrained test there is no peak deviator stress, as shown in Fig. 11.10(b). The critical state strength of the sample at F_{pd} is less than it would have been if were fully undrained and reached the critical state at F_u. The undrained strength is the maximum strength in a fully undrained test which is at F_u. If a slip plane forms with local drainage the peak undrained strength is the yield stress at Y. In Fig. 11.12(c) the actual critical state point for a nominally undrained test in which there is some partial drainage is at F_{pd} but you would plot it at F_{um} at the measured value of p' and at a specific volume equal to the initial specific volume. If a number of similar tests were carried out the measured critical state line would be plotted as the broken line in Fig. 11.12(c), which is below the true critical state line.

A triaxial test which is fully drained has a peak strength at Y and a critical state strength at F_d. The specific volume in a slip plane is at F_d in Fig. 11.12(c) but the average measured specific volume is smaller and is at F_{dm} in Fig. 11.12(c). The actual critical state point for a drained triaxial test is at F_d but you would plot it at F_{dm} in Fig. 11.12(c) at the measured values of p' and specific volume. Again, if a number of similar tests were carried out the measured critical state line would be plotted as the broken line in Fig. 11.12(c) which is below the true critical state line.

Further analyses of local drainage in slip planes and its influence on measured soil behaviour are given by Atkinson (2000). The important thing to remember is once slip planes appear in a soil sample you can no longer trust the results. Undrained tests are no longer fully undrained due to local drainage and in drained tests specific volumes in slip planes are larger than the average for the whole sample. Both lead to erroneous interpretations of the results.

11.8   Summary

1.   The initial state of soil, before shearing, is fixed by the appropriate normal compression and swelling lines and the final state is fixed by the critical state line. The path between the initial and final states is governed by the loading (i.e. drained or undrained) and by the state boundary surface.
2.   There is an important distinction to be made between the behaviour of soils on the wet side of critical (which compress on drained loading, or where the pore pressures rise on undrained loading) and soils on the dry side of critical (which dilate on shearing, or where the pore pressures fall).
3.   In the simple idealization the behaviour is taken to be elastic when the state is inside the state boundary surface. Yielding and plastic straining occur as the state moves on the state boundary surface.
4.   There are relationships between stress ratio and dilation for states on the state boundary surface on the wet side and on the dry side of the critical state. These relationships provide a means of determining the critical state of soil from tests in which the sample did not reach the critical state.
5.   Overconsolidated soils, on the dry side of critical, which soften on shearing beyond the peak, often develop strong slip surfaces where intense shearing and volume changes are concentrated in a very thin region of material. In this case measurements made at the boundaries of a test sample become unreliable.

Worked examples

Example 11.1: Determination of state path and yielding   A soil has the parameters $M = 0.98$, $\lambda = 0.20$, $\kappa = 0.05$ and $N = 3.25$. A constant volume section of the state boundary surface is a semi-circle passing through the origin. Samples were isotropically compressed and swelled in a stress path triaxial cell to different stresses but the same initial specific volume $v_0 = 1.97$; the initial stresses were: sample A, $p'_0 = 600$ kPa, sample B, $p'_0 = 400$ kPa, sample C, $p'_0 = 150$ kPa (sample A was normally consolidated). The samples were tested undrained by increasing q with p held constant.

The state paths are shown in Fig. 11.13. When the state is inside the state boundary surface the behaviour is elastic and shearing and volumetric effects are decoupled; hence $\delta p' = 0$ for undrained loading. The states of the samples after compression and swelling, at their yield points and at failure at their critical states, shown in Table 11.1, were found by scaling from the diagram.

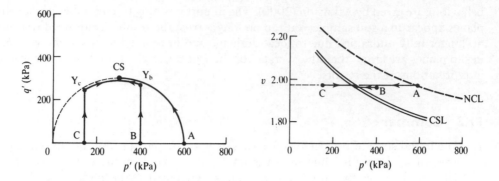

Figure 11.13 State paths in undrained tests – Example 11.1.

Table 11.1 Yield and critical states in undrained tests – Example 11.1

Sample	Initial state		Yield point			Critical state		
	p'_0 (kPa)	v_0	q'_y (kPa)	p'_y (kPa)	v_y	q'_f (kPa)	p'_f (kPa)	v_f
A	600	1.97	0	600	1.97	294	300	1.97
B	400	1.97	280	400	1.97	294	300	1.97
C	150	1.97	260	150	1.97	294	300	1.97

Example 11.2: Determination of state path and yielding   Three further samples D, E and F of the soil described in Example 11.1 were prepared at the same initial state as samples A, B and C. Each sample was tested drained following a stress path with increasing q' with p' held constant.

The state paths are shown in Fig. 11.14. When the state is inside the state boundary surface the behaviour is elastic and shearing and volumetric effects are decoupled; hence $\delta v = 0$ for constant p' stress path tests. The states of the samples after compression and

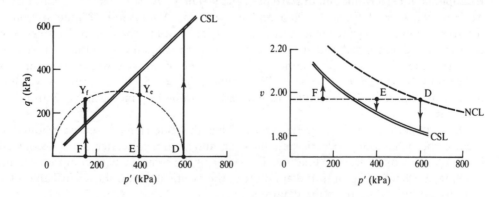

Figure 11.14 State paths in drained tests – Example 11.2.

Table 11.2 Yield and critical states in drained tests – Example 11.2

Sample	Initial state		Yield point			Critical state		
	p_0' (kPa)	v_0	q_y' (kPa)	p_y' (kPa)	v_y	q_f' (kPa)	p_f' (kPa)	v_f
D	600	1.97	0	600	1.97	588	600	1.83
E	400	1.97	280	400	1.97	392	400	1.90
F	150	1.97	260	150	1.97	147	150	2.10

swelling, at their yield points and at failure at their critical states, shown in Table 11.2, were found by scaling from the diagram.

Example 11.3: Calculation of undrained stress path   A soil has the parameters $M = 0.98$, $\lambda = 0.20$, $\kappa = 0.05$ and $N = 3.25$, but the shape of the state boundary surface is unknown. A sample is isotropically normally compressed in a triaxial apparatus to $p_0' = 600$ kPa and tested undrained by increasing the axial stress with the total mean stress p held constant. It is observed that the change of pore pressure can be approximated by $\Delta u = \Delta q'/300$.

The test result are given in Table 11.3 for equal increments of q'. An undrained stress path defining a constant volume section of the state boundary surface or the wet side of critical is shown in Fig. 11.15.

Table 11.3 Results of undrained test – Example 11.3

q' (kPa)	p (kPa)	u (kPa)	p' (kPa)
0	600	0	600
50	600	8	592
100	600	33	567
150	600	75	525
200	600	133	467
250	600	208	392
300	600	300	300

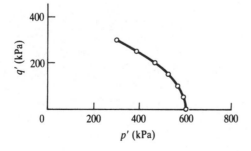

Figure 11.15 Undrained stress path – Example 11.3.

Reference

Atkinson, J. H. (2000) 40th Rankine Lecture. Non-linear stiffness in routine design, *Geotechnique*, Vol. 50, No. 5, pp. 487–508.

Further reading

Atkinson, J. H. and P. L. Bransby (1978) *The Mechanics of Soils*, McGraw-Hill, London.
Muir Wood, D. M. (1991) *Soil Behaviour and Critical State Soil Mechanics*, Cambridge University Press, Cambridge.
Schofield, A. N. and C. P. Wroth (1968) *Critical State Soil Mechanics*, McGraw-Hill, London.

Chapter 12

Cam clay

12.1 Introduction

Figure 11.7 shows a simple state boundary surface for soil; to develop a simple theoretical model for the stress–strain behaviour of soil this could be taken to be a yield surface. Yield curves are the lines of intersection of elastic walls with the yield surface as shown in Fig. 11.9 and these could be taken to be plastic potentials. We could then use the ideas of yielding, hardening and normality set out in Chapter 3 to derive a set of constitutive equations for soil. All that is required is a mathematical expression for the shape of the boundary surface.

Suitable equations for the state boundary surface could be obtained by fitting expressions to laboratory test data, by purely theoretical consideration of the mechanics of granular materials or by a combination of these. A very simple and neat theoretical equation was obtained by research workers in the University of Cambridge during the 1960s and this will be described here. Over the years many others have tried to improve on the original Cambridge equation and while some have succeeded in obtaining better agreement with experimental observations the simplicity and elegance of the original is inevitably lost. What I am going to do in this chapter is to describe the original simple theoretical model to get across the basic techniques involved in constructing constitutive equations for soil. Anyone seriously interested in applying these techniques in practice will need to study the more complex, and more realistic, soil models.

12.2 Basic features of the Cam clay models

The Cambridge theories are known under the umbrella term of Cam clay. The first model described by Schofield and Wroth (1968) is known as original Cam clay and a second model described by Roscoe and Burland (1968) is known as modified Cam clay. All the theories within the Cam clay family are basically similar. Soil is taken to be frictional with logarithmic compression. The state boundary surface is taken as a yield surface and as a plastic potential surface, and hardening is related to the plastic volumetric strains. The principle differences between the various members of the Cam clay family are in the precise equations used to describe the yield curves. For example, in original Cam clay they are logarithmic spirals while in modified Cam clay they are ellipses.

The term Cam clay was coined by the Cambridge research workers because the river in Cambridge is called the Cam. Do not misunderstand this. You cannot go to Cambridge and dig up any Cam clay; it is simply the name of a theoretical model or a set of equations. The status of Cam clay is like the status of elasticity. You cannot find any elasticity anywhere; what you can find is steel or copper which behave in a way very like the theory of elasticity, at least over small strains. In the same way you cannot find any Cam clay; what you can find are reconstituted (and some intact) soils that behave in a way very like the theoretical model called Cam clay.

12.3   State boundary surface for ordinary Cam clay

The basic equation for the state boundary surface for ordinary Cam clay is

$$\frac{q'}{Mp'} + \left(\frac{\lambda}{\lambda - \kappa}\right) \ln p' - \left(\frac{\Gamma - v}{\lambda - \kappa}\right) = 1 \tag{12.1}$$

The basic parameters M, λ, κ and Γ have been described in earlier chapters. They are all material parameters and depend only on the nature of the soil grains.

This equation defines the state boundary surface shown in Fig. 12.1. The surface meets the $v : p'$ plane along the isotropic normal compression line where $q' = 0$ and $v = N - \lambda \ln p'$ and hence, substituting in Eq. (12.1),

$$N - \Gamma = \lambda - \kappa \tag{12.2}$$

The curves shown in Fig. 12.1 are constant volume sections and undrained stress paths. The equation for an undrained stress path can be obtained from Eq. (12.1) with

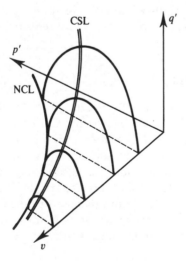

Figure 12.1 State boundary surface for ordinary Cam clay.

$v = \Gamma - \lambda \ln p'_{\mathrm{c}}$, where p'_{c} is the stress at the intersection of the constant volume section and the critical state line and is

$$\frac{q'}{Mp'} + \left(\frac{\lambda}{\lambda - \kappa}\right) \ln\left(\frac{p'}{p_{\mathrm{c}}}\right) = 1 \qquad (12.3)$$

A yield curve is the intersection of an elastic wall given by $v = v_\kappa - \kappa \ln p'$ with the state boundary surface. At the critical state line the specific volume is v_{c} and the mean stress is p'_{c}, as shown in Fig. 12.2(b) where $v_{\mathrm{c}} = v_\kappa - \kappa \ln p'_{\mathrm{c}} = \Gamma - \lambda \ln p'_{\mathrm{c}}$. Eliminating v and v_κ, the equation for the yield curve shown in Fig. 12.2(a) is

$$\frac{q'}{Mp'} + \ln\left(\frac{p'}{p_{\mathrm{c}}}\right) = 1 \qquad (12.4)$$

Note that the equations of the constant volume section, or undrained stress path, and the yield curve are different except for the special case of a soil with $\kappa = 0$. From Eq. (12.4), with $q' = 0$, the yield stress p'_{y} is related to the critical state stress p'_{c} on the same yield curve by

$$\frac{p'_{\mathrm{y}}}{p'_{\mathrm{c}}} = \exp(1) = 2.72 \qquad (12.5)$$

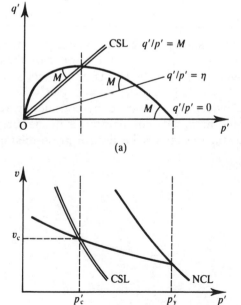

(a)

(b)

Figure 12.2 Yield curve for ordinary Cam clay.

Differentiating Eq. (12.4) we get

$$\frac{dq'}{dp'} = \frac{q'}{p'} - M \tag{12.6}$$

which is simply another way of writing an equation for a yield curve. Equation (12.6) shows that the logarithmic spiral curve has the very simple property that the gradient dq'/dp' is related to the gradient q'/p' of the radius from the origin.

Equation (12.3) describes a constant volume section of the state boundary surface and it was obtained essentially by normalizing the stresses in Eq. (12.1) with respect to the critical pressure p'_c. Alternatively the stresses could be normalized with respect to v_λ which was defined in Sec. 9.9 and is given by

$$v_\lambda = v + \lambda \ln p' \tag{12.7}$$

Substituting Eq. (12.7) into Eq. (12.1) we get

$$\left(\frac{q'}{Mp'}\right) = 1 + \left(\frac{\Gamma - v_\lambda}{\lambda - \kappa}\right) \tag{12.8}$$

At the critical state point $q' = Mp'$ and $v_\lambda = \Gamma$ and at the normal compression line $q' = 0$ and $v_\lambda = N$. The state parameter S_v defined in Sec. 9.9 is $\Gamma - v_\lambda$ and Eq. (12.7) becomes

$$\left(\frac{q'}{Mp'}\right) = 1 + \left(\frac{S_v}{\lambda - \kappa}\right) \tag{12.9}$$

and this is shown in Fig. 12.3. At the critical state point $S_v = 0$ and at the normal compression line $v_\lambda = N$ and $S_v = \Gamma - N$. For values of $S_v > 0$ the soil is on the dry side of critical and $q > Mp'$ and for values of $S_v < 0$ the soil is on the wet side of critical and $q < Mp'$.

Figure 12.3 is similar to Fig. 10.18(c) which was obtained by considering relationships between the peak stress ratio and the initial state parameter in shear tests. Figure 12.3 shows that the simple Cam clay theory extends the ideas of stress–dilatancy

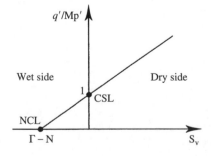

Figure 12.3 State boundary surface for Cam clay.

into the whole range of states, both on the dry side and on the wet side of the critical state.

12.4 Calculation of plastic strains

The yield curve is taken to be a plastic potential so that the vector of plastic strain increment $\delta\varepsilon^{\mathrm{p}}$ is normal to the curve, as shown in Fig. 12.4. If two lines are orthogonal the product of their gradients is -1, so

$$\frac{\mathrm{d}q'}{\mathrm{d}p'} \cdot \frac{\mathrm{d}\varepsilon_{\mathrm{s}}^{\mathrm{p}}}{\mathrm{d}\varepsilon_{\mathrm{v}}^{\mathrm{p}}} = -1 \tag{12.10}$$

and, from Eq. (12.6), the plastic strain increments are given by

$$\frac{\delta\varepsilon_{\mathrm{v}}^{\mathrm{p}}}{\delta\varepsilon_{\mathrm{s}}^{\mathrm{p}}} = M - \frac{q'}{p'} \tag{12.11}$$

At the critical state when $q'/p' = M$ we have $\delta\varepsilon_{\mathrm{v}}^{\mathrm{p}} = 0$. On the wet side $q'/p' < M$ and so $\delta\varepsilon_{\mathrm{v}}^{\mathrm{p}}$ is positive (i.e. compressive), while on the dry side $q'/p' > M$ and so $\delta\varepsilon_{\mathrm{v}}^{\mathrm{p}}$ is negative (i.e. dilative), as shown in Fig. 12.5.

Notice that Eq. (12.11) is almost the same as Eq. (10.20); the only difference is that Eq. (10.20) gives total strains while Eq. (12.11) gives the plastic strains. Equation (10.20) was obtained by analogy with the work done by friction and dilation and the derivation was for peak states on the dry side. The similarity between Eqs. (10.20) and (12.11) demonstrates that the basis of ordinary Cam clay is an equivalent work equation, but now extended to the wet side as well as the dry side. A more rigorous derivation of ordinary Cam clay from work principles was given by Schofield and Wroth (1968).

12.5 Yielding and hardening

As the state moves on the state boundary surface from one yield curve to another there will be yielding and hardening (or softening if the state is on the dry side) and,

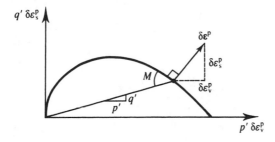

Figure 12.4 Plastic potential and plastic strains for Cam clay.

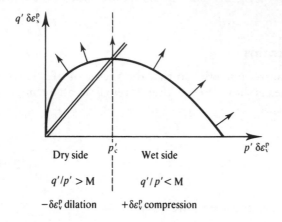

Figure 12.5 Vectors of plastic strain for Cam clay.

in Cam clay, the change of the yield stress is related to the plastic volume change. Figure 12.6 shows an increment of loading A → B on the wet side of critical, and the state moves from one yield curve to a larger one with an increase in yield stress and a reduction in volume. The increment of loading C → D on the dry side is associated with a decrease in yield stress and an increase in volume. Equation (12.1) can be rewritten as

$$v = \Gamma + \lambda - \kappa - \lambda \ln p' - \frac{(\lambda - \kappa)q'}{Mp'} \qquad (12.12)$$

Differentiating, dividing by v and noting that $\delta\varepsilon_v = -\delta v/v$, we have

$$\delta\varepsilon_v = \left(\frac{\lambda - \kappa}{vp'M}\right)\delta q' + \left[\frac{\lambda}{vp'} - \frac{(\lambda - \kappa)\eta'}{vp'M}\right]\delta p' \qquad (12.13)$$

If we now subtract the elastic volumetric strains given by Eq. (11.3), the plastic volumetric strains are

$$\delta\varepsilon_v^p = \left(\frac{\lambda - \kappa}{vp'M}\right)\left[\delta q' + (M - \eta')\delta p'\right] \qquad (12.14)$$

and, from Eq. (12.11), the plastic shear strains are

$$\delta\varepsilon_s^p = \left(\frac{\lambda - \kappa}{vp'M}\right)\left(\frac{\delta q'}{M - \eta'} + \delta p'\right) \qquad (12.15)$$

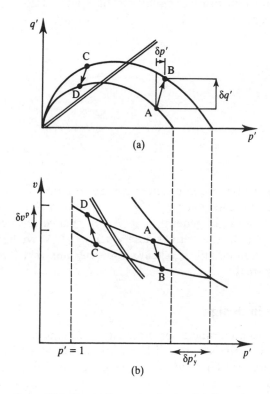

Figure 12.6 Hardening and softening for Cam clay.

12.6 Complete constitutive equations for ordinary Cam clay

The complete constitutive equations for Cam clay are obtained simply by adding the elastic strains given by Eqs. (11.3) and (11.4) to the plastic strains given by Eqs. (12.14) and (12.15) to obtain

$$\delta\varepsilon_s = \frac{1}{vp'}\left\{\left[\frac{\lambda-\kappa}{M(M-\eta')}+\frac{g}{3}\right]\delta q' + \left[\frac{\lambda-\kappa}{M}\right]\delta p'\right\} \qquad (12.16)$$

$$\delta\varepsilon_v = \frac{1}{vp'}\left\{\left[\frac{\lambda-\kappa}{M}\right]\delta q' + \left[\frac{\lambda-\kappa}{M}(M-\eta')+\kappa\right]\delta p'\right\} \qquad (12.17)$$

These apply for states that are on the state boundary surface; for states inside the boundary surface the elastic strains given by Eqs. (11.3) and (11.4) can be recovered by putting $\lambda = \kappa$ into Eqs. (12.16) and (12.17).

Equations (12.16) and (12.17) are constitutive equations like Eq. (3.26) and components of the compliance matrix are

$$C_{11} = \frac{1}{vp'} \left[\frac{\lambda - \kappa}{M(M - \eta')} + \frac{g}{3} \right] \tag{12.18}$$

$$C_{22} = \frac{1}{vp'} \left[\frac{\lambda - \kappa}{M}(M - \eta') + \kappa \right] \tag{12.19}$$

$$C_{12} = C_{21} = \frac{1}{vp'} \left[\frac{\lambda - \kappa}{M} \right] \tag{12.20}$$

These demonstrate that in Cam clay the basic compliances contain the intrinsic soil parameters M, λ, κ and g and the current state given by v, p' and $\eta' = q'/p'$. Thus, in Cam clay, the behaviour is non-linear since, in general, v, p' and q' change during a loading path. Notice that towards failure at the critical state when $\eta' \to M$ we have $C_{11} \to \infty$ and $C_{22} \to 0$. Thus, near ultimate failure, shear strains become very large while volumetric strains become very small.

12.7   Applications of Cam clay in design

Although Eqs. (12.16) and (12.17) are a complete set of constitutive equations for soil there is still quite a lot of further analysis required before they can be used for detailed design. For example, they are written in terms of shearing and volumetric effects, but for calculations they need to be rewritten in terms of the normal and shear stresses and strains on horizontal and vertical planes in the ground and possibly in three dimensions.

Ordinary Cam clay has the advantage that with yield curves as logarithmic spirals the algebra is relatively simple. Although it describes the main features of soil behaviour qualitatively there are a number of detailed aspects where it is not so good. Another model, modified Cam clay, is based on yield curves that are ellipses; this is described in detail by Muir Wood (1991).

The Cam clay equations can be implemented in finite element and similar numerical analyses as described by Britto and Gunn (1987). Be warned though: these analyses are quite complex and difficult to do properly. If you are interested in making use of these advanced techniques you are advised to start by working with people who have previous experience.

12.8   Summary

1. Cam clay is a theoretical model for soil behaviour: it includes strength and stress–strain behaviour within a single, relatively simple set of equations.
2. Cam clay combines the theories of critical state soil mechanics and the idea of a state boundary surface with the theories of plasticity, including yielding, hardening and plastic flow.
3. There are different versions of Cam clay, depending on the precise equation for the state boundary surface.

Worked examples

Example 12.1: Calculation of strains for overconsolidated Cam clay   A soil has the parameters $M = 0.98, \lambda = 0.20$ and $\kappa = g = 0.05$ and its behaviour can be represented by the Cam clay model. A sample is isotropically compressed in a stress path triaxial cell to $p' = 300$ kPa and swelled to $p_0' = 200$ kPa where the specific volume is $v_0 = 2.13$. It is then subjected to a drained test in which $\delta q' = \delta p' = 10$ kPa.

The strains are given by Eqs. (12.16) and (12.17) with $\lambda = \kappa$, since the state of the overconsolidated sample is inside the state boundary surface. Hence,

$$\delta\varepsilon_s = \frac{g}{3vp'}\delta q' = \frac{0.05 \times 10 \times 100}{2.13 \times 200 \times 3} = 0.04\%$$

$$\delta\varepsilon_v = \frac{\kappa}{vp'}\delta q' = \frac{0.05 \times 10 \times 100}{2.13 \times 200} = 0.11\%$$

Example 12.2: Calculation of strains for normally consolidated Cam clay   A second sample of the soil in Example 12.1 was isotropically compressed to $p_0' = 200$ kPa where the specific volume was $v_0 = 2.19$. It was then subjected to a drained test in which $\delta q' = \delta p' = 10$ kPa.

The strains are given by Eqs. (12.16) and (12.17), with the initial state $p' = 200$ kPa, $v = 2.19$ and $\eta' = 0$ corresponding to isotropic compression. The compliances given by Eqs. (12.18) to (12.20) are

$$C_{11} = \frac{1}{200 \times 2.19}\left(\frac{0.15}{0.98^2} + \frac{0.05}{3}\right) = 0.39 \times 10^{-3}\,\text{m}^2/\text{kN}$$

$$C_{22} = \frac{1}{200 \times 2.19}(0.15 + 0.05) = 0.46 \times 10^{-3}\,\text{m}^2/\text{kN}$$

$$C_{12} = \frac{1}{200 \times 2.19}\left(\frac{0.15}{0.98}\right) = 0.35 \times 10^{-3}\,\text{m}^2/\text{kN}$$

and, hence,

$$\delta\varepsilon_s = (C_{11}\delta q' - C_{12}\delta p') \times 100 = 0.74\%$$

$$\delta\varepsilon_v = (C_{12}\delta q' + C_{22}\delta p') \times 100 = 0.81\%$$

References

Britto, A. M. and M. J. Gunn (1987) *Critical State Soil Mechanics via Finite Elements*, Ellis Horwood, Chichester.

Muir Wood, D. M. (1991) *Soil Behaviour and Critical State Soil Mechanics*, Cambridge University Press, Cambridge.

Roscoe, K. H. And J. B. Burland (1968) 'On the generalised stress–strain behaviour of "wet" clay', in *Engineering Plasticity*, J. Heyman and F. A. Leckie (eds), Cambridge University Press, Cambridge.

Schofield, A. N. and C. P. Wroth (1968) *Critical State Soil Mechanics*, McGraw-Hill, London.

Further reading

Atkinson, J. H. and P. L. Bransby (1978) *The Mechanics of Soils*, McGraw-Hill, London.
Muir Wood, D. M. (1991) *Soil Behaviour and Critical State Soil Mechanics*, Cambridge University Press, Cambridge.
Schofield, A. N. and C. P. Wroth (1968) *Critical State Soil Mechanics*, McGraw-Hill, London.

Chapter 13

Stiffness of soil

13.1 Introduction

Stiffness relates increments of stress and increments of strain. A knowledge of soil stiffness is required to calculate ground movements and to obtain solutions to problems of soil–structure interaction, such as loads on retaining walls. Often simple analyses are carried out assuming that soil is linear and elastic and solutions for foundations will be considered in Chapter 22. However, it is recognized that soil strains are often significantly inelastic and more complicated elasto-plastic models such as Cam clay (see Chapter 12) have been developed to model the stress–strain behaviour of soil.

The stress–strain behaviour of soil is actually more complex than that given by the simple Cam clay model, particularly at small strains and for states inside the state boundary surface where, in the simple theory, the strains are elastic. A detailed treatment of soil stiffness is beyond the scope of this book. What I am going to do in this chapter is simply describe the essential features of the stress–strain behaviour of soil as an introduction to further studies.

13.2 Cam clay and soil stiffness

In Chapter 12 the basic ideas of the classical theories of elasticity and plasticity were combined with the basic soil mechanics theories of friction and logarithmic compression into a general model known as Cam clay. A set of non-linear constitutive equations was obtained in terms of the intrinsic soil parameters $\lambda, M, \Gamma, \kappa$ and g, together with parameters describing the current state and the loading history.

The basic equations for Cam clay for states on the state boundary surface (Eqs. 12.16 and 12.17) contain elastic and plastic components of straining, while for states inside the state boundary surface the basic equations (Eqs. 11.3 and 11.4) contain only elastic strains. It turns out that the ordinary Cam clay equations are reasonably good for states on the state boundary surface but the basic Cam clay theories are rather poor for states inside the state boundary surface where soil behaviour is not elastic and recoverable.

The consequences of this for geotechnical design are illustrated in Fig. 13.1. This shows two soils subjected to exactly the same loading paths A $\to$ B and C $\to$ D. The soil which starts from A is lightly overconsolidated; it yields at Y when the state reaches the state boundary surface and then it moves along Y $\to$ B on the state boundary surface with elastic and plastic strains. The soil which starts from C is heavily

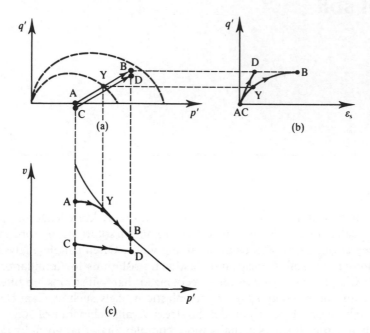

Figure 13.1 Compression of lightly and heavily overconsolidated soils.

overconsolidated, the state does not reach the state boundary surface and in Cam clay the strains are taken to be elastic throughout the loading path C → D. The stress–strain curves are shown in Fig. 13.1(b) and these correspond to the volume changes shown in Fig. 13.1(c).

For lightly overconsolidated soils following the path A → B in Fig. 13.1, the greater proportion of the strains occur along Y → B as the state moves on the state boundary surface and only a small proportion occurs along A → Y, where the soil is inside the boundary surface. For these soils we can use the Cam clay or similar theories to calculate ground movements since the significant errors which occur in the calculations of the elastic strains along A → Y will be relatively small compared with the total strains for the whole path A → B. For heavily overconsolidated soils, on the other hand, the state remains inside the state boundary surface for the whole path C → D and the errors in the strains calculated using the Cam clay theories will be relatively large.

13.3   Stiffness–strain relationships for soil

From Eqs. (3.21) to (3.25) a general set of constitutive equations can be written as

$$\begin{Bmatrix} \delta q' \\ \delta p' \end{Bmatrix} = \begin{bmatrix} 3G' & J' \\ J' & K' \end{bmatrix} \begin{Bmatrix} \delta \varepsilon_s \\ \delta \varepsilon_v \end{Bmatrix} \tag{13.1}$$

where G' is the shear modulus, K' is the bulk modulus and J' are moduli that couple shear and volumetric effects. For undrained loading for which $\delta \varepsilon_v = 0$, we have

$$\frac{dq'}{d\varepsilon_s} = 3G' \tag{13.2}$$

$$\frac{dp'}{d\varepsilon_s} = J' \tag{13.3}$$

and, for isotropic compression for which $\delta \varepsilon_s = 0$, we have

$$\frac{dp'}{d\varepsilon_v} = K' \tag{13.4}$$

$$\frac{dq'}{d\varepsilon_v} = J' \tag{13.5}$$

Notice that for undrained loading Eq. (13.2) also defines the undrained shear modulus G_u and hence

$$G_u = G' \tag{13.6}$$

Figure 13.2 shows the general characteristics of shearing and compression stress–strain curves for undrained shearing and isotropic compression tests with stages of loading, unloading and reloading. In Fig. 13.2(a) the gradient of the curve is the shear modulus $3G'$ and in Fig. 13.2(b) the gradient is the bulk modulus K'; we could obtain similar curves and evaluate J_1' and J_2' by plotting $\delta q'$ against $\delta \varepsilon_v$ and $\delta p'$ against $\delta \varepsilon_s$. In Fig. 13.2 the soil had been unloaded from B and Q and so the initial states C and R are inside the state boundary surface and the soil yields at D and S.

In Fig. 13.2 the stress–strain lines CDE and RST look non-linear, but it is difficult to see exactly how the soil is behaving, especially for small increments at the start of the reloading. The principal features of stress–strain curves can be seen more clearly by examining how stiffness changes as loading progresses. Tangent moduli are more important than secant moduli but it is not easy to calculate tangent moduli from test

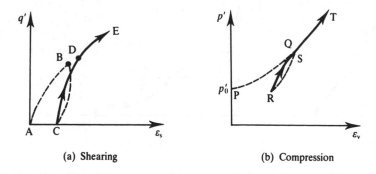

(a) Shearing   (b) Compression

Figure 13.2 Shearing and compression of soils.

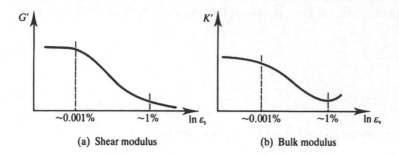

(a) Shear modulus   (b) Bulk modulus

Figure 13.3 Characteristic stiffness–strain curves for soil.

data and secant moduli are often used. The progress of the test can be described by the change of deviator stress q' for the shear modulus G' and the change of mean effective stress p' for bulk modulus K'. Alternatively the progress of the test can be described by the shear strain ε_s for shear modulus or the volumetric strain ε_v for bulk modulus. If strains are used they are usually plotted to a logarithmic scale.

Figure 13.3 shows the general shapes of stiffness–strain curves for a typical soil. (Surprisingly, the general shape applies for normally consolidated soils as well as for lightly and heavily overconsolidated soil and the consequences of this will be considered later.) The curves for shear and bulk modulus are basically similar at strains less than 1% or so. The tangent shear modulus becomes zero at the critical state while at large strains the tangent bulk modulus increases as the specific volume decreases. If the soil has a peak the tangent shear modulus is zero at the peak and is negative as the soil weakens from the peak towards the critical state.

This stress–strain behaviour is significantly different from that given by the simple Cam clay theory described in Chapter 12. Figure 13.4 illustrates characteristic stress–strain behaviour observed in laboratory tests and given by Cam clay. For the drained constant p' loading path O → Y → A in Fig. 13.4(a), the state reaches the state boundary surface at Y and travels on the boundary surface along Y → A. For Cam clay the behaviour is taken to be elastic along O → Y and, since p' and v remain constant for the particular loading path considered, the shear modulus $G' = vp'/g$ remains constant,

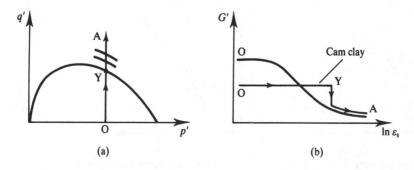

(a)   (b)

Figure 13.4 Characteristic stress–strain behaviour for soil observed in laboratory tests and given by the Cam clay theories.

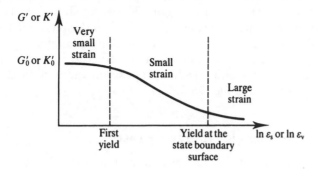

Figure 13.5 Characteristic ranges of soil stiffness.

as shown in Fig. 13.4(b). When the state reaches the state boundary surface at Y, yield occurs and the stiffness drops sharply to the value given by the full Cam clay expression in Eq. (12.16). Figure 13.4(b) indicates that after yield the behaviour observed in laboratory tests will be very like that given by the Cam clay theories (with suitable values for the soil parameters), but before yield the observed stiffness–strain behaviour is very different in character from that given by Cam clay.

The principal features of soil stiffness are illustrated in Fig. 13.5. There are three regions, as indicated, where the behaviour is different. For very small strains, smaller than some value corresponding to the first yield (usually of the order of 0.001%), the stiffness is approximately constant and the stress–strain behaviour is linear. For large strains, where the state has reached the state boundary surface (usually greater than about 1%), the behaviour is elasto-plastic and the Cam clay theories are quite good. In the intermediate, small strain, range the stiffness changes rapidly with strain and the behaviour is highly non-linear.

13.4   Strains in the ground

In most geotechnical structures that are designed to restrict ground movements, such as foundations and retaining walls, the strains in the ground are usually quite small. Figure 13.6 illustrates a stiff retaining wall and a foundation. The outward movement

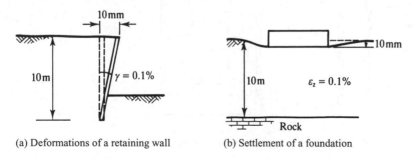

(a) Deformations of a retaining wall   (b) Settlement of a foundation

Figure 13.6 Strains in the ground near typical geotechnical structures.

of the top of the wall and the settlement of the foundation are both 10 mm and these would be acceptable displacements in many designs. The mean shear strains in the ground near the wall and the volumetric strains below the foundation are 0.1%. In practice there will be local strains greater than these, especially near the edge of the foundation, and the strains will decay to zero far from the structures. This means that in the ground soil stiffness will vary continuously with position and with loading throughout most of the range illustrated in Fig. 13.5.

13.5   Measurement of soil stiffness in laboratory tests

The best method for investigating soil stiffness and evaluating stiffness parameters is to conduct stress path triaxial tests in the laboratory using one of the hydraulic triaxial cells described in Sec. 7.9. This apparatus permits tests to be carried out in which the initial state and the loading path can be controlled. The principal problem arises in the measurement of the small and very small strains required to investigate the whole of the characteristic stiffness strain curves shown in Fig. 13.5. To examine the whole of the stiffness–strain curve it is necessary to measure strains less than 0.001%; if the length of the sample is about 100 mm you will need to measure displacements smaller than 0.001 mm or 1 μm.

The problem is not so much with the resolution and accuracy of the dial gauges, displacement transducers and volume gauges used to measure axial and volumetric strains in triaxial tests, as with the errors that occur due to compliance, or movement, in the apparatus. (Do not forget the distinction between accuracy and resolution discussed in Chapter 7.) Figure 13.7 illustrates a conventional triaxial test; the axial displacement ΔL is measured using a displacement transducer or dial gauge mounted on the loading ram and the volume change is measured from the volume of pore water entering or leaving the sample through the drainage leads. Errors arise due to (1) axial

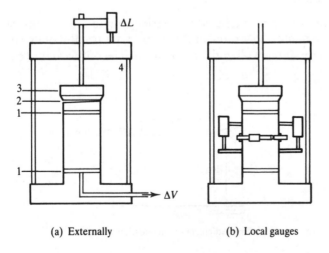

(a) Externally   (b) Local gauges

Figure 13.7 Measurement of strains in triaxial tests.

displacements at the ends of the sample, (2) displacements where the loading ram joins the top platen, (3) movements in the load cell and (4) movements in the cell.

The errors that can arise due to the compliances illustrated in Fig. 13.7(a) can be very significant and can easily swamp the required measurements of small strains. In conventional triaxial tests the measured axial strains are unreliable at strains smaller than about 0.1% irrespective of the resolution and accuracy of the transducer or dial gauge. If a hydraulic triaxial cell is used and if very careful measurements are made of the displacements in the apparatus, it is possible to obtain reliable measurements of axial and volumetric strains smaller than 0.01%. One way to improve the accuracy of measurements of strain in triaxial tests is to use gauges inside the cell mounted directly on the sample, as shown in Fig. 13.7(b). Using these kinds of instruments strains smaller than 0.001% can be measured reliably.

It is very difficult to measure the stiffness of soil at very small strains (i.e. less than about 0.001%) in triaxial tests by direct observations of strains. The simplest method is to calculate the shear modulus from the velocity of dynamic waves. The very small strain shear modulus G'_0 is given by

$$G'_0 = \frac{\gamma V_s^2}{g} \tag{13.7}$$

where V_s is the velocity of shear waves through the sample, γ is the unit weight of the soil and $g = 9.81$ m/s^2. Shear waves can be generated and their velocity measured directly using shear elements set into the top and bottom platens or from resonant frequencies in torsional shearing. The equipment and techniques for making these measurements are rather specialized and if you need to determine G'_0 you will need help; it is enough now to know that the techniques are available.

Note that in these dynamic tests the rates of loading are very large and saturated soil will be undrained. This does not matter for measurement of shear modulus since, for shearing alone, $G' = G_u$. The undrained bulk modulus of saturated soil is theoretically infinite (since $\delta \varepsilon_v = 0$ for undrained loading) and so we cannot easily determine the small strain bulk modulus K'_0 of saturated soil from the velocity of compression waves.

Figure 13.8 summarizes the principal features of the application and measurement of soil stiffness over a wide range of strain. In the field, strains in the ground near retaining walls and below foundations are relatively small and are usually less than 1%, except in small regions near the edges of foundations. Stiffness cannot be measured reliably in ordinary triaxial tests at strains less than 0.1% unless special procedures are followed, so the ordinary triaxial test is not much good for measuring soil stiffness in the range of practical interest. Stiffness at small strains can be measured reliably using local gauges attached to the sample and the shear modulus at very small strain G'_0 can be obtained from measurements of shear wave velocity.

13.6   Stiffness of soil at small and very small strains

At large strains (i.e. greater than about 1%) the state of lightly or heavily overconsolidated soil will have reached the state boundary surface and the stiffness parameters in Eq. (13.1) depend on the current state (v, p' and η') as given by Eqs. (12.18) to (12.20). For states inside the state boundary surface, at small and very small strains,

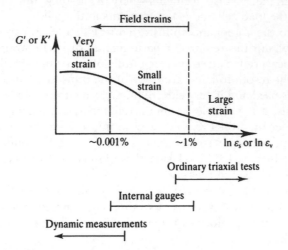

Figure 13.8 Characteristic ranges of stiffness in the field and in laboratory tests.

soil stiffness is highly non-linear, but we might expect that the stiffness at a particular strain will also depend on the current state and on the history.

(a)   Stiffness at very small strain

In dynamic tests used to measure G'_0, samples are vibrated at a constant state at strains less than about 0.001%. The damping is a negligible and at these very small strains soil is assumed to be linear and elastic. (If a typical value for G'_0 is 100 MPa then a strain $\delta\varepsilon_s = 0.001\%$ corresponds to an increment of stress $\delta q'$ of only 3 kPa.)

The general relationship between G'_0 and the current state is of the form

$$\frac{G'_0}{p'_r} = A\left(\frac{p'}{p'_r}\right)^n Y_p^m \qquad (13.8)$$

where p'_r is a reference pressure included to make Eq. (13.8) dimensionless Y_p is the current yield stress ratio defined in Sec. 8.3 and A, m and n depend on the nature of the soil (Viggiani and Atkinson, 1995). Notice that in Eq. (13.8) the value of G'_0 is related to p' and Y_p without the specific volume or voids ratio. This is possible because v, p' and Y_p are not independent as discussed in Sec. 8.3 and so v is included in the parameters p' and Y_p. Alternatively, G'_0 could be related to v and $Y_{p'}$. The value of the exponent n is generally in the range 0.5 to 1.0 and typical values for m are in the range 0.2 to 0.3. Equation (13.8) can be rewritten as

$$\ln\left(\frac{G'_0}{p'_r}\right) = \ln A + m \ln Y_p + n \ln\left(\frac{p'}{p'_r}\right) \qquad (13.9)$$

Plotting data from a set of tests carried out at different values of p' and Y_p, as shown in Fig. 13.9, provides a convenient method for evaluating the parameters A, m and n.

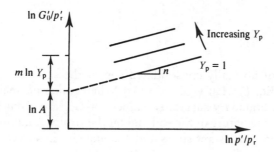

Figure 13.9 Typical variation of very small strain stiffness of soil with stress and overconsolidation.

(b)   Stiffness at small strain

The general relationships between shear modulus G' and strain, state and history for small strains in the range 0.001% to 1% are illustrated in Fig. 13.10 and the same general relationships hold for the other stiffness parameters. The value of G'/p' depends on strain (because of the non-linearity) and on $\ln Y_p$ and, at a

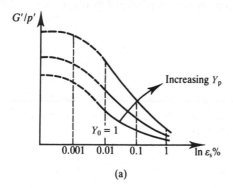

(a)

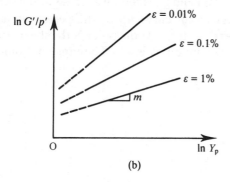

(b)

Figure 13.10 Typical variation of small strain stiffness of soil with strain, stress and overconsolidation.

particular strain,

$$\frac{G'}{p'} = AY_p^m \tag{13.10}$$

where $A = G'_{nc}/p'$ is the stiffness of normally consolidated soil at the same strain. (Notice that Eq. (13.8) reduces to Eq. (13.10) when $n = 1$.) When soil is normally consolidated its state is on the state boundary surface so values for G'_{nc} are given by Eq. (12.18). Values for m depend on the nature of the soil and on the strain. A number of other factors, such as a rest period at constant stress and a change in the direction of the stress path between successive loading stages, also effect soil stiffness, but the rate of loading has virtually no effect provided that the soil is either fully drained or fully undrained.

13.7   Rigidity and non-linearity

It is clear that the stress–strain behaviour of soil is highly non-linear over most of the range of practical interest in ground engineering and this non-linear behaviour should be taken into account in design. There are some relatively simple parameters which can be used to describe how non-linear a soil is (Atkinson, 2000).

For materials which are linear-elastic and perfectly plastic rigidity R was defined in Sec. 3.6 as

$$R = \frac{E'}{q'_f} \tag{13.11}$$

where E' is Young's modulus and q'_f is strength expressed as the diameter of the Mohr circle at failure. For soils, which are highly non-linear and which have a peak and a critical state strength, rigidity can be defined as

$$R = \frac{E'_0}{q'_p} \tag{13.12}$$

where E'_0 is Young's modulus at very small strain and q'_p is the shear stress at the peak.

Figure 13.11(a) shows a non-linear stress–strain curve up to the peak state and Fig. 13.11(b) is the corresponding stiffness–strain curve. At the point X the shear stress is q'_x, the strain is ε_x and the tangent Young's modulus is E'_x. The tangent Young's modulus is given by

$$E'_t = \frac{dq'}{d\varepsilon} \tag{13.13}$$

and hence

$$q'_x = \int_0^{q'_x} dq' = \int_0^{\varepsilon_x} E'_t d\varepsilon \tag{13.14}$$

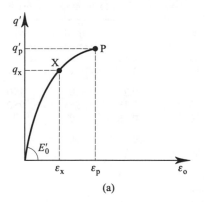
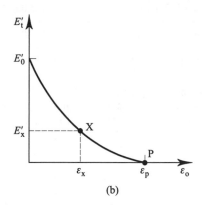

(a)   (b)

Figure 13.11 Relationships between stress–strain curves and strength.

In Eq. (13.14) the term on the right hand side is the area below the stiffness–strain curve from the origin to the point X and q'_x is the stress at the point X. It follows that the peak strength is equal to the area below the stiffness–strain curve from the origin to ε_p. This places strong restrictions on the permissible shapes of non-linear stress–strain and stiffness–strain curves.

Figure 13.12(a) shows a typical stress–strain curve up to the peak state and Fig. 13.12(b) is the corresponding stiffness–strain curve: the stiffness decreases from an initial value E'_0 to zero at the peak. In Fig. 13.12(a) there is a reference strain ε_r defined as

$$\varepsilon_r = \frac{q'_p}{E'_0} \tag{13.15}$$

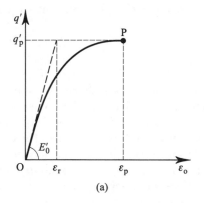
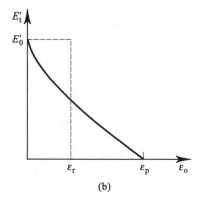

(a)   (b)

Figure 13.12 Degree of non-linearity of soil.

Table 13.1 Typical values for rigidity and degree of non-linearity of some common materials

Material	E_0 MPa	Strength MPa	Rigidity	ε_r %	ε_p %	n_1
Concrete	28,000	40	700	0.15	0.35	2
Glass	70,000	1000	70	1.5	1.5	1
Mild steel	210,000	430	500	0.2	30	150
Copper	120,000	200	600	0.15	35	250
Aluminium	70,000	100	700	0.15	10	70
Rubber	10	20	0.5	200	800	4
Timber	10,000	20	500	0.2	5	25
Soft soil	100	0.05	2000	0.05	10	200
Stiff soil	300	0.3	1000	0.1	1	10

(Comparing Eqs. (13.12) and (13.15) $\varepsilon_r = 1/R$) the strain at the peak ε_p is larger than the reference strain and the degree of non-linearity n_1 is defined as

$$n_1 = \frac{\varepsilon_p}{\varepsilon_r} \tag{13.16}$$

Note that, as shown above, the peak strength q'_p is equal to the area below the stiffness–strain curve from the origin to the peak state. As a consequence, the non-linear stress–strain and stiffness–strain curves, shown as solid lines in Fig. 13.12, must have shapes such that the area below the solid and broken stiffness–strain curves in Fig. 13.13(b) are the same.

Table 13.1 gives typical values for rigidity and degree of non-linearity for some common materials and for characteristic soft and stiff soils. (Some of the data in Table 13.1 were given also in Table 3.1.) For soils, the degree of non-linearity varies from about 10 to about 200. This is a large range, almost as big as for all other materials. Notice that the rigidity of soil is relatively large, largely because of the relatively low strength compared with other materials and the rigidity of soft soil is *larger* than the rigidity of stiff soil which is a surprising result. The reasons for this were discussed by Atkinson (2000).

13.8   Numerical modelling of soil stiffness

Equations (13.8) and (13.10) are convenient expressions relating the shear modulus to the current state and to the stress history and there will be similar expressions for the bulk modulus K'. However, to be of practical use for design, soil behaviour must be represented by mathematical expressions similar to those developed for Cam clay in Chapter 12, although these are likely to be more complex to take account of the non-linear behaviour for states inside the boundary surface. One possibility is to regard soil behaviour inside the state boundary surface as essentially elastic, but non-linear, and to use curve-fitting techniques to obtain an empirical expression relating shear modulus G' and bulk modulus K' to strain. This is the approach followed by

Duncan and Chang (1970) and by Jardine *et al.* (1991). This method requires complex laboratory tests in which the stress paths mimic the *in situ* paths and numerical analyses that should stop and restart at each change in the direction of a stress path. An alternative approach is to regard soil behaviour in the small strain region as inelastic, with yielding and hardening with moving yield surfaces inside the state boundary surface. One approach is to adapt the Cam clay models by including additional yield surfaces (e.g. Mroz, Norris and Zienkiewics, 1979; Atkinson and Stallebrass, 1991). In these models the parameters remain the fundamental parameters required by Cam clay together with additional parameters that describe the relative sizes of the additional yield surfaces.

At small strains in the region 0.001 to 1% the general relationships between shear modulus, strain and stress shown in Fig. 13.10 are similar for normally consolidated and overconsolidated soils. Furthermore, unloading and reloading loops, like those illustrated in Fig. 13.2, result in substantial irrecoverable strains. These observations indicate that the basic rules governing stiffness of overconsolidated soils at small strains are similar to those for normally consolidated soil which, as we have seen, are essentially elasto-plastic and not purely elastic as assumed in the Cam clay theories.

All this is really quite advanced and any further discussion of developments in theories for soil stiffness at small strain is clearly beyond the scope of this book.

13.9 Summary

1. The stress–strain behaviour of soil is highly non-linear over the whole range of loading except at very small strains less than about 0.001%. These are three ranges of behaviour:

    (a) very small strain (usually less than 0.001%),
    (b) small strain,
    (c) large strains (for states on the state boundary surface).

2. For states on the state boundary surface the strains are relatively large and can be modelled reasonably using Cam clay or a similar elasto-plastic model.
3. For very small strains the stress–strain behaviour is approximately linear and the shear modulus is given by

$$\frac{G_0'}{p_r'} = A\left(\frac{p'}{p_r'}\right)^n Y_p^m \tag{13.8}$$

    where A, m and n depend on the nature of the soil.
4. For small strains the soil is highly non-linear: at a particular strain the shear modulus is given by

$$\frac{G'}{p'} = AY_p^m \tag{13.10}$$

    where A and m depend both on the nature of the soil and on the strain.

References

Atkinson, J. H. and S. E. Stallebrass (1991) 'A model for recent history and non-linearity in the stress–strain behaviour of overconsolidated soil', *Proceedings of 7th IACMAG '91, Cairns*, pp. 555–560.

Duncan, J. M. and C. Y. Chang (1970) 'Non-linear analysis of stress and strain in soils', *ASCE, J. of the Soil Mechanics and Foundation Engng Div.*, 96, SM5, 1629–1653.

Jardine, R. J., D. M. Potts, H. D. St. John and D. W. Hight (1991) 'Some applications of a non-linear ground model', *Proceedings 10th European Conference on Soil Mechanics and Foundation Engineering, Florence*, Vol. 1, pp. 223–228.

Mroz, Z., V. A. Norris and O. C. Zienkiewics (1979) 'Application of an anisotropic hardening model in the analysis of elasto-plastic deformation of soils', *Geotechnique*, 29, 1, 1–34.

Viggiani, G. and J. H. Atkinson (1995) Stiffness of fine-grained soil at very small strains. *Geotechnique*, Vol. 45, No. 1, pp. 149–154.

Further reading

Atkinson, J. H. (2000) 40th Rankine Lecture. Non-linear stiffness in routine design, *Geotechnique*, Vol. 50, No. 5, pp. 487–508.

Atkinson, J. H. and G. Sallfors (1991) 'Experimental determination of stress–strain–time characteristics in laboratory and *in situ* tests', *Proceedings of 10th ECSMFE, Florence*, Vol. 3, pp. 915–956.

Wroth, C. P. and G. T. Houlsby (1985) 'Soil mechanics – property characterization and analysis procedures', *Proceedings of 11th ICSMFE, San Francisco*, pp. 1–55.

Chapter 14

Steady state seepage

14.1 Groundwater conditions

You know that water flows downhill and you have probably studied the flow of water in pipes and open channels in courses on hydraulics. Water also flows through soils in much the same way but the flow is retarded as it flows past the grains. Theories for groundwater flow are covered in courses in hydraulics and all I will do here is consider the topics essential for geotechnical engineering. There are essentially three separate conditions for groundwater in geotechnical engineering and simple examples of these are illustrated in Fig. 14.1.

(a) Hydrostatic states

This condition, illustrated in Fig. 14.1(a), was discussed in Sec. 6.3. If the water table, or phreatic surface, is level there is no flow. Pore pressures are hydrostatic, are given by $u = \gamma_w h_w$ and this is the same whether there are soil grains or not.

(b) Steady state seepage

If the phreatic surface is not level, as in Fig. 14.1(b), water will flow along flowlines such as ABC. At any point, such as at A, the pore pressures will be $u = \gamma_w h_w$, where h_w is the height of water in a standpipe. Note that the level of water in the pipe does not necessarily define the phreatic surface (see Sec. 14.5). Notice also that in Fig. 14.1(b) the flow is apparently uphill from A to B and that the pore pressure at C is greater than that at B.

The basic rule for the flow of water through a single element of soil is Darcy's law, which was introduced in Sec. 6.10 in connection with relative rates of loading and drainage. In this chapter we will extend Darcy's law to cover seepage through a whole region of soil. The essential feature of steady state seepage is that neither the pore pressures nor the rates of flow change with time. Since effective stresses remain constant, the soil grains can be taken to be stationary as water flows through the pore channels.

(c) Consolidating soil

When pore pressures change with time effective stresses and soil volumes also change with time. This process, which couples Darcy's seepage theory with soil compression

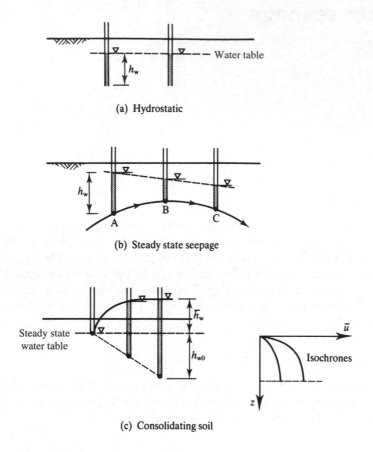

(a) Hydrostatic

(b) Steady state seepage

(c) Consolidating soil

Figure 14.1 Groundwater conditions.

and swelling, is known as consolidation. This is covered in Sec. 6.9 and Chapter 15. During consolidation, pore pressures are the sum of the steady state pore pressures $u_0 = \gamma_w h_{w0}$ and the excess pore pressure $\bar{u} = \gamma_w \bar{h}_w$, as shown in Fig. 14.1(c). Graphs of excess pore pressure $\bar{u}$ at given times are called isochrones. Remember consolidation works both ways. If the excess pore pressure is positive water is squeezed out and the soil compresses. If they are negative water is sucked in and the soil swells.

14.2   Practical problems of groundwater flow

Any child who has dug a hole in the beach or constructed a small soil dam across a stream has soon recognized the importance of groundwater in ground engineering. It is impossible to excavate much below the groundwater table and dams soon fail by downstream erosion, even if they are not overtopped first. The hole can only be continued if water is pumped from the excavation, and possibly from the surrounding ground as well, and engineers will need to determine the quantities of water to be pumped. They will also be interested in the quantities of water leaking from water storage dams.

It is common knowledge that landslides occur most frequently after periods of rainfall when pore pressures in the ground are highest. (Remember that this has nothing to do with water lubricating soil because the critical state friction angle ϕ'_c is the same for dry and saturated soil as discussed in Chapter 9.) We have already seen that soil strength and stiffness are governed by the effective stresses which depend on the pore pressures as well as on the total stresses, so that calculation of pore pressures in soil with steady state seepage will be an essential component of geotechnical design calculations.

Figure 14.2 illustrates two typical cases of steady state seepage in geotechnical problems. In both cases water flows from regions of high water level to regions of low water level along flowlines such as ABC: notice that in Fig. 14.2(a) the water flows upwards from B to C. In Fig. 14.2(a) the flow is confined because the top flowline PQRS is confined by the impermeable concrete dam. In Fig. 14.2(b) the flow is unconfined and there is a phreatic surface, which is also the top flowline TU. In both cases we will be interested in calculating both the rates of leakage below or through the dams and the distributions of pore pressures.

In Fig. 14.2(a) water flows upwards in the region of C, where the flowline emerges at the downstream ground surface. If the seepage velocities are large, soil grains may be disturbed and washed away. If this should happen the erosion would seriously jeopardize the stability of the dam. The same thing might happen to the dam in Fig. 14.2(b) if the downstream drain is inadequate so that the top flowline TU emerges from the downstream face of the dam. After overtopping this is the most common cause of failure of dams made by children at the seaside.

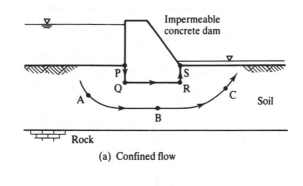

(a) Confined flow

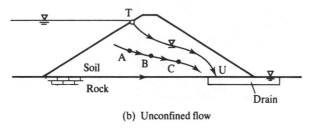

(b) Unconfined flow

Figure 14.2 Problems in groundwater flow.

14.3   Essentials of steady state seepage

Darcy's law governing flow of water through soil is very like Ohm's law for the flow of electricity through a conducting material, and an electrical flow model can be used to solve problems in groundwater seepage. In both cases a potential causes a current to flow against a resistance so that electrical conductivity is analogous to permeability. We have already seen that hydraulic potential is not the same as pore pressure and it is necessary to include a term to take account of elevation.

To define hydraulic potential it is necessary to have a datum as in Fig. 14.3(a). Since it is only changes of potential that matter the datum could be anywhere, but it is best to put it low down to avoid negative values of potential. From Fig. 14.3(a), the potential at A is

$$P = h_w + z = \frac{u}{\gamma_w} + z \tag{14.1}$$

(Note that this is simply Bernoulli's expression for total head since, in groundwater seepage, the velocity terms are small compared with the pressure and elevation terms.)

In Fig. 14.3(b) the points A and B are δs apart on the same flowline and the hydraulic gradient between A and B is

$$i = -\frac{\delta P}{\delta s} \tag{14.2}$$

The negative sign is introduced into Eq. (14.2) so that the hydraulic gradient is positive in the direction of flow. (Note that in Fig. 6.12 and in Eq. (6.19) the hydraulic potential and the hydraulic gradient were defined in terms of h_w only. This was allowable in that case because the flowlines in Fig. 6.12 were horizontal and so the z term in Eq. (14.1) remains constant. From now on we will work with potentials and hydraulic gradients using Eqs. (14.1) and (14.2), taking account of pore pressure and elevation terms.)

Figure 14.4 shows part of a flownet with two flowlines AB and CD at an average distance δb apart. The points A and C have the same potential and so do the

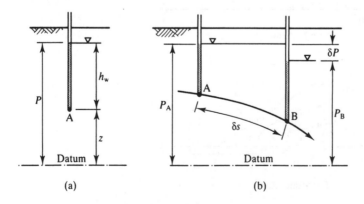

Figure 14.3 Pore pressure and potential.

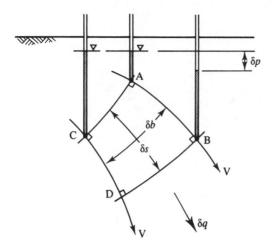

Figure 14.4 Flowlines and equipotentials.

points B and D. The lines AD and BD are called equipotentials (because they are lines of equal potential) and the average distance between them is δs. Flowlines and equipotentials intersect at 90° as shown. (The proof of this is given in textbooks on hydraulics.)

Figure 14.4 represents two-dimensional seepage through isotropic soil in which the value of k is the same in all directions and through a slice of unit thickness normal to the page; all the discussion in this chapter is for two-dimensional seepage through isotropic soil. The rate of flow (in cubic metres per second) between the two flowlines is δq and the mean seepage velocity (in metres per second) is

$$V = \frac{\delta q}{\delta b} \tag{14.3}$$

Darcy's law states that

$$V = ki \tag{14.4}$$

where k is the coefficient of permeability which has the units of velocity. Typical values of k for soils were given in Sec. 6.10. Remember that for coarse-grained soils $k > 10^{-2}$ m/s while for fine-grained soils $k < 10^{-8}$ m/s; these very large differences mean that coarse-grained soils with high permeability can act as drains while fine-grained soils with very low permeability can be used as nearly watertight barriers.

Notice that the seepage velocity V given by Eq. (14.4) is not the velocity of a drop of water as it seeps through the pore spaces. From Fig. 14.5 the velocity of the drop of water is $V_w = \delta q/\delta w$, where δw is the area occupied by the pore spaces in an area

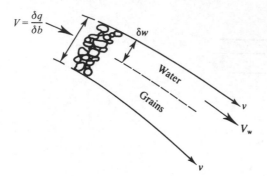

$$V = \frac{\delta q}{\delta b}$$

Figure 14.5 Difference between seepage and flow velocities.

of soil δb and

$$\frac{V}{V_w} = \frac{\delta w}{\delta b} = 1 - \frac{1}{v} \tag{14.5}$$

where v is the specific volume. This means that if you use dye or a tracer to examine groundwater flow you will measure V_w, which is not the same as the velocity given by Darcy's law in Eq. (14.4).

14.4   Flow through a simple flownet

Figure 14.4 shows the conditions of steady state seepage through a single element bounded by two flowlines and two equipotentials. The rate of flow through the element is given by Eqs. (14.3) and (14.4) as

$$\delta q = \delta bki \tag{14.6}$$

If we can assign a value of potential to an equipotential we could calculate the pore pressures from Eq. (14.1), if necessary interpolating between the equipotentials. The flow through a whole region and the pore pressures throughout the region can be found by considering an assembly of elements called a flownet.

Figure 14.6 shows part of a simple flownet. The flowlines and equipotentials intersect orthogonally, and if $\delta s = \delta b$ the flownet is square. There are four flowlines and so the number of flow channels, N_f, is three. The total rate of flow through the region is Δq and, making use of Eq. (14.6),

$$\Delta q = N_f\, \delta q = N_f\, \delta bki \tag{14.7}$$

Because the flowlines are straight and parallel, the seepage velocity, and hence the hydraulic gradient, is constant and so the equipotentials are equally spaced as shown.

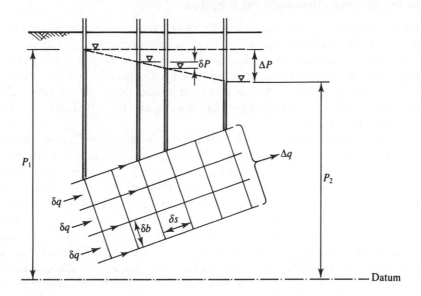

Figure 14.6 Flow through a simple flownet.

There are six equipotentials and so the number of equipotential drops in the square flownet, N_d, is five; therefore, from Eq. (14.2),

$$i = -\frac{\delta P}{\delta s} = -\frac{\Delta P}{N_d \, \delta s} \qquad (14.8)$$

Hence, from Eqs. (14.7) and (14.8) the rate of flow through the whole flownet is

$$\Delta q = -k\frac{N_f}{N_d}\Delta P \qquad (14.9)$$

where ΔP is the change of potential across the whole flownet.

Although Eq. (14.9) was derived for the simple flownet in Fig. 14.6 with straight flowlines and equipotentials, it is applicable to any flownet with curved elements provided that the elements are 'square' in the sense that the flowlines and equipotentials intersect orthogonally and the mean dimensions of each element are the same (i.e. $\delta s = \delta b$). Notice that the ratio N_f/N_d depends only on the geometry of the boundary of the flownet so that in Fig. 14.6 we could have $N_f = 6$ and $N_d = 10$ by halving the size of each element. If the values of potential, P_1 and P_2, at the inflow and outflow boundaries are known the values of potential can be found at any equipotential (because the drop in potential is the same across any element) and the pore pressure at any point within the flownet can be calculated from Eq. (14.1).

14.5   Flownet for two-dimensional seepage

A solution to any problem in two-dimensional steady state seepage can be found by drawing a flownet with curvilinear squares. This must be a proper scale drawing with the correct boundary conditions. The solution gives the rate of flow from Eq. (14.9) and the distribution of pore pressure from Eq. (14.1). Techniques for constructing flownets by sketching, by electrical models and by numerical analysis are covered in textbooks on hydraulics. All I will do here is find solutions to two simple cases to illustrate the general principles.

In Fig. 14.7 water seeps from a river into a trench supported by walls and which is pumped dry. The geometry is symmetric about the centre-line. The flow is confined so there is no phreatic surface. If a standpipe is placed with its tip just at the ground level, such as at G or at C, water will rise to the river level or the pumped level: therefore AG is an equipotential with value P_1 and similarly CF is an equipotential with value P_2. Any impermeable boundary, such as the wall and the rock surface, must be a flowline and so is the axis of symmetry; therefore, ABC and DEF are flowlines because flowlines cannot cross. A roughly sketched flownet is shown in Fig. 14.7(b). This satisfies the boundary conditions in Fig. 14.7(a); flowlines and equipotentials are orthogonal and each element is more or less 'square' with approximately equal length and breadth. For this flownet the total number of flow channels is $N_f = 8$ (i.e. four on each side of the centre-line) and the number of equipotential drops is $N_d = 10$.

In Fig. 14.8 water seeps through a soil embankment dam to a drain in the down-stream toe. The flow is unconfined and there is a phreatic surface in a position approximately as shown by the broken line. If a standpipe is placed with its tip anywhere on the upstream face, water will rise to the reservoir level so AB is an equipotential with value P_1. Similarly, the drain CD is an equipotential with value P_2.

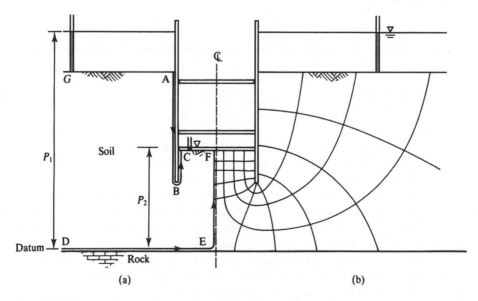

(a)   (b)

Figure 14.7 Flownet for steady state flow into a trench excavation.

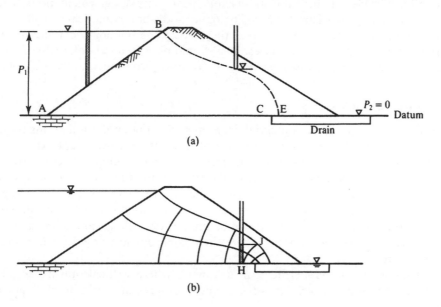

Figure 14.8 Flownet for steady state seepage through a dam.

The top of the impermeable rock AC is a flowline. The phreatic surface BE is not precisely located by the geometry of the dam alone but its position will be fixed by the geometry of the flownet. The phreatic surface is a flowline and, on the phreatic surface, the pore pressure is zero. A roughly sketched flownet is shown in Fig. 14.8(b). Again this satisfies the boundary conditions, flowlines and equipotential are orthogonal and each element is more or less 'square'. Notice that the equipotentials intersect the phreatic surface at equal vertical intervals (because $u = 0$ along the phreatic surface). For this flownet, $N_f = 2$ and $N_d = 5$.

The level of water in a standpipe does not necessarily rise to the phreatic surface. In Fig. 14.8(b) the tip of the standpipe is on the equipotential HJ. If the tip of a standpipe is on the phreatic surface at J the water remains at J and so the level of water in any standpipe on HJ must be at the level of J. For the standpipe at H the water rises not to the phreatic surface but to the level of J as shown.

The flownets can be used to calculate the rates of leakage into the trench excavation and through the dam using Eq. (14.9). Note that this contains the coefficient of permeability k and the accuracy of the solution will depend more on how well you can determine a value for k than on how well you can draw a flownet. The flownets can also be used to calculate pore pressures. You will need these to calculate the loads on the walls and props in Fig. 14.7 and the stability of the dam slopes in Fig. 14.8, but to calculate pore pressures the flownet must be accurately drawn. Notice that the geometry of a flownet and the pore pressures are independent of the value of coefficient of permeability k.

The flownets shown in Figs. 14.7 and 14.8 were sketched by me very quickly using a soft pencil and a good eraser. They are a bit rough – not all the elements are properly 'square' and sometimes the flownets and equipotential do not intersect exactly orthogonally – but they are probably good enough for many design calculations. They could be improved by use of an electrical analogue model or a numerical analysis. The important thing about my flownets is that they satisfy the boundary conditions and there are no fundamental inconsistencies. You should now go to a book on hydraulics and study flownets for other cases, particularly for flow into drains, wells and slots.

14.6   Piping and erosion

As water flows through soil the potential drops and the drag on the soil grains results in an increase in effective stress in the direct of flow. If the flow is upwards, these seepage stresses act against the self-weight stresses and the resultant effective stresses reduce. This condition occurs in the base of the excavation in Fig. 14.7. If the upward flow is large the condition could occur where the effective stresses and the strength become zero, and this would clearly have very serious consequences for the stability of an excavation. This condition is known as piping, or boiling, and is the cause of quicksand: natural quicksands occur where there is an upward flow of water under artesian pressure.

Figure 14.9 shows the last element in a flownet where vertical upward seepage emerges at the ground surface. The stresses and pore pressures at a depth δs in the ground are

$$\sigma_v = \gamma\,\delta s + \gamma_w h_w \tag{14.10}$$

$$u = \gamma_w(\delta P + h_w + \delta s) \tag{14.11}$$

Hence, making use of Eq. (14.2)

$$\sigma'_v = \gamma_w\,\delta s\left[\left(\frac{\gamma}{\gamma_w} - 1\right) - i\right] \tag{14.12}$$

and the vertical effective stress σ'_v reduces with increasing i. If $\sigma'_v = 0$ the critical hydraulic gradient i_c is

$$i_c = \frac{\gamma}{\gamma_w} - 1 \tag{14.13}$$

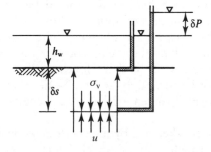

Figure 14.9 Critical hydraulic gradient.

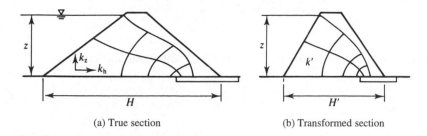

(a) True section   (b) Transformed section

Figure 14.10 Flownets for anisotropic soil.

For many soils γ is approximately 20 kN/m^3 and i_c is approximately unity. Piping or boiling will generally only occur for upward seepage towards the ground surface, as shown in Fig. 14.9. You can create piping in the apparatus shown in Fig. 6.7 by extending the standpipe and filling it to a height above ground level that is about twice the depth of the model.

14.7   Seepage through anisotropic soils

Many soils are layered either because they were naturally deposited in changing depositional environments or because they were compacted in layers, with the result that the permeability for horizontal flow k_h is often considerably greater than the permeability for vertical flow k_z. In this case the flownet is not square and flowlines and equipotentials do not intersect orthogonally, as shown in Fig. 14.10(a).

The flownet can, however, be made to be square by transforming the horizontal axis to H' and the mean coefficient of permeability to k', where

$$H' = \sqrt{\frac{k_z}{k_h}} H \tag{14.14}$$

$$k' = \sqrt{k_z k_h} \tag{14.15}$$

as shown in Fig. 14.10(b). The theoretical derivations for these transformations are beyond the scope of this book and are given in textbooks on hydraulics and groundwater flow.

14.8   Summary

1. For steady state seepage pore water pressures u at point are given by the potential P:

$$P = h_w + z = \frac{u}{\gamma_w} + z \tag{14.1}$$

where z is the elevation of the point above an arbitrary datum.

2. Seepage of water through soil is governed by Darcy's law

$$V = ki \tag{14.4}$$

where V is the seepage velocity and i is the hydraulic gradient given by

$$i = -\frac{\delta P}{\delta s} \tag{14.2}$$

3. Steady state seepage through a region of soil is described by a square flownet consisting of an orthogonal net of flowlines and equipotential. Pore pressures can be calculated from equipotentials. The total rate of flow through a flow through a flownet is given by

$$\Delta q = -k\frac{N_f}{N_d}\Delta P \tag{14.9}$$

where ΔP is the change of potential across the whole flownet. Flownets can be obtained by sketching orthogonal nets that satisfy the boundary conditions.

4. Seepage towards the ground surface or towards a slope may cause instabilities due to piping or erosion.

    For upward seepage towards the ground surface the critical hydraulic gradient when $\sigma_v' = 0$ is given by

$$i_c = \frac{\gamma}{\gamma_w} - 1 \tag{14.13}$$

Worked examples

Example 14.1: Confined flow   Figure 14.11 illustrates flow towards a long (out of the page) land drain through a layer of soil with permeability $k = 10^{-6}$ m/s sandwiched between clay and rock, both of which may be considered to be impermeable. The water level in the drain is 1 m below the water table which is at ground level 9 m away.

The phreatic surface is above the top of the soil and the flow is confined by the impermeable clay. A simple square flownet is shown in Fig. 14.11 in which $N_f = 3$

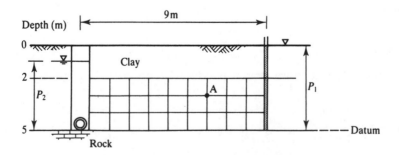

Figure 14.11 Confined flow towards a drain – Example 14.1.

and $N_d = 9$. Taking the datum for potential at the rock level, $P_1 = 5$ m and $P_2 = 4$ m and, from Eq. (14.9) the rate of flow into the drain (from one side) per unit length out of the page is

$$\Delta q = -k \frac{N_f}{N_d} \Delta P = -10^{-6} \times \tfrac{9}{3} \times (4 - 5) = 3 \times 10^{-6} \, \text{m}^3/\text{s}$$

At the point A the elevations is $z_a = 2$ m and the potential is

$$P_a = P_1 - \tfrac{3}{9} \Delta P = 5 - \tfrac{3}{9}(5 - 4) = 4.67 \, \text{m}$$

Hence, from Eq. (14.1), the pore pressure at A is

$$u_a = \gamma_w(P_a - z_a) = 9.81 \times (4.67 - 2) = 26 \, \text{kPa}$$

Example 14.2: Unconfined flow   Figure 14.12 illustrates leakage from a canal into a nearby river. (Both the river bank and the canal bank are supported by sheet piles that leak.) The coefficient of permeability of the soil is 10^{-6} m/s.

The phreatic surface joins the water levels in the river and canal and the flow is unconfined. From the flownet sketched $N_f = 3$ and $N_d = 7$ and, taking the datum for potential at the bed of the canal, $P_1 = 4$ m and $P_2 = 2$ m. Hence, from Eq. (14.9) the rate of leakage per unit length of the page is

$$\Delta q = -k \frac{N_f}{N_d} \Delta P = -10^6 \times \tfrac{7}{3} \times (2 - 4) \approx 5 \times 10^{-6} \, \text{m/s}$$

At the point A, scaling from the diagram, the elevation is $z_a = 1.73$ m and the potential is

$$P_a = P_1 - \tfrac{5}{7} \Delta P = 4 - \tfrac{5}{7}(4 - 2) = 2.57 \, \text{m}$$

Hence, from Eq. (14.1), the pore pressure at A is

$$u_a = \gamma_w(P_a - z_a) = 9.81 \times (2.57 - 1.73) = 8.2 \, \text{kPa}$$

Notice that water in a standpipe at A rises to the level where the equipotential through A meets the phreatic surface.

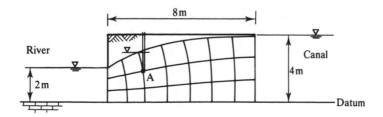

Figure 14.12 Unconfined flow – Example 14.2.

Further reading

Atkinson, J. H. and P. L. Bransby (1978) *The Mechanics of Soils*, McGraw-Hill, London.
Cedergren, H. R. (1997) *Seepage, Drainage and Flownets*, 3rd Edition, Wiley, New York.
Powers, J. P. (1992) *Construction Dewatering*, Wiley, New York.
Somerville, S. H. (1986) *Control of Ground Water for Temporary Works*, CIRIA, Report 113, London.

Consolidation

15.1   Basic mechanism of consolidation

In Sec. 6.9 we saw that, in general, any undrained loading or unloading will create excess pore pressures $\bar{u}$ in the region of the loading. These excess pore pressures may be positive or negative with respect to the long term steady state pore pressures u_∞ and they give rise to hydraulic gradients that cause seepage flow. These seepage flows lead to volume changes that, in turn, are associated with the changes of effective stress as the excess pore pressures dissipate. As the excess pore pressure diminish the hydraulic gradients and rates of flow also diminish, so that the volume changes continue at a reducing rate. After a long time the seepage and volume changes will stop when the excess pore pressures and hydraulic gradients become zero and the pore pressures reach their steady state values.

The coupling of seepage due to hydraulic gradients with compression or swelling due to the resulting seepage flow and changes of effective stress is known as consolidation. This process accounts for settlement of foundations with time, progressive softening of soil in excavations and other similar effects. In order to calculate the rate at which excess pore pressures reduce it is necessary to develop a simple theory for consolidation.

A general theory for three-dimensional consolidation is quite complicated and here I will consider a simpler theory for one-dimensional consolidation in which all seepage flow and soil strains are vertical and there is no radial seepage or strain. This is relevant to conditions in an oedometer test (see Sec. 7.6), as shown in Fig. 15.1(a), and in the ground below a wide foundation on a relatively thin layer of soil, as shown in Fig. 15.1(b). In both cases the seepage of water from within the body of the soil is vertical and upwards towards a surface drainage layer where the steady state pore pressure is always $u_0 = u_\infty$ and the excess pore pressure is always zero.

15.2   Theory for one-dimensional consolidation

Figure 15.2 shows an element in a consolidating soil. (Here all dimensions increase positively downwards to avoid difficulties with signs.) In a time interval δt the thickness changes by δh. The flow of water through the element is one-dimensional and the rates of flow in through the top and out through the bottom are q and $q + \delta q$ respectively.

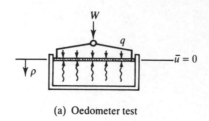

(a) Oedometer test

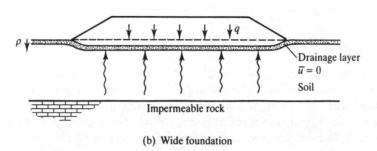

(b) Wide foundation

Figure 15.1 Examples of one-dimensional consolidation.

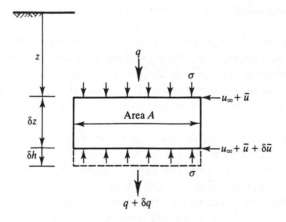

Figure 15.2 One-dimensional consolidation.

From the definition of the coefficient of compressibility m_v given by Eq. (8.9),

$$\delta h = -m_v \delta z \delta \sigma' \tag{15.1}$$

The theory requires that m_v remains constant and so it is valid only for relatively small increments of stress. Since the soil grains are incompressible an equation of continuity relates the change of volume of the element to the change of flow through it:

$$A\,\delta h = -\delta q \delta t \tag{15.2}$$

Combining Eqs. (15.1) and (15.2) and in the limit noting that q and σ' are both functions of z and t,

$$\frac{\partial q}{\partial z} = A m_v \frac{\partial \sigma'}{\partial t} \tag{15.3}$$

The rate of seepage flow is given by Darcy's law as

$$V = \frac{q}{A} = ki \tag{15.4}$$

where V is the seepage velocity and the hydraulic gradient i is

$$i = \frac{1}{\gamma_w} \frac{\delta \bar{u}}{\delta z} \tag{15.5}$$

From Eqs. (15.4) and (15.5) and in the limit,

$$\frac{\partial q}{\partial z} = -\frac{Ak}{\gamma_w} \frac{\partial}{\partial z} \left(\frac{\partial \bar{u}}{\partial z} \right) = -\frac{Ak}{\gamma_w} \frac{\partial^2 \bar{u}}{\partial z^2} \tag{15.6}$$

and, from Eqs. (15.3) and (15.6),

$$\frac{k}{m_v \gamma_w} \frac{\partial^2 \bar{u}}{\partial z^2} = -\frac{\partial \sigma'}{\partial t} \tag{15.7}$$

The effective stress is given by $\sigma' = \sigma - (u_\infty + \bar{u})$ and, noting that u_∞ remains constant,

$$\frac{\partial \sigma'}{\partial t} = \frac{\partial \sigma}{\partial t} - \frac{\partial \bar{u}}{\partial t} \tag{15.8}$$

The simple and common case is where consolidation takes place after an increment of undrained loading or unloading so that the total stress remains constant during the consolidation. Then, from Eqs. (15.7) and (15.8) with $\partial \sigma / \partial t = 0$,

$$c_v \frac{\partial^2 \bar{u}}{\partial z^2} = \frac{\partial \bar{u}}{\partial t} \tag{15.9}$$

where

$$c_v = \frac{k}{m_v \gamma_w} \tag{15.10}$$

The parameter c_v is known as the coefficient of consolidation and has the units of square metres per year. Values of c_v depend on both the permeability k and on the compressibility m_v, both of which vary greatly for different soils.

Equation (15.9) is the basic equation for one-dimensional consolidation. Solutions will give the variations of excess pore pressure $\bar{u}$ with depth z and with time t. Note that consolidation theory deals with excess pore pressure $\bar{u}$ and not with absolute pore pressures.

15.3   Isochrones

Solutions to Eq. (15.9) can be represented graphically by plotting the variation of $\bar{u}$ with depth at given times. The resulting family of curves are called isochrones. A simple way to visualize isochrones is to imagine a set of standpipes inserted into the consolidating soil below a rapidly constructed embankment as shown in Fig. 15.3(a).

Before construction water rises in the standpipes to the steady state water table in the drain at the surface where the initial and long term pore pressures are $u_0 = u_\infty$. Undrained construction of the embankment adds a total stress $\Delta\sigma$ at the surface, which gives rise to initial excess pore pressures $\bar{u}_i = \Delta\sigma$ throughout the soil. The initial excess pore pressures registered by the standpipes are uniform with depth and water rises to the same height in all the pipes, as shown by the broken (initial) line in Fig. 15.3(a). The corresponding isochrone for $t = 0$ is shown in Fig. 15.3(b). (Notice that because $\gamma \approx 2\gamma_w$ the standpipes must project well above the maximum height of the embankment.)

At a time shortly after construction excess pore pressure at the top of the soil near the drain will have reduced to zero and excess pore pressures will have reduced elsewhere, so the variation of the levels of water in the standpipes is similar to that shown by the curved broken line. This broken line gives the shape of the isochrone at a particular time. After a very long time all the excess pore pressures have dissipated and the levels of water in the standpipes are at the long term steady state groundwater table; the isochrone for $t = \infty$ is the final broken line.

Figure 15.3(b) shows a set of isochrones for the one-dimensional consolidation illustrated in Fig. 15.3(a) plotted as $\bar{u}$ against depth z. Each isochrone corresponds to a particular time: for $t = 0, \bar{u}_i = \Delta\sigma$ at all depths and at $t = \infty, \bar{u}_\infty = 0$.

15.4   Properties of isochrones

Isochrones must satisfy the one-dimensional consolidation equation together with the drainage boundary conditions, and these requirements impose conditions on the geometry and properties of isochrones. Consolidation, with dissipation to a drain at the surface, as shown in Fig. 15.3, starts near the surface and progresses down through

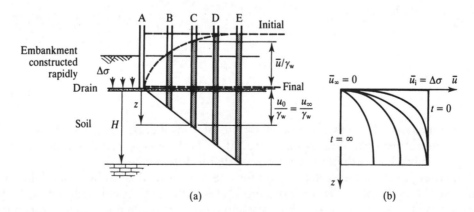

Figure 15.3 Isochrones for one-dimensional consolidation.

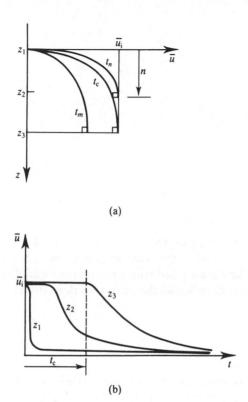

Figure 15.4 Dissipation of excess pore pressure during consolidation.

the soil. At relatively small times, such as t_n in Fig. 15.4, consolidation is limited to the upper levels only and below a depth n the excess pore pressures have not fallen. At large times, such as t_m, consolidation is occurring throughout the layer. There is a critical time t_c when excess pore pressures at the base first start to dissipate; the isochrone for t_c is shown in Fig. 15.4(a). Figure 15.4(b) illustrates the dissipation of excess pore pressure at the different depths indicated in Fig. 15.4(a). Near the surface, at a depth z_1, the excess pore pressures dissipate very rapidly but near the base, at a depth z_3, the excess pore pressures remain at $\bar{u}_i$ until the critical time t_c.

The gradient of an isochrone is related to the hydraulic gradient by

$$\frac{\partial \bar{u}}{\partial z} = -\gamma_w i \tag{15.11}$$

and from Darcy's law the seepage velocity is

$$V = -\frac{k}{\gamma_w} \frac{\partial \bar{u}}{\partial z} \tag{15.12}$$

By inspection of isochrones in Fig. 15.4(a) the gradients of the isochrones, and hence the seepage velocities, increase towards the surface. At the base of an isochrone there

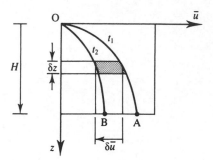

Figure 15.5 Area between two isochrones.

is no seepage flow, either because it represents the limit of consolidation for t_n or because of the impermeable boundary for t_m, and so the isochrones must be vertical at the base, as shown in Fig. 15.4(a). Since soil grains and water are incompressible the velocity of the upward seepage at any level must equal the rate of settlements at that level and

$$\frac{\partial \rho}{\partial t} = \frac{k}{\gamma_w} \frac{\partial \overline{u}}{\partial z} \qquad (15.13)$$

The movement of isochrones represents changes of excess pore pressure and changes of effective stress. Figure 15.5 shows isochrones for t_1 and t_2. From Eq. (15.1) the change of thickness δh of the thin slice δz is given by $\delta h = -m_v \delta z \delta \sigma'$. If the total stress remains constant, $\delta \sigma' = -\delta \overline{u}$ and

$$\delta h = m_v \delta z \delta \overline{u} \qquad (15.14)$$

where $\delta z \delta \overline{u}$ is the shaded area in Fig. 15.5. Summing the changes of thickness for all thin slices in the depth z, the change of surface settlement between the times t_1 and t_2 is given by

$$\delta \rho = m_v \times \text{area OAB} \qquad (15.15)$$

Hence the settlement of a consolidating layer in a given time is given by m_v times the area swept by the isochrone during the time interval.

15.5   Solution for one-dimensional consolidation by parabolic isochrones

Simple and reasonably accurate solutions for the rate of settlement for one-dimensional consolidation can be obtained by assuming that the general shapes of the isochrones in Fig. 15.4(a) can be approximated by parabolas. It is necessary to treat the cases $t < t_c$ and $t > t_c$ separately; the ideas behind each analysis are the same but the algebra differs slightly.

(a) $t = t_n < t_c$

Figure 15.6(a) shows an isochrone for time t_n; the slope is vertical at N and no consolidation has occurred below a depth n. From Eq. (15.15) (noting that the area below a parabola is $\frac{1}{3} \times$ base $\times$ height), the surface settlement is given by

$$\Delta \rho_t = m_v \times \text{area AEN} = \frac{1}{3} m_v n \Delta \sigma \tag{15.16}$$

Differentiating Eq. (15.16) and noting that m_v and $\Delta \sigma$ are assumed to be constants during consolidation, the rate of settlement is given by

$$\frac{\mathrm{d}\rho_t}{\mathrm{d}t} = \frac{1}{3} m_v \, \Delta \sigma \frac{\mathrm{d}n}{\mathrm{d}t} \tag{15.17}$$

The rate of surface settlement is also related to the gradient of the isochrone at A. From Eq. (15.13) and noting that from the geometry of a parabola the gradient at A is $2\Delta\sigma/n$, we have

$$\frac{\mathrm{d}\rho_t}{\mathrm{d}t} = \frac{k}{\gamma_w} \frac{2\Delta\sigma}{n} \tag{15.18}$$

Hence, equating the rates of surface settlement from Eqs. (15.17) and (15.18),

$$n\frac{\mathrm{d}n}{\mathrm{d}t} = 6\frac{k}{m_v \gamma_w} = 6c_v \tag{15.19}$$

and, integrating with the boundary condition $n = 0$ at $t = 0$,

$$n = \sqrt{12 c_v t} \tag{15.20}$$

Equation (15.20) gives the rate at which the effects of consolidation progress into the soil from the drainage boundary; no dissipation of excess pore pressure will occur at

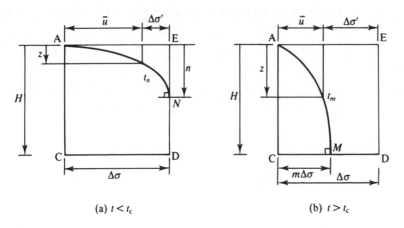

(a) $t < t_c$ (b) $t > t_c$

Figure 15.6 Geometry of parabolic isochrones.

depths greater than n. Using Eq. (15.20) and the geometry of a parabola it is possible to calculate the excess pore pressure at any depth and at any time $t < t_c$.

In practice, the most important thing to calculate is the surface settlement $\Delta\rho_t$ after a time $t < t_c$; and this is found by substituting for n into Eq. (15.16), giving

$$\Delta\rho_t = \frac{1}{3}m_v\Delta\sigma\sqrt{12c_v t} \tag{15.21}$$

The final surface settlement $\Delta\rho_\infty$ will occur after a long time when all excess pore pressures have dissipated and $\Delta\sigma' = \Delta\sigma$. Hence, from Eq. (15.1),

$$\Delta\rho_\infty = m_v H\Delta\sigma \tag{15.22}$$

Combining Eqs. (15.21) and (15.22),

$$\frac{\Delta\rho_t}{\Delta\rho_\infty} = \frac{2}{\sqrt{3}}\sqrt{\frac{c_v t}{H^2}} \tag{15.23}$$

Equation (15.23) may be written in terms of a dimensionless degree of consolidation U_t and a dimensionless time factor T_v given by

$$U_t = \frac{\Delta\rho_t}{\Delta\rho_\infty} \tag{15.24}$$

$$T_v = \frac{c_v t}{H^2} \tag{15.25}$$

and the general solution becomes

$$U_t = \frac{2}{\sqrt{3}}\sqrt{T_v} \tag{15.26}$$

This solution is valid until the point N in Fig. 15.6(a) reached D when $t = t_c$; at this instant $n = H = \sqrt{12c_v t}$ so $T_v = \frac{1}{12}$ and $U_t = 0.33$. For $t > t_c$ the isochrone no longer touches ED and a new analysis is required.

(b) $t = t_m > t_c$

Figure 15.6(b) shows an isochrone for t_m; it intersects the base orthogonally at M where $\bar{u} = m\Delta\sigma$. Making use of the geometry of a parabola and proceeding as before,

$$\Delta\rho_t = m_v\Delta\sigma H\left(1 - \frac{2}{3}m\right) \tag{15.27}$$

$$\frac{d\rho_t}{dt} = -\frac{2}{3}m_v\Delta\sigma H\frac{dm}{dt} = \frac{k}{\gamma_w}\frac{2m\Delta\sigma}{H} \tag{15.28}$$

$$m\frac{dm}{dt} = -\frac{3c_v}{H^2} = -\frac{1}{t}3T_v \tag{15.29}$$

Integrating Eq. (15.29) between the limits $m = 1$ and $T_v = \frac{1}{12}$ at $t = t_c$ and $m = 0$ at $t = \infty$, we have

$$m = \exp\left(\tfrac{1}{4} - 3T_v\right) \tag{15.30}$$

Equation (15.30), together with the geometry of a parabola, may be used to calculate the excess pore pressure at any depth and at any time $t > t_c$. Proceeding as before, the surface settlement and the degree of consolidation are given by

$$\Delta\rho_t = m_v H \Delta\sigma \left[1 - \tfrac{2}{3}\exp\left(\tfrac{1}{4} - 3T_v\right)\right] \tag{15.31}$$

$$U_t = 1 - \tfrac{2}{3}\exp\left(\tfrac{1}{4} - 3T_v\right) \tag{15.32}$$

The complete solution for one-dimensional consolidation with parabolic isochrones consists of Eq. (15.26) for $T_v < \frac{1}{12}$ and Eq. (15.32) for $T_v > \frac{1}{12}$, as shown in Fig. 15.7. For most practical purposes consolidation can be taken to be completed at $T_v = 1$. Excess pore pressures can be found from the geometry of the parabolic isochrones shown in Fig. 15.6 with values for n and m calculated from Eqs. (15.20) and (15.30) respectively.

Notice that in all the examples discussed so far drainage has been one-way to the upper surface and the base was impermeable, as illustrated in Fig. 15.8(a). Often in practice and in laboratory tests the drainage is two-way to drains at the top and bottom, as illustrated in Fig. 15.8(b). In this case the soil consolidates as two symmetric halves, each with one-way drainage, and the rate of consolidation is governed by H^2. We can avoid ambiguity by redefining H as the maximum drainage path; thus H in Eq. (15.25) is the longest direct path taken by a drop of water as it is squeezed from the soil.

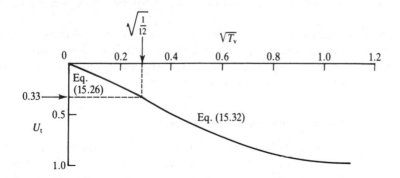

Figure 15.7 Solution for consolidation from parabolic isochrones.

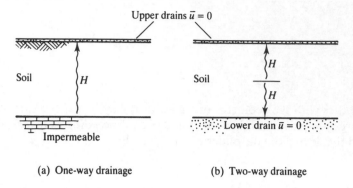

Figure 15.8 Boundary drainage conditions for one-dimensional consolidation.

15.6   Other consolidation solutions

The solutions obtained using parabolic isochrones are simple and illustrative but are restricted to the case of one-dimensional consolidation where the initial excess pore pressure $\bar{u}_i$ is the same everywhere. Other solutions are available for other cases.

The one-dimensional consolidation equation can be solved analytically and the solution is in the form of a Fourier series (Taylor, 1948). The degree of consolidation is given by

$$U_t = 1 - \sum_{m=0}^{\infty} \frac{2}{M^2} \exp(-M^2 T_v) \tag{15.33}$$

where $M = \frac{1}{2}\pi(2m+1)$. For values of U_t not greater than about 0.6, Eq. (15.33) can be approximated to

$$U_t = \frac{2}{\sqrt{\pi}} \sqrt{T_v} \tag{15.34}$$

which is close to Eq. (15.26) which is the solution using parabolic isochrones for small times.

The solutions will be slightly different if the initial excess pore pressures are not everywhere the same. The two common cases are where the initial excess pore pressures increase or decrease linearly with depth. Relationships between U_t and $\sqrt{T_v}$ for three cases of initial excess pore pressure are shown in Fig. 15.9.

15.7   Determination of c_v from oedometer tests

The results of a single stage of consolidation of a sample in an oedometer test may be used to estimate a value for the coefficient of consolidation of a soil. Since the time factor T_v is a function of c_v, we cannot immediately plot experimental results of U_t against T_v. However, if the test is continued until consolidation is complete, we may

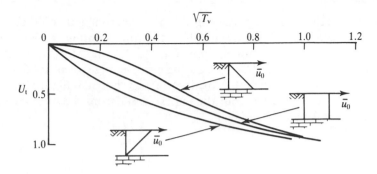

Figure 15.9 Solutions for one-dimensional consolidation.

find the final settlement ρ_∞ and, hence, the degree of consolidation at any time, and thus plot U_t against time t. If the experimental U_t against t curve can be fitted to a theoretical U_t against T_v curve, a relationship between t and T_v may be obtained and c_v found from Eq. (15.25). Two alternative curve-fitting approximations are available.

(a) A√(time) method

This method makes use of the observation that settlement against √(time) curves have an initial portion that may be approximated by a straight line, and this straight line can be fitted to Eq. (15.34). Figure 15.10(a) shows the results of a single stage of consolidation of a sample of clay in an oedometer test plotted as U_t against $\sqrt{t}$. The slope of the initially linear part of the curve is given by $\sqrt{t_1}$, as shown in Fig. 15.10(a). The experimental curve and the curve in Fig. 15.10(a) fit when $U_t = 1$ and $t = t_1$ in Eq. (15.25). Hence,

$$\sqrt{T_v} = \sqrt{\frac{C_v t_1}{H^2}} = \frac{\sqrt{3}}{2} \qquad (15.35)$$

$$c_v = \frac{3H^2}{4t_1} \qquad (15.36)$$

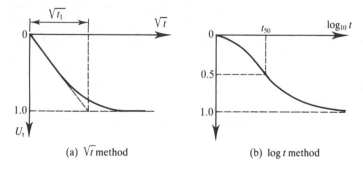

(a) √t method   (b) log t method

Figure 15.10 Determination of c_v from oedometer test results by curve fitting.

(b) A $\log_{10}$ (time) method

As an alternative, it is sometimes more convenient to fit the experimental and theoretical consolidation curves at $U_t = 0.5$, i.e. when half of the consolidation is complete. The value of T_v for $U_t = 0.5$ may be found from Eq. (15.33) and is $T_v = 0.196$. To estimate a value for t_{50}, the time for $U_t = 0.5$ during a single stage of consolidation in an oedometer test, it is convenient to plot U_t against $\log t$ as shown in Fig. 15.10(b). The value for t_{50} may be read directly from the experimental consolidation curve. Theoretical and experimental curves fit when

$$T_v = \frac{c_v t_{50}}{H^2} = 0.196 \tag{15.37}$$

$$c_v = 0.196 \left(\frac{H^2}{t_{50}} \right) \tag{15.38}$$

Note that U_t cannot be calculated until the final settlement $\Delta\rho_\infty$ has been found. Ideally, settlement–time curves would approach horizontal asymptotes as illustrated in Fig. 15.10 and it would not be difficult to estimate a value for $\Delta\rho_\infty$. For most experimental settlement–time curves, however, these horizontal asymptotes are not clearly defined and, moreover, there is often an initial settlement which is observed immediately after the loading increment has been applied. For most practical cases it is necessary to estimate a value for $\Delta\rho_\infty$ by means of special constructions. A construction for estimating $\Delta\rho_\infty$ from a plot of $\Delta\rho_t$ against $\sqrt{t}$ was proposed by Taylor and a construction for estimating $\Delta\rho_\infty$ from a plot of $\Delta\rho_t$ against $\log_{10} t$ was proposed by Casagrande; both constructions are described by Taylor (1948).

15.8 Continuous loading and consolidation

If the loading in a test which is supposed to be drained is applied too quickly excess pore pressures will occur but there will also be some drainage, so the loading is neither fully drained nor fully undrained. This is, of course, what happens in the ground, but solutions of general problems of coupled loading and drainage are very difficult. There are, however, relatively simple solutions for coupled one-dimensional loading and these form the basis of continuous loading consolidation tests (Atkinson and Davison, 1990).

Figure 15.11(a) shows a continuous loading one-dimensional compression test with a drain at the top and an impermeable boundary at the bottom. At a particular instant in the test the total stress is σ, the settlement is ρ and the pore pressures at the top and bottom of the sample are u_0 and u_b, so the excess pore pressure at the base is $\overline{u}_b = u_b - u_0$. The shaded are in Fig. 15.11(b) is $\sigma'H$, where σ' is the mean vertical effective stress and the isochrone is taken to be parabolic. Figure 15.11(c) shows the variations of total stress σ, settlement ρ and pore pressures u_0 and u_b, all of which must be measured during the test.

From Eqs. (15.7) and (15.8) the basic equation for coupled loading and consolidation is

$$c_v \frac{\partial^2 \overline{u}}{\partial z^2} = \frac{\partial \overline{u}}{\partial t} - \frac{\partial \sigma}{\partial t} = -\frac{\partial \sigma'}{\partial t} \tag{15.39}$$

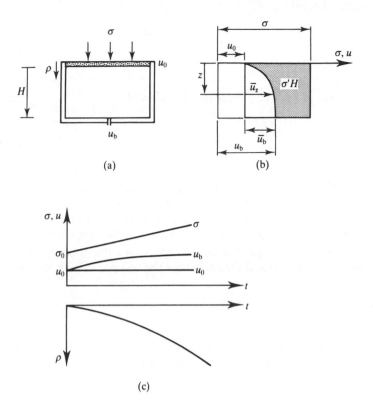

Figure 15.11 Behaviour of soil during continuous loading consolidation tests.

If the rate of loading is sufficiently slow so that $\bar{u}_b$ is relatively small compared with $\sigma - u_0$, then the mean effective stress can be approximated by $\sigma' = \sigma - u_0$. From the definition of the coefficient of compressibility m_v given by Eq. (8.11),

$$m_v = -\frac{1}{H}\frac{dH}{d\sigma'} \tag{15.40}$$

If the isochrone is a parabola then the excess pore pressure at any depth z is given by

$$\bar{u}_z = \bar{u}_b\left(\frac{2z}{H} - \frac{z^2}{H^2}\right) \tag{15.41}$$

Differentiating twice,

$$\frac{d^2\bar{u}}{dz^2} = -\frac{2\bar{u}_b}{H^2} \tag{15.42}$$

and, substituting into Eq. (15.39),

$$c_v = \frac{H^2}{2\bar{u}_b} \frac{d\sigma'}{dt} \tag{15.43}$$

Then, from Eqs. (15.40) and (15.43) together with Eq. (15.10),

$$k = \frac{\gamma_w H}{2\bar{u}_b} \frac{dH}{dt} \tag{15.44}$$

The compression, consolidation and permeability parameters, m_v, c_v and k, can be evaluated from any one-dimensional continuous loading test in terms of the current values of sample thickness H and the excess pore pressure at the undrained face $\bar{u}_b$ and the gradients $d\sigma'/dH$, $d\sigma'/dt$ and dH/dt. In a test in which the sample dimensions and pore pressures are recorded at frequent intervals, values for the gradients may be determined by a numerical procedure and the values for the soil parameters calculated at equally frequent intervals.

15.9 Summary

1. Consolidation occurs when excess pore pressures dissipate, usually at constant total stress. This results in compression or swelling as the effective stresses change.
2. The basic equation of one-dimensional consolidation is

$$c_v \frac{\partial^2 \bar{u}}{\partial z^2} = \frac{\partial \bar{u}}{\partial t} \tag{15.9}$$

where the coefficient of consolidation is $c_v = k/m_v\gamma_w$, which has the units of square metres per year. Values of c_v can be determined from results of oedometer tests.
3. Solutions to Eq. (15.9) are represented by isochrones, which show the variation of excess pore pressure with time throughout the consolidating layer. Simple solutions for one-dimensional consolidation can be found, assuming that the isochrones are parabolas.
4. Standard solutions for consolidation settlements are given in terms of the degree of consolidation and the time factor:

$$U_t = \frac{\Delta\rho_t}{\Delta\rho_\infty} \tag{15.24}$$

$$T_v = \frac{c_v t}{H^2} \tag{15.25}$$

Relationships between U_t and T_v depend on the distribution of the initial excess pore pressures and the drainage geometry.

Worked examples

Example 15.1: Interpretation of oedometer test results   The first two columns of Table 15.1 contain data from a single increment of an oedometer test in which the total vertical stress was raised from $\sigma = 90$ kPa to $\sigma = 300$ kPa. At $t = 0$ the sample was 20 mm thick and it was allowed to drain from the top and from the bottom.

For two-way drainage the drainage path is $H = 10$ mm. The degree of consolidation U_t is given by Eq. (15.24), taking the final settlement as $\Delta\rho_\infty = 1.920$ mm corresponding to $t = 24$ h.

(a) $\sqrt{t}$ method. Figure 15.12(a) shows U_t plotted against $\sqrt{t}$. Scaling from the diagram, $\sqrt{t_1} = 4.6$ and hence $t_1 = 21.2$ min. From Eq. (15.36),

$$c_v = \frac{3H^2}{4t_1} = \frac{3 \times (10 \times 10^{-3})^2}{4 \times 21.2} \times 60 \times 24 \times 365 = 1.9 \, \text{m}^2/\text{year}$$

(b)   Log t method. Figure 15.12(b) shows U_t plotted against $\log t$. From the figure, $\log t_{50} = 0.70$ and $t_{50} = 5.01$ min. From Eq. (15.38),

$$c_v = \frac{0.196H^2}{t_{50}} = \frac{0.196 \times (10 \times 10^{-3})^2}{5.01} \times 60 \times 24 \times 365 = 2.1 \, \text{m}^2/\text{year}$$

The mean value for the coefficient of consolidation from the two methods is $c_v = 2 \, \text{m}^2/\text{year}$.

During the increment the vertical effective stress changes from $\sigma' = 90$ kPa at the start to $\sigma' = 300$ kPa at the end. The vertical strain is $\Delta\varepsilon_z = 1.920/20 = 0.096$ and from Eq. (8.9) the coefficient of compressibility is

$$m_v = \frac{\Delta\varepsilon_z}{\Delta\sigma'_z} = \frac{0.096}{300 - 90} = 4.6 \times 10^{-4} \, \text{m}^2/\text{kN}$$

Table 15.1 Results of an oedometer test – Example 15.1

Time (min)	Settlement $\Delta\rho_t$ (mm)	U_t	$\sqrt{t}$ (min$^{1/2}$)	$\log t$
0	0	0	0	–
0.25	0.206	0.107	0.5	−0.602
1	0.414	0.216	1	0
2.25	0.624	0.325	1.5	0.352
4	0.829	0.432	2	0.602
9	1.233	0.642	3	0.954
16	1.497	0.780	4	1.204
25	1.685	0.878	5	1.398
36	1.807	0.941	6	1.556
49	1.872	0.975	7	1.690
24 h	1.920	1.000	–	–

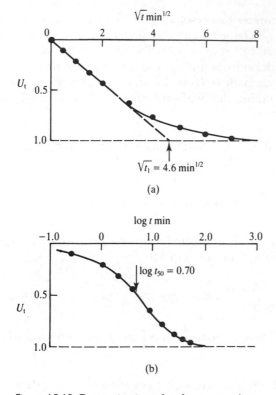

Figure 15.12 Determination of c_v from an oedometer test – Example 15.1.

From Eq. (15.10) the coefficient of permeability is given by

$$k = c_v m_v \gamma_w = \frac{2.0 \times 4.6 \times 10^{-4} \times 9.81}{60^2 \times 24 \times 365} = 2.9 \times 10^{-10}\,\text{m/s}$$

Example 15.2: Settlement of an oedometer sample   In a stage of an oedometer test the total stress was raised by 100 kPa. The sample was initially 20 mm thick and it was drained from both ends. The properties of the soil were $c_v = 2\,\text{m}^2/\text{year}$ and $m_v = 5 \times 10^{-4}\,\text{m}^2/\text{kN}$.

From Eq. (15.1) the final settlement, after consolidation is complete, was

$$\rho = m_v z \Delta \sigma_z' = 5 \times 10^{-4} \times 20 \times 100 = 1.0\,\text{mm}$$

(a)   The time factor at which the settlement will be 0.25 mm (i.e. when $U_t = 0.25$) is given by Eq. (15.26):

$$T_v = \frac{3U_t^2}{4} = \frac{3 \times 0.25^2}{4} = 0.05$$

From Eq. (15.25), taking $H = 10\,\mathrm{mm}$ for two-way drainage, the time when the settlement is 0.25 mm is

$$t = \frac{T_v H^2}{c_v} = \frac{0.05 \times (10 \times 10^{-3})^2}{2} \times 60 \times 24 \times 365 = 1.3\,\mathrm{min}$$

(b)   After 3 min the time factor and degree of consolidation are

$$T_v = \frac{c_v t}{H^2} = \frac{2 \times 3}{(10 \times 10^{-3})^2 \times 60 \times 24 \times 365} = 0.11$$

$$U_t = \frac{2}{\sqrt{3}}\sqrt{T_v} = \frac{2 \times \sqrt{0.11}}{\sqrt{3}} = 0.39$$

and the settlement is

$$\rho_t = \rho_\infty U_t = 1.0 \times 0.39 = 0.39\,\mathrm{mm}$$

References

Atkinson, J. H. and L. R. Davison (1990) 'Continuous loading oedometer tests', *Q. J. Engng Geol.*, 23, 347–355.

Taylor, D. W. (1948) *Fundamentals of Soil Mechanics*, Wiley, New York.

Chapter 16

Natural soils

16.1 Characteristics of natural soils

In previous chapters I described the basic mechanics of soils and in later chapters these simple theories for soil behaviour will be used to investigate the performance of soil structures such as slopes, retaining walls and foundations. The behaviour described and the theories developed are largely idealizations for the behaviour of reconstituted soils, but natural soils differ from reconstituted soils in a number of important aspects.

A reconstituted soil is manufactured in the laboratory by mixing it with water to form a slurry at a very high water content, pouring it into a mould and loading and unloading it by one-dimensional compression or consolidation in a test apparatus to the required initial state. Coarse grains are not bonded to each other; fine grains, especially clays, may be very slightly bonded due to small surface forces. If the soil is well graded the grains of different sizes are distributed randomly. The sample is tested soon after it has been manufactured. Soil properties measured in tests on reconstituted samples depend only on the nature of the grains and they are material parameters. Natural soils have features which are known as *structure*, which is a combination of *fabric* and *bonding* and these features arise from their formation and age.

Natural soils are either sedimented, usually through water, or they are residual soils which are the end products of weathering in situ. In reconstituted soil the grains are distributed randomly: in structured soil, fabric is the way in which grains are arranged in a non-random way. It includes layers or lenses of poorly graded soil within a body of soil with different grading and collections or flocs of fine grained soil which appear as larger grains. (Go and look carefully at freshly excavated slopes in soils and you will almost always be able to see layering; occasionally you can find thick beds of nearly uniform clay deposited in an unchanging environment but these are rare.) Soil fabric is the non-random arrangement of soil grains and it is created largely during deposition.

Natural soils are compressed and swelled by further deposition and erosion of over-lying sediment, by loading from glaciers and ice sheets and by changes in groundwater. With geological time all soils age and their properties change. Most natural soils are very old. London Clay is about 60 million years old and even recent glacial soils are over 10,000 years old. Occasionally you may come across natural soils like Mississippi delta muds or the soils of the Fens of East Anglia, which are only decades or centuries old, but these are the exception.

Ageing gives rise to bonding and to other physical and chemical changes. Bonding is the way in which soil grains are attached to one another. It can arise from natural attractions of very fine grains or from deposition of salts from solution in the ground-water. Weathering weakens bonding and may change the chemical composition of the grains. There may be continuing deformation due to creep at constant effective stress and there may be changes in the chemistry of the pore water. All these, and other, events contribute to features in natural soils which are not all present in reconstituted samples.

It is very difficult to discover the true behaviour of natural soils. The obvious way is to recover undisturbed samples from the ground and test them in the laboratory but, unfortunately, the process of recovering the sample from the ground and installing it in the test apparatus will probably alter its behaviour. There is no possibility of recovering and testing a truly undisturbed sample; the best we can do is to take and test an intact sample with the very minimum of disturbance. If the correct procedures for sampling and testing are followed the behaviour of an intact sample will be very close to the behaviour of the soil in the ground, but it is essential to follow the correct procedures.

This book deals with the basic, simple theories of soil mechanics relevant to recon-stituted soils and a detailed discussion of all the effects and consequences of structure in natural soils is beyond its scope. It is, however, important to note these effects, which is the purpose of this chapter. The important thing is to consider the behaviour of your intact samples of natural soils within the basic simple framework developed for reconstituted soils.

16.2 One-dimensional compression and swelling of soils in the ground

The behaviour of soils during one-dimensional compression and swelling in laboratory tests was discussed in Sec. 8.5 and similar behaviour will occur during deposition and erosion of soil in the ground. Figure 16.1(a) illustrates a soil element below a ground level which rises and falls due to deposition and erosion and Fig. 16.1(b) shows the resulting changes of effective stress and water content. So far I have considered volume and volume changes in terms of the specific volume v or the voids ratio e, but in this chapter I shall consider water content w, as this is a commonly measured and often quoted parameter. Water content, specific volume and voids ratio are simply related (see Sec. 5.5 and, for saturated soil, $e = wG_s$. At points A and B the soil is normally consolidated and at C it is overconsolidated. Notice that although the vertical stresses at A and C are similar the water contents are very different. Figure 16.1(c) illustrates the changes of vertical and horizontal effective stresses during deposition and erosion. These can be related by a coefficient of earth pressure at rest, K_0, given as

$$K_0 = \frac{\sigma_h'}{\sigma_z'} \tag{16.1}$$

For normally consolidated and lightly overconsolidated soils $\sigma_h' < \sigma_z'$ and $K_0 < 1$, while for heavily overconsolidated soils $\sigma_h' > \sigma_z'$ and $K_0 > 1$. An approximation often

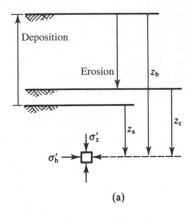

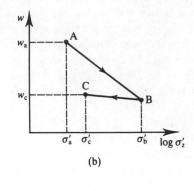

(b)

(a)

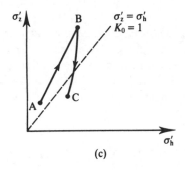

(c)

Figure 16.1 One-dimensional consolidation and swelling of soil in the ground due to deposition and erosion.

used to estimate K_0 is

$$K_0 = K_{0nc}\sqrt{Y_0} \qquad (16.2)$$

where Y_0 is the yield stress ratio defined in Sec. 8.3 and $K_{0nc} = 1 - \sin\phi'_c$ is the value of K_0 for normally consolidated soil.

In previous chapters I showed that many aspects of soil behaviour (but not the critical states) depend on the state which arises as a consequence of the history of loading and unloading. This means that reconstituted samples should be compressed and swelled one-dimensionally in the apparatus before shearing and intact samples of natural soil should be recompressed to the estimated state in the ground.

The state of an element of soil in the ground depends on the current stresses (i.e. on the depth) and on the overconsolidation (i.e. on the current depth and on the depth of erosion). Figure 16.2 illustrates the variations of water content with depth for a deposit which is lightly eroded (i.e. the depth of erosion z_e is small) or heavily eroded (i.e. the depth of erosion is large). For the lightly eroded soil the difference between the water contents at A and B is relatively large, while for the heavily eroded soil

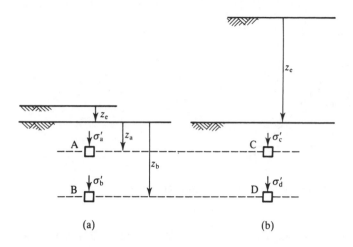

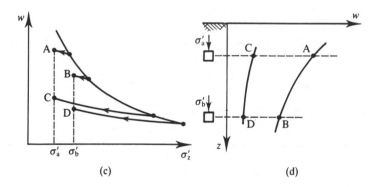

Figure 16.2 Variations of water content in the ground in normally consolidated and overconsolidated soils.

the difference between the water contents at C and D is much smaller and the water contents themselves are smaller. For the heavily eroded soil the smaller variation of water content with depth is a result of the very large maximum past stress.

16.3   Changes of state and yield stress ratio in natural soils

The effects of structure can be regarded as changes of state and changes of the state boundary surface and both give rise to changes in the yield stress ratio Y_0. Yield stress ratio is essentially the distance of the current state from the state boundary surface and it was defined in Sec. 8.3. In the simple theories of soil mechanics the state can only change by loading on the state boundary surface and unloading to a new state inside the surface. Figure 16.3 shows the state of a soil at A inside the state boundary surface

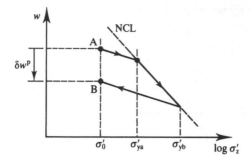

Figure 16.3 Change of yield stress ratio due to deposition and erosion.

and it is similar to Fig. 16.1 (b). The yield stress ratio Y_0 is defined in Eq. 8.12 as

$$Y_0 = \frac{\sigma'_y}{\sigma'} \tag{16.3}$$

The state at A can move to B only by loading along the normal compression line which is part of the state boundary surface. The recoverable plastic water content change δw_p is associated with a change of the yield stress from σ'_{ya} to σ'_{yb} and a change in yield stress ratio.

There are other ways in which the state and yield stress ratio of a natural soil can change and I will discuss some of these in the following sections. Some involve changes of the water content and others involve changes of the state boundary surface.

Notice that the overconsolidation ratio R_0 of the soil at A is smaller than that at B because the maximum past stress σ'_{ya} is smaller than the maximum past stress σ'_{yb} and the current stresses are the same. In this case the overconsolidation ratio R_0 is the same as the yield stress ratio Y_0.

16.4   Effects of volume changes due to vibration or creep

If coarse grained soils are vibrated at constant effective stress they will compress and there will be irrecoverable plastic volume changes and changes in the yield stress ratio. Figure 16.4(a) shows the state path A → B corresponding to compression by vibration at constant effective stress; the yield stress has increased from σ'_{ya} to σ'_{yb} with a consequent increase in the yield stress ratio. Figure 16.4(b) is the corresponding state path normalised with respect to the critical stress σ'_c; this assumes that both σ'_z and σ'_h remain constant so the change of the state is due to the increase of σ'_c as the volume decreases.

If soil, or any other material, is left under a constant effective stress there will be continued deformations which are due to creep. In soils, creep is greatest in fine grained soils. The effects of creep in fine grained soil illustrated in Fig. 16.5(a) are similar to those due to compression by vibration of coarse grained soil except vibration

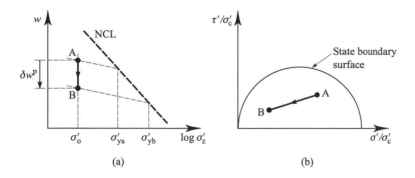

Figure 16.4 Change of yield stress ratio due to vibration of coarse grained soil.

compression occurs more or less instantaneously while creep occurs slowly and at a rate which diminishes with time. The basic constitutive equation for creep given in Sec. 3.11 is of the form

$$\delta w = C_\alpha \ln \left(\frac{t}{t_0} \right) \tag{16.4}$$

And so the water content decreases with the logarithm of time, as illustrated in Fig. 16.5(b).

During compression at constant effective stress by vibration or creep shown in Figs. 16.4 and 16.5 the yield stress ratio increases because the yield stress has increased from σ'_{ya} to σ'_{yb}. Notice that the overconsolidation ratio of the soil at A is the same as that at B because neither the current stress nor the maximum past stress have changed. This demonstrates that yield stress ratio is a more fundamental measure of soil state than overconsolidation ratio.

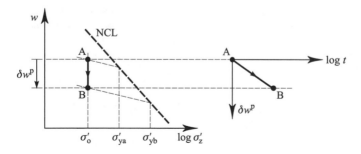

Figure 16.5 Change of yield stress ratio due to creep of fine grained soil.

16.5   Influence of layering in sedimented soils

Natural sedimentation of soil through water or wind is often episodic and in a natural soil there will be layers each representing different episodes of deposition. Each layer has coarser grains at the bottom, deposited first, and finer grains at the top, deposited last. (This feature is common in sedimented soils and is called graded bedding.)

Overall the soil is well graded and at a particular effective stress a normally consolidated reconstituted sample will have a relatively low water content w_r in Fig. 16.6(a). In the natural sedimented soil each part of each layer is poorly graded and at the same effective stress a normally consolidated natural sample which contains several layers will have a relatively high water content w_s at S in Fig. 16.6(b). The slopes of both normal compression lines are assumed to be approximately the same.

For a state at A in Fig. 16.6(a) there are two yield stresses, one σ'_{ys} corresponding to sedimented bedded soil and the other σ'_{yr} corresponding to reconstituted soil and there are two possible yield stress ratios. Figure 16.6(a) illustrates compression of sedimented and reconstituted samples of the same soil starting from the same state at A. They have different yield stresses and they travel down different normal compression lines.

In theory the critical state lines for both sedimented and reconstituted samples are the same because, by definition, it is the ultimate state reached after very large deformations when the sedimented sample has become completely reconstituted by shear straining. (In practice it is difficult to achieve these critical states in laboratory tests but they may be reached in natural landslides.) The critical stress σ'_c and the state parameters S_v and S_σ are the same for both sedimented and reconstituted samples of the same soil.

16.6   Influence of bonding and weathering

During vibration or creep compression only the water content of soil changes but during bonding and weathering both the water content and the state boundary surface may change. A detailed discussion of the effects of bonding and weathering is beyond the scope of this book and all I can do here is outline the basic features: for more

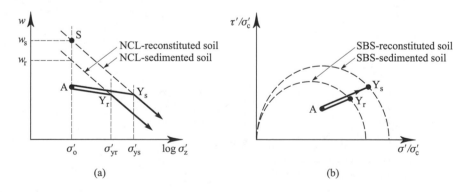

Figure 16.6 Compression of sedimented and reconstituted soil.

detailed discussion there are papers by Lerouil and Vaughan (1990) and Coop and Atkinson (1993).

The principal mechanism of bonding is by deposition of additional material, often calcium carbonate, from the groundwater. This has the dual effect of reducing the water content and shifting the state boundary surface. However, the critical state corresponds to relatively large straining when the soil is essentially reconstituted and to reach these states bonding must fracture. This means that the critical states of bonded and unbonded soil will be about the same and the main influence of bonding will be on the position of the state boundary surface and on yield and peak strength.

Figure 16.7(a) shows normal compression lines for bonded soil and for the same soil after it has been reconstituted. Soil is sedimented at a state near A. The path A $\rightarrow$ B represents a reduction in water content due to deposition of bonding material at constant stress. At the same time the normal compression line moves to the right as the yield stress increases due to bonding. The path B $\rightarrow$ Y_b $\rightarrow$ C represents compression of initially bonded soil and part of this is outside the normal compression line for reconstituted soil. The yield point Y_b lies on the state boundary surface for bonded soil but. after yield, the state moves towards the line for reconstituted soil as the cementing breaks with continuing strain. Notice the relatively large compression from Y_b to C as the cementing fractures.

There is, however, only one critical state line so values of the normalising parameter σ'_c can be obtained unambiguously. (This is the principal reason for selecting σ'_c as the normalizing parameter rather than the equivalent stress on the normal compression line σ'_e; see Sec. 9.6.)

Figure 16.7(b) illustrates the state path A $\rightarrow$ B $\rightarrow$ Y_b $\rightarrow$ C corresponding to the loading path in Fig. 16.7(a) with stresses normalized with respect to σ'_c. (The path is for loading with constant stress ratio.) Part of the path lies outside the state boundary surface for reconstituted soil and the yield point Y_b lies on the state boundary surface for bonded soil. The states at A and C lie at the same point on the state boundary surface for reconstituted soil and the state B is overconsolidated. The distance between the state boundary surfaces for bonded and reconstituted soil depends principally on the strength and amount of the cementing.

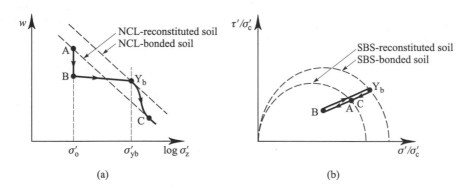

(a)   (b)

Figure 16.7 Behaviour of bonded soil.

Weathering is essentially the opposite of development of bonding. It involves physical and chemical changes at approximately constant effective stress with increase or decrease of water content. The effect of weathering is normally to decrease the strength of bonding and to reduce the yield stress ratio as the state boundary surface collapses towards the state boundary surface for reconstituted soil.

16.7   Changes in pore water salinity

The fabric of a fine grained soil is influenced by the salinity of the water through which it was deposited. Clays deposited through sea water are often flocculated and have higher water contents than the same clay deposited through fresh water. If the original saline pore water subsequently loses salinity the state boundary surface becomes smaller.

Figure 16.8(a) shows normal compression lines for the same soil sedimented in saline water and in fresh water. Soil deposited in saline water is normally consolidated at A and, with time, becomes lightly overconsolidated at B due to creep. If the soil is loaded it will yield at Y_s for soil in saline water where the yield stress is σ'_{ys} and the state will move down the normal compression line for saline water. Notice that $\sigma'_{ys} > \sigma'_b$ and so the yield stress ratio at B is greater than 1.

If the pore water becomes fresh, while the state is at B the normal compression line moves to the left. However, the original fabric, created during deposition in saline water is preserved. If the soil is loaded it will yield at Y_s for soil in saline water where the yield stress is σ'_{ys} but it will then move towards C on the normal compression line for soil in fresh water. The relatively large compression from Y_s to C is similar to that for bonded soil shown in Fig. 16.7. Notice that $\sigma'_{yf} < \sigma'_b$ and so the yield stress ratio at B is less than 1 and B is outside the current normal compression line.

Figure 16.8(b) illustrates the state path A $\rightarrow$ B $\rightarrow$ Y_s $\rightarrow$ C corresponding to the loading path in Fig. 16.8(a) with stresses normalized with respect to the critical stress σ'_c for soil in fresh water. The soil with fresh pore water but with fabric created in saline water yields at Y_s and ends at C on the state boundary surface for the soil with fresh water. The state at B is outside the state boundary surface for soil with fresh pore water.

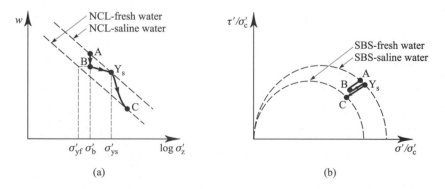

Figure 16.8 Influence of changes of pore water salinity.

16.8   Summary

1.   The state of a soil in the ground is determined primarily by the history of deposition and erosion, but it may be altered subsequently by the various processes of ageing.
2.   The principal processes of ageing are creep, cementing, weathering and changes in the salinity of the pore water.
3.   Ageing may change either the current state or the position of the state boundary surface.

References

Coop, M. A. and J. H. Atkinson (1993) 'Mechanics of cemented carbonate sands', *Geotechnique*, 43, 2, 53–67.

Leroueil, S. and P. R. Vaughan, (1990) 'The general and congruent effects of structure in natural soils and weak rocks', *Geotechnique*, 50, 3, 467–488.

Sides, G. and L. Barden (1970) The microstructure of dispersed and flocculated samples of kaolinite, illite and montmorillonite, *Can. Geot. Jnl.*, 8, 391–400.

Further reading

Mitchell, J. K. and K. Sogo (2005) *Fundamentals of soil behaviour*, 3rd Edition, Wiley, New York.

Chapter 17

Ground investigations

17.1 Introduction

Engineers designing structures and machines normally choose materials and specify their strength and stiffness and they often combine materials to make composites (e.g. steel and concrete in reinforced concrete). Similarly, highway engineers can specify the soils and rocks to be used in the construction of roads. Geotechnical engineers, on the other hand, cannot choose and must work with the materials in the ground. They must therefore determine what there is in the ground and the engineering properties of the ground, and this is the purpose of ground investigations.

The basic techniques of ground investigation are drilling, sampling and testing, *in situ* and in the laboratory, but these must be complemented by geological information and a sound appreciation of the relevant soil mechanics principles. Consequently, it is in the area of ground investigation that geology and engineering combine and where engineering geologists and geotechnical engineers cooperate.

Ground investigation is, of course, far too big a topic to be covered in one short chapter and all I will do here is outline the basic issues as a starting point for further study. The detailed techniques vary from country to country, and from region to region, and depend both on the local ground conditions, on historical precedents, on contractural procedures and on the available equipment and expertise. As with laboratory testing, procedures for ground investigations are covered by national standards and codes of practice; in the United Kingdom this is BS 5930:1999. You should look up the standards covering the region where you work in to see what they contain. Detailed descriptions of the current practices in the United Kingdom are given by Clayton, Simons and Matthews (1995).

17.2 Objectives of ground investigations

When you look at the face of a cliff or an excavation you see a section of the ground and when you look at a site you have to imagine what an excavation would reveal. A major part of a ground investigation is to construct a three-dimensional picture of the positions of all the important soil and rock layers within the site that may be influenced by, or may influence, the proposed construction. Of equal importance is the necessity to sort out and identify the groundwater conditions. In distinguishing the important soil and rock layers engineering classifications based on the nature and state of the soils (see Chapter 5) should be used rather than the geological classifications, which are based on age (see Chapter 4).

There is no simple answer to the problem of how many holes should be drilled and to what depths and how many tests should be carried out. Most of the standards and codes of practice make various recommendations, but really you should do enough investigation to satisfy everybody that safe and economical works can be designed and constructed. There will inevitably be uncertainties and these will require conservatism in design which will lead to additional costs of construction. There is a balance to be struck between costs of more investigations and savings in construction: it is a matter of apportioning risk.

Figure 17.1 illustrates a very simplified section along the centre-line of a road. (Notice that the horizontal and vertical scales are not the same.) The ground conditions revealed by drilling and other methods have been greatly idealized so that a number of characteristic layers have been identified and the boundaries between them drawn as smooth lines. The actual soils in the ground within any one layer are likely to be variable, horizontally and vertically, and their boundaries irregular. Something like Fig. 17.1 is about the best you can do with a reasonable investigation. Notice that Fig. 17.1 is a section along the centre-line of the road and to complete the investigation you should be able to draw cross-sections and sections on either side of the road.

The section shown in Fig. 17.1 is similar to that shown in Fig. 4.4(c) and I have already discussed the sequence of geological events and processes that formed this sequence of deposits. Certain features of the nature and state of the various layers can be estimated from consideration of their depositional environment and subsequent geological history. The grading and mineralogy of the soft clay and the stiff clay are the same (so they have the same nature), but their water contents are different (so they have different states); the soft clay is normally consolidated or lightly overconsolidated while the stiff clay is heavily overconsolidated.

For each of the principal strata in Fig. 17.1 you will need to determine representative parameters for strength, stiffness and water seepage flow (i.e. permeability). These will

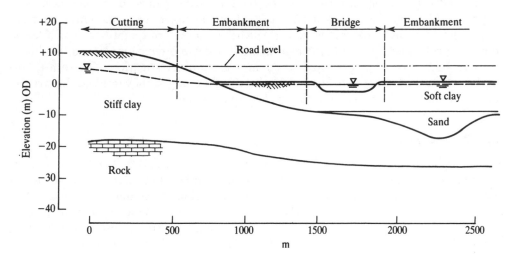

Figure 17.1 A Simple geotechnical cross-section.

be selected from the results of laboratory and *in situ* tests. These parameters may be constant for a particular layer or they may vary with depth; generally we expect strength and stiffness to increase with depth. The question of which parameters to determine depends on the ground and on the structure to be designed and constructed. There are relationships between the engineering properties of soils, such as strength and stiffness, and their nature and state. These issues are discussed in Chapter 18.

After any ground investigation you should know the following for each of the principal strata:

1. Its engineering description and classification in terms of the nature (grading and plasticity) and state (stress and specific volume or overconsolidation).
2. The positions of the boundaries between the different strata (i.e. you should be able to draw sections like that in Fig. 17.1 in any direction).
3. The geological environment when the soil was deposited and the history of subsequent deposition, erosion, weathering and ageing.
4. Descriptions of visible features of structure and fabric (e.g. layering, fissuring and jointing).
5. Representative values for the parameters for strength, stiffness and permeability relevant to the design and construction of the works.

You should also be sure that you know all about the groundwater. A very experienced ground engineer once said to me that he would not start an excavation until he knew exactly what he was digging into and what the groundwater conditions were; this is very good advice.

17.3   Planning and doing investigations

You cannot really plan an entire ground investigation because you do not know what is there before you start and so you cannot select the best methods or decide how much to do. A ground investigation must, therefore, be carried out in stages: each stage can be planned with existing information and the knowledge gained from one stage will assist with planning the next. Currently in the United Kingdom a ground investigation is often let as a single contract with a specification and bill of quantities, which leads to major problems in planning the investigation and can often cause later difficulties.

There should be three principal stages in a ground investigation. (These are not rigid demarcations. There is often overlap between the stages; they need not be strictly sequential and one or other may have to be expanded later.)

(a)   Desk studies

This consists of study of all the information that you can find existing on paper or electronically. The major sources are topographical and geological maps and sections, geological reports and local authority records. Other sources include air photographs, historical archives and reports on earlier site investigations at the site or at nearby sites. Experienced geotechnical engineers and engineering geologists can often decipher the principal ground conditions from the desk study, so leading to well-planned later stages.

(b)   Preliminary investigations

Preliminary investigations are carried out at the site, rather than in the office, but they do not yet involve major expenditure on drilling, sampling and testing. The purposes are, firstly, to confirm or revise the findings of the desk study and, secondly, to add further information. This additional information will come from detailed engineering geological mapping, and this is best done by engineers and geologists working together or by experienced engineering geologists. Preliminary investigations may also involve some limited sub-surface exploration by trial pits, probing or exploratory drilling and geophysical sensing using seismic, electrical resistivity and other methods.

(c)   Detailed investigations

Detailed investigations consist of drilling, sampling and laboratory and *in situ* testing. They may also involve more detailed geological mapping, groundwater and chemical studies and other appropriate investigations necessary for the works. This is where the bulk of the expenditure is incurred and planning of the detailed investigations should set out to discover the required facts in the most efficient way. This will require some foreknowledge which can be gained from the desk study and preliminary investigations.

17.4   Test pitting, drilling and sampling

The standard method of ground investigation is excavation and sampling supplemented by *in situ* and laboratory testing. The excavations are usually done by drilling but also by opening test pits.

(a)   Test pitting

A test pit is an excavation that a geotechnical engineer or engineering geologist can enter to examine the soil profile *in situ*. Pits can be excavated by large drilling machines of the kind used for boring piles, by an excavator or by hand digging. Remember that any excavation in soil with vertical or steep sides is basically unstable and must be supported before anyone enters it.

(b)   Drilling

Drill holes can be advanced into the ground using a number of different techniques; the principal kinds are illustrated in Fig. 17.2. Augers may be drilled to shallow depths by hand and large diameter augers can be drilled by machines used also for installation of bored piles (see Chapter 23). Wash boring is used in sands and gravels and rotary drilling is used mainly in rocks. Light percussion drilling is widely used in the United Kingdom and you can very often see the typical tripod rigs at work.

In some soils, particularly stiff clays and in rocks, boreholes will remain open unsupported, but in soft clays and particularly in coarse-grained soils the hole will need to be cased to maintain stability. Boreholes should normally be kept full of water, or bentonite mud, to prevent disturbance below the bottom of the hole.

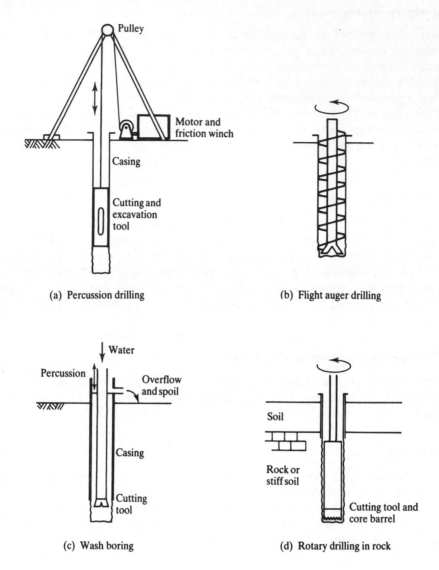

Figure 17.2 Methods for ground investigation drilling (schematic).

(c) Sampling

Samples obtained from test pits or boreholes may be disturbed or intact. (Samples are often called disturbed or undisturbed but, as no soil sample is ever truly undisturbed, the word intact can be used for samples taken with minimum disturbance.) Disturbed samples are used principally for description and classification. They may be reconstituted to determine the properties of the soil as described in Chapter 16. Intact samples may be cut from the base or sides of test pits using saws or knives or taken in tubes

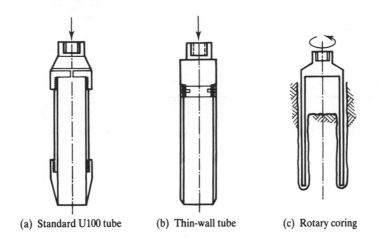

(a) Standard U100 tube   (b) Thin-wall tube   (c) Rotary coring

Figure 17.3 Methods of sampling in boreholes (schematic).

pushed into the bottom of a borehole. There are many different tube samples; two designs used in the United Kingdom are shown in Fig. 17.3(a) and (b).

The tube sampler most often used in practice in the United Kingdom is the U100 illustrated in Fig. 17.3(a). The tube, nominally 100 mm in diameter, is screwed to a cutting shoe and a sampler head. The thickness of the wall of the cutting head is 6 to 7 mm, which is relatively large. A thin wall sample tube like that illustrated in Fig. 17.3(b) has a wall thickness of 1 to 2 mm and a cutting edge formed by machining. Both samplers are capable of taking samples in many soft and stiff clays. Intact samples may be recovered by coring (see Fig. 17.3c), where a rotary drill cuts an annulus around the core sample. In the past this method was used exclusively for rocks but is now also used in stiff clays.

17.5   *In situ* testing

Laboratory tests to determine soil strength, stiffness and permeability are described in Chapter 7, but there are also a number of *in situ* tests. These can be grouped into probing tests, loading tests and permeability tests.

(a)   *Probing tests*

In these tests a tool, usually cone-shaped, is hammered or pushed into the ground and the resistance to penetration recorded. This gives some measure of the strength and stiffness of the ground. In the standard penetration test (SPT) shown in Fig. 17.4(a) a solid cone or thick-wall tube is hammered, with a standardized blow, into the bottom of a borehole. The result is given as N, the number of blows to achieve a standard penetration; values increase from about 1 to more than 50 with increasing relative density or overconsolidation ratio.

In the static cone, or Dutch cone, penetration test shown in Fig. 17.4(b) the instrument is steadily pushed into the ground from the surface and the resistance recorded

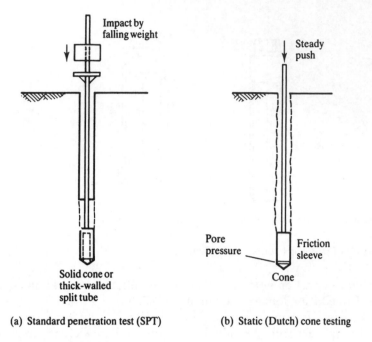

Figure 17.4 Probing tests (schematic).

continuously. Most static cone penetrometers have a sleeve behind the cone which measures a frictional or shearing resistance. Some modern cones, known as piezocones, also measure pore pressures generated at the tip or shoulder of the cone. Methods for interpretation of static cone tests were given by Meigh (1987). Many of these depend on empirical correlations between test observations and soil characteristics.

(b) Loading tests

In these tests an instrument loads the soil in a controlled manner and stresses and deformations are observed. The ultimate load, when the deformations are large, is related to the strength of the soil and the load–deformation behaviour is related to soil stiffness. Plate loading tests illustrated in Fig. 17.5(a) may be carried out near the ground surface or at the bottom of a borehole and measurements are made of the load on the plate *F* and its settlement ρ. Simple analysis of plate tests are rather like the methods used for design of foundations, discussed in Chapter 22.

The shear vane test, illustrated in Fig. 17.5(b), is used to measure the undrained strength s_u. A vane with four blades is pushed into the ground from the surface or from the bottom of a borehole. The vane is rotated and the torque *T* measured. At the ultimate state the shear stress on the cylinder of soil containing the vane is given by

$$T = \frac{1}{2}\pi D^2 H \left(1 + \frac{1}{3}\frac{D}{H}\right) s_u \tag{17.1}$$

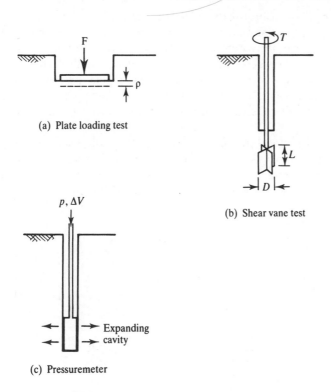

(a) Plate loading test

(b) Shear vane test

(c) Pressuremeter

Figure 17.5 *In situ* loading tests.

and a value for s_u can be calculated from the measured torque. If the rotation is continued for several revolutions the strength will drop to the residual (see Sec. 9.2).

Pressuremeter tests are illustrated in Fig. 17.5(c). A flexible cylinder is inflated and the pressures and volume changes measured. The best pressuremeters measure radial displacements directly (instead of volume changes) and some measure pore pressures as well. Pressuremeters may be installed in pre-drilled boreholes or self-boring devices drill themselves into the ground with less disturbance. Results of pressuremeter tests are used to calculate both soil strength, stiffness and the *in situ* horizontal stress σ_h. Methods for analysis of pressuremeter tests are described by Mair and Wood (1987).

17.6   Investigating groundwater and permeability

Whatever else you do in a ground investigation you must be sure to define the groundwater conditions. This will include determining the current steady state pore pressures and the final steady state pore pressures after construction. If the works involve a seepage flow of water, either steady state or during consolidation, you will need values of the coefficient of permeability.

Pore pressures can be measured by observing the level of water in a standpipe (see Sec. 14.1) in a borehole. Notice that if you drill a borehole into saturated clay with a groundwater table, or phreatic surface, near the ground surface the hole will remain dry for a considerable time. The reason for this is that if the clay has low permeability it will take a very long time for sufficient water to flow from the ground to fill the borehole. This means you can only determine pore pressures, and groundwater conditions, from observation in boreholes in soils with relatively high permeability. For clays and soils with low permeability you will need to use special piezometers (i.e. instruments to measure pore pressures). In the final analysis the groundwater conditions must be reasonable and self-consistent and compatible with the soils and the regional hydrogeology.

Values for the coefficient of permeability k can be found from the results of *in situ* pumping tests. For coarse-grained soils steady state conditions will be reached quickly. Figure 17.6(a) illustrates steady state flow towards a pumped well. The potential at a radius r is P and, from Darcy's law (see Chapter 14), the rate of flow q is

$$q = Aki = 2\pi r P k \frac{dP}{dr} \tag{17.2}$$

or

$$\frac{dr}{r} = \frac{2\pi k}{q} P \, dP \tag{17.3}$$

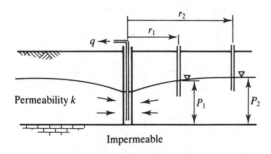

(a) Well pumping tests

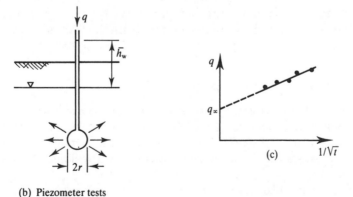

(b) Piezometer tests

Figure 17.6 In situ permeability tests.

(Notice that the hydraulic gradient at the phreatic surface is strictly dP/ds, but dP/dr is a sufficiently good approximation.) Integrating Eq. (17.3) between P_1 at r_1 and P_2 at r_2 we have

$$\ln\left(\frac{r_2}{r_1}\right) = \frac{\pi k}{q}\left(P_2^2 - P_1^2\right) \tag{17.4}$$

Hence k can be obtained from observations of the pumping rate q and water levels in standpipes at a number of different radii.

For fine-grained soils steady state seepage will not be reached quickly and during a reasonable test period there will be simultaneous steady state flow and consolidation or swelling. Figure 17.6(b) illustrates a flow from a spherical cavity radius r with a constant excess pore pressure $\bar{u} = \gamma_w \bar{h}_w$. The rate of flow at any time t is given by

$$q = 4\pi r k \bar{h}_w \left(1 + \frac{r}{\sqrt{\pi c_s t}}\right) \tag{17.5}$$

where c_s is the coefficient of consolidation for spherical consolidation. (This is similar to c_v for one-dimensional flow, discussed in Chapter 15) A condition of steady state flow would be reached after infinite time and, with $t = \infty$ in Eq. (17.5),

$$q_\infty = 4\pi r k \bar{h}_w \tag{17.6}$$

A value of q_∞ can be found by plotting q against $1/\sqrt{t}$, as shown in Fig. 17.5(c), and extrapolating. Hence a value for k can be obtained from Eq. (17.6). If the cavity is not spherical the term $4\pi r$ must be replaced by an intake factor F which depends on the geometry.

17.7   Ground investigation reports

The findings of ground investigations are recorded in two different kinds of reports.

(a)   Factual reports

These simply describe the procedures and findings without comment or interpretation. The report will contain text describing what was done, how, where and by whom. It will summarize the factual findings of the desk study, the field investigations and the *in situ* and laboratory tests.

The basic information from the drilling and sampling operations is contained in borehole logs. (Similar logs contain information from test pits.) A typical borehole log is shown in Fig. 17.7, this is idealized and simplified to illustrate the principle features which should be recorded. The top panel gives the date, time, place, method of drilling and other basic information. The legend is a pictorial representation of the principal strata with a word description alongside. To the left are depths and levels. To the right are columns for sample recovery, groundwater observations and *in situ* tests. Borehole logs prepared by different ground investigation companies differ in detail but should contain at least this basic information. The borehole log in Fig. 17.7 is for a borehole

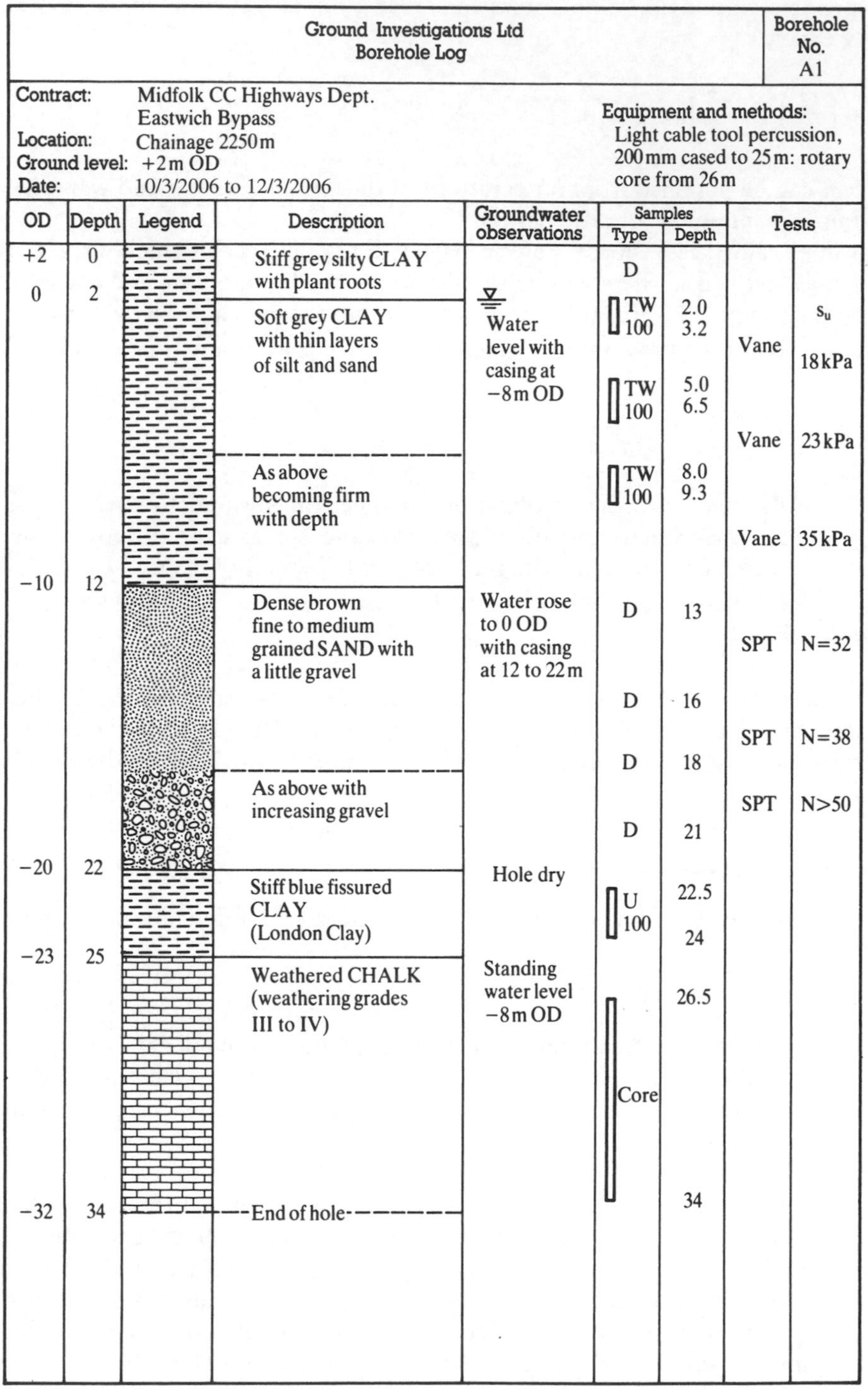

Ground Investigations Ltd Borehole Log	Borehole No. A1

Contract:	Midfolk CC Highways Dept. Eastwich Bypass	Equipment and methods: Light cable tool percussion, 200 mm cased to 25 m: rotary core from 26 m
Location:	Chainage 2250 m	
Ground level:	+2 m OD	
Date:	10/3/2006 to 12/3/2006	

OD	Depth	Legend	Description	Groundwater observations	Samples Type	Samples Depth	Tests	Tests
+2	0		Stiff grey silty CLAY with plant roots		D			
0	2		Soft grey CLAY with thin layers of silt and sand	Water level with casing at −8 m OD	TW 100	2.0 3.2	Vane	s_u 18 kPa
					TW 100	5.0 6.5	Vane	23 kPa
			As above becoming firm with depth		TW 100	8.0 9.3	Vane	35 kPa
−10	12		Dense brown fine to medium grained SAND with a little gravel	Water rose to 0 OD with casing at 12 to 22 m	D	13	SPT	N=32
					D	16	SPT	N=38
					D	18	SPT	N>50
			As above with increasing gravel		D	21		
−20	22		Stiff blue fissured CLAY (London Clay)	Hole dry	U 100	22.5 24		
−23	25		Weathered CHALK (weathering grades III to IV)	Standing water level −8 m OD		26.5		
					Core			
−32	34		End of hole			34		

Figure 17.7 Borehole log.

drilled at chainage 2250 m on the section in Fig. 17.1. (How many more boreholes would you need to drill before you could draw the section in Fig. 17.1, given some idea of the basic geology of the site?)

(b) Interpretive reports

An interpretive report will contain all the information in a factual report or it may refer to a separate factual report, but it will contain geological and engineering interpretations of the results of the investigations. An interpretive report should contain detailed engineering geological maps and sections giving a comprehensive three-dimensional picture of the engineering geology and hydrogeology of the site. For each of the principal soil and rock strata identified the interpretive report should give values for the parameters for strength, stiffness and permeability that will be used in the design. (These should relate to the requirements for the design of the individual structures in the scheme and the methods of analysis proposed.)

17.8 Summary

1. In any geotechnical engineering activity investigations are required to determine the ground conditions. The objectives are to locate and identify all the principal soil and rock strata, estimate design values for their strengths and stiffnesses and determine the groundwater conditions.
2. Ground investigations should, ideally, be carried out in stages, involving desk studies, preliminary investigations and detailed investigations. Detailed investigations consist of test pitting, drilling and sampling, laboratory testing and *in situ* testing.
3. Often reasonable estimates can be made of the state and the undrained strength of soil in the ground from the geological history of deposition, erosion and groundwater changes (see Chapter 18). These estimates are, however, likely to be substantially modified by structure and ageing (see Chapter 16).
4. The results of a ground investigation may be contained either in a factual report or in an interpretive report. The principal component of a factual report is the borehole logs which record all the details of each borehole: it will also record the procedures and results of the laboratory and *in situ* tests. An interpretive report should contain, in addition, cross-sections of the site showing all the principal soil and rock strata, recommended values for all the required design parameters, and, possibly, outline designs.
5. On completion of an investigation you should be able to provide, at least, the following information:

    (a) Cross-sections and plans showing the location of each of the principal strata and the groundwater conditions.
    (b) A list of the principal strata. This should include, for each stratum: descriptions of the nature and state of the soil or rock based on classification tests; the geological name a description of the depositional environments and the subsequent geological events.
    (c) A full description of the groundwater conditions.

(d) Values for the soil parameters required for the design: these would include the strength, stiffness and permeability (or consolidation) parameters appropriate to the ground conditions and the works.

(e) Statements about the uncertainties (because you can never know everything about the ground from the results of a few boreholes and tests).

6. The simple relationships linking the intrinsic parameters and soil profiles with soil classification tests and geological history are useful, particularly for preliminary design studies. However, we do not yet know enough about the fundamental mechanical properties of soils to select final design parameters from classification tests alone, so engineers must always conduct thorough ground investigations, including detailed laboratory and *in situ* testing.

References

Clayton, C. R. I., N. E. Simons and M. C. Matthews (1995) *Site Investigation*, Blackwell Science.
Mair, R. J. and D. M. Wood (1987) 'Pressuremeter testing', in *C1RIA Ground Engineering Report*, Butterworth, London.
Meigh, A. C. (1987) Cone penetration testing', *CIRIA Ground Engineering Report*, Butterworth, London.

Further reading

BS 1377 (1990) *Methods of Test for Soils for Civil Engineering Purposes*, British Standards Institution, London.
BS 5930 (1999) *Code of Practice for Site Investigations*, British Standards Institution, London.
Clayton, C. R. I., N. E. Simons and M. C. Matthews (1995) *Site Investigation*, Blackwell Science.
Weltman, A. J. and J. M. Head (1993) *Site Investigation Manual*, CIRIA, London.

Chapter 18

Soil parameters for design

18.1 Introduction

In previous chapters I described simple theories for the mechanics of soils covering strength, stiffness and permeability. In later chapters I will describe analyses for the behaviour of slopes, foundations, retaining walls and tunnels. These analyses may be empirical, they may be solved by hand calculation or they may be solved by complex numerical analyses but in any case they will require input of numerical values for design parameters for applied loads and soil parameters. They will also need input of factors of safety or load factors and these, too, are design parameters. The soil parameters and factors required are different for slopes, for foundations and for retaining walls and they are different for different methods of analysis. I will discuss the parameters required for each type of problem in the appropriate chapter; here I will discuss the general principles for selection of design parameters and factors for simple analyses.

Complex numerical analyses using the finite element or similar methods require numerical models for soil behaviour, like Cam clay described in Chapter 12. The parameters required for Cam clay are the simple critical state parameters but other numerical models often require special parameters and I will not discuss these here.

Soils are collections of mineral or clay grains which are packed together loosely or densely; they may also have a structure which is a combination of fabric and bonding, as described in Chapter 16. There are some soil parameters which depend only on the nature of the grains; these are called *material parameters*. There are some soil parameters which depend also on the state; these are called *state dependent parameters*. Soil parameters may be modified by structure.

Soil parameters can be measured in the laboratory tests described in Chapter 7 or in in situ tests described in Chapter 17. They may also be estimated from classification and descriptions of the nature and state of the soil. Natural soils in the ground vary over relatively short distances. Even a set of tests on identical reconstituted samples will give different results due to small variations in the test conditions. Consequently, any set of determinations of a single parameter will show a statistical variation and, in selection of design values, it is necessary to take account of these variations.

18.2 Principles of design

Engineers are required to design structures which are safe, serviceable and economical: they must not fall down, they must not move too much and they should not

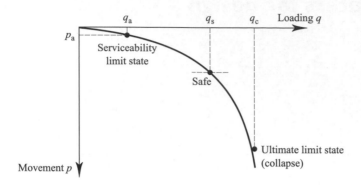

Figure 18.1 Loading and movement of structures.

be too expensive. Different analyses are needed to meet these criteria even for a particular type of structure.

Figure 18.1 shows the relationship between loading and movement of a structure; it would apply equally for a slope or a foundation, for a building frame or for a machine part. There is an ultimate limit state where the load is q_c, the movements are very large and the structure is collapsing. There is a safe state where the load is q_s, the movements are relatively large but the structure has not collapsed. There is a factor of safety given by

$$q_s = \frac{1}{F_s} q_c \qquad (18.1)$$

There are a few cases, usually involving excavation, where the soil is required to fail and then, for a factor of safety, the applied loads must be greater than the failure loads. It is no good if a machine digging an excavation or a tunnel is not powerful enough.

There is a serviceability limit state where the allowable movements ρ_a are small. There is a load factor given by

$$q_a = L_f q_c \qquad (18.2)$$

such that the allowable bearing pressure q_a causes movements that are acceptably small. Values for factors of safety and for load factors are design parameters and should be chosen by the designer. Typical values for different structures will be discussed in later chapters. Notice that values for a factor of safety will be greater than 1.0 while values for a load factor will be in the range 0 to 1.0.

For structures, such as slopes in rural locations where relatively large movements do no damage, the design is controlled by the ultimate limit state with a factor of safety. The important soil parameter is its strength. For soils we have to decide between the peak, critical state and residual strengths. Factors are there to ensure that the design is not too close to its ultimate limit state.

For structures, such as foundations, the design is controlled by the serviceability limit state. The structure must not move too much and, if it has not moved much, it is unlikely to fall down. In geotechnical engineering there are two quite different ways to design structures to limit movements. One is the load factor method where a load factor L_f is applied to the bearing capacity q_c, as shown in Fig. 18.1. The other method for design of foundations is the stiffness method in which movements are calculated from soil stiffness and applied loads. Whatever factors are applied are there to account for uncertainties. I will deal with each method and the appropriate parameters and factors when I deal with slopes, foundations and retaining walls in later chapters.

18.3   Description and classification

As a start you should always carefully describe the soil and classify it, as discussed in Chapter 5. The most important things to classify are the nature of the soil and particularly its grading (is it coarse grained or fine grained?) and the state (is it loose or dense?). You can observe the nature of the soil in disturbed samples but measurements of state will require good undisturbed samples.

It is important to distinguish between coarse grained and fine grained soils so you can determine whether analyses will be drained (effective stress analyses) or undrained (total stress analyses). There are no hard and fast rules; it is a matter of engineering judgment. You should draw grading curves and look particularly at the grain size where the curve is at the 35% fraction. For coarse grained soils you should examine the grains you can see and describe their shape (rounded, angular, elongated, flaky) and their surface texture (rough or smooth). For fine grained soils you should measure the Atterberg limits (liquid limit and plastic limit). These all describe the nature of the grains and lead to estimates of material parameters.

You should try to determine the state of the soil. This is not so easy because state is a combination of water content and stress and it should be related to the critical state. In a sample in your hand the total stresses are zero and the effective stresses are governed by the suctions which the soil can sustain and the effective stresses in the ground near a loaded structure may be quite different. You certainly should determine the water content and unit weight and from these calculate the specific volume, as described in Sec. 5.5. If you have measured the Atterberg limits you should calculate the liquidity index. If the soil is coarse grained you should determine its maximum and minimum specific volumes and calculate the relative density.

You should investigate the structure of the soil in undisturbed samples or in the ground. Look for bedding and for cracks. You should always put small samples into a glass of water and examine bonding or dispersion, as described in Sec. 9.11.

18.4   Drained or undrained or consolidation: total or effective stress parameters

For soils it is essential to separate drained analyses which are done using effective stresses and pore pressures from undrained analyses which are done using total stresses. In Chapter 6 I distinguished carefully between loading (or unloading) events which were drained from those which were undrained. During an event which is drained there is no change of pore pressure or the pore pressures change from one steady state to another, both of which are determinable. During an event which is undrained there is no change of water content but pore pressures change and are unknown; in saturated soil there is no change of volume.

Whether a particular loading or unloading event is drained or undrained depends on the rate of loading and the rate at which drainage can occur; this depends on, among other things, the permeability. Table 6.2 gives typical values for the coefficient of permeability for soils with different gradings and Table 6.3 gives durations of typical engineering constructions. In both cases the variations are very large.

Drained loading: effective stress parameters

If the soil is coarse grained and the permeability is relatively large, or if the rate of loading is relatively slow, drainage will occur during the period of construction. This case is called drained. Pore pressures are in equilibrium and can be found from the position of the water table or from a steady state seepage flownet, as described in Chapter 14. Since pore pressures are known effective stresses can be determined. These analyses are known as effective stress analyses and the parameters are known as effective stress parameters. Typical effective stress parameters include the critical state friction angle ϕ_c' and the bulk modulus K'.

Undrained loading: total stress parameters

If the soil is fine grained and the permeability is relatively small or if the rate of loading is relatively fast there will be no drainage during the period of construction. This case is called undrained. Pore pressures are unknown but, in saturated soil, there is no change of volume. Since pore pressures are unknown only total stresses can be determined. These analyses are known as total stress analyses and the parameters are known as total stress parameters. Typical total stress parameters include the undrained strength s_u and the undrained Young's Modulus E_u. In saturated soil there is no volume change so the undrained bulk modulus K_u must be infinite and, from Eq. (3.31), the undrained Poisson's ratio ν_u must be 0.5.

Consolidation

If the loading is undrained pore pressures in the vicinity of the structure change but pore pressures far from the structure do not. As a result there will be hydraulic gradients and seepage of water until the pore pressures are everywhere in equilibrium. This will cause changes of effective stress and ground movements. This process is consolidation (see Chapter 15) and you will need to obtain parameters for analyses of movements due to consolidation.

In most cases in practice the loading or unloading will be neither fully drained nor fully undrained and there will be some drainage and some pore pressure changes during construction. There are complex coupled analyses which account for partial drainage, but for routine geotechnical design it is necessary to consider the soil to be either fully drained, in which case the analyses will be effective stress, or fully undrained, in which case they will be total stress. In some instances it may be necessary to consider both cases.

18.5   Ultimate limit state: critical state and residual strengths and factors of safety

The first fundamental requirement of engineering design is to examine the ultimate limit state and to demonstrate that there is an adequate margin of safety against failure. These analyses do not consider movements.

Figure 18.2 illustrates the typical behaviour of soil in a drained shear test and it is similar to Figs. 9.1 and 9.3 in Chapter 9. The strength of the soil is the shear stress mobilized under different conditions. There is a peak strength at relatively small strains of the order of 1% where there are volumetric strains. There is a critical state strength at strains of the order of 10% where the soil continues to distort at constant stress and constant volume. For clay soils there is a residual strength at large displacement. For soils which do not contain significant quantities of clay the residual strength is the same as the critical state strength. Figure 18.2 illustrates typical soil behaviour in drained tests. There will be similar relationships between shear stress, pore pressure and strain

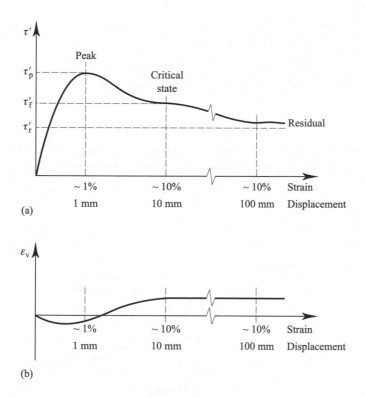

Figure 18.2 Behaviour of soils during shearing.

or displacement in undrained tests, at least up to the critical state. The question is which of these strengths – peak, critical state or residual – should be used as design parameters.

Residual strength

After very large displacements the shear stress which can be mobilized is the residual strength τ_r'. For drained loading this is given by

$$\tau_r' = (\sigma - u) \tan \phi_r' \tag{18.3}$$

where the residual friction angle ϕ_r' is a material parameter.

Figure 18.3(a) shows a pile driven into the ground. Some of the load on the pile is transferred into the ground by shear stresses between the side of the pile and the soil. Driving the pile into the ground has generated large displacements between the pile and the soil and so the limiting shear stress between the side of the pile and the ground is governed by the residual strength. It will also be limited by shearing on the interface between the soil and the pile if this is smaller. Figure 18.3(b) shows an old landslide. The near-surface soil has slid down-slope along a clearly defined slip surface. In analyses of the stability of the slope the limiting shear stress across this slip surface is governed by the residual strength.

Since the residual strength of a clay soil is often very much smaller than its critical state and peak strengths it is essential to discover the presence of pre-existing slip surfaces in the ground. This requires careful and detailed ground investigation. Old landslides can often be very difficult to detect as surface changes over geological time-scales, together with vegetation, hide the evidence of the movements.

Critical state strength

The critical state strength is reached after relatively large strains of the order of 10%, as illustrated in Fig. 18.2. Very much larger strains and displacements are required to

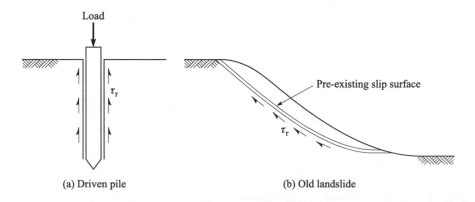

Figure 18.3 Cases where the limiting shear stress is the residual strength.

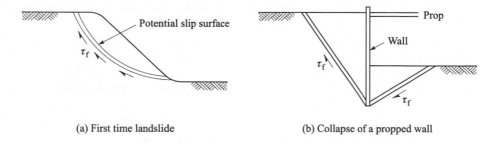

(a) First time landslide

(b) Collapse of a propped wall

Figure 18.4 Cases where the critical state strength should be used for design.

reduce the strength towards the residual. Consequently, for most purposes the critical state strength is the worst that needs to be considered in design. The residual strength should be considered for design in clay soils only if there are pre-existing slip surfaces or if very large displacements are expected.

After strains of the order of 10% or so the shear stress which can be mobilized is the critical state strength τ'_f. For drained loading this is given by

$$\tau'_f = (\sigma - u) \tan \phi'_c \tag{18.4}$$

where the critical friction angle ϕ'_c is a material parameter. For undrained loading it is given by

$$\tau_f = s_u \tag{18.5}$$

where the critical undrained strength s_u depends on the water content and so it is a state dependent parameter.

Figure 18.4 illustrates two cases in which the critical state strength should be used in design. In Fig. 18.4(a) the shear stress on the potential slip surface in the slope is governed by the critical state strength because, even in a stable slope, there will probably be strains in the ground larger than 1%. In Fig. 18.4(b) the ultimate limit state of a stiff propped retaining wall should be investigated using the critical state strength.

Factor of safety

For a safe design it is necessary to ensure that there is an adequate margin against collapse and this is done by applying a factor of safety F_s to the strength. The safe shear stresses τ'_s or τ_s in the soil are given by:

$$\tau'_s = \frac{\tau'}{F_s} = (\sigma - u)\frac{\tan \phi'}{F_s} \tag{18.6}$$

$$\tau_s = \frac{\tau}{F_s} = \frac{s_u}{F_s} \tag{18.7}$$

where ϕ' and s_u are the appropriate residual or critical state strengths. The values chosen for the factors of safety depend on many things, including uncertainties in determination of strength and the consequences of failure.

Notice that in Eqs. (18.3) and (18.4) the strength depends not only on the friction angle but also on the total stress (which depends on the unit weight γ and on the external loads from foundations) and on the pore pressure. Some engineers apply partial factors to each of these to reflect the different levels of uncertainty in their determination.

18.6   Serviceability limit state: peak strength with a load factor

A further fundamental requirement of engineering design is to examine the serviceability limit state and to demonstrate that the movements will not exceed some limit determined by the design team. In geotechnical engineering there are two principal methods for examining serviceability limit states. One is to apply a load factor to a collapse analysis and, again, the question is whether the collapse analysis should be done with the peak or with the critical state strength.

The critical state strength is clearly not appropriate for serviceability limit state design because this would mean that you would design the same structure for dense and loose sand or for normally consolidated and overconsolidated clay and this is clearly illogical. A dense sand is stiffer than a loose sand and, for the same movement, it can have a larger load and the same for overconsolidated and normally consolidated clays.

Figure 18.5(a) shows the behaviour of two samples of the same soil in a triaxial compression test: up to the peak the behaviour would be similar for drained and undrained tests. (Note that in a triaxial test $q = (\sigma'_a - \sigma'_r)$ and the same symbol q is used for loading.) The behaviour shown in Fig. 18.5(a) is similar to that shown in Fig. 10.15. Sample 2 is further from the critical state line, it is more heavily overconsolidated and its state parameter is larger than that of sample 1. Both samples reach a peak strength

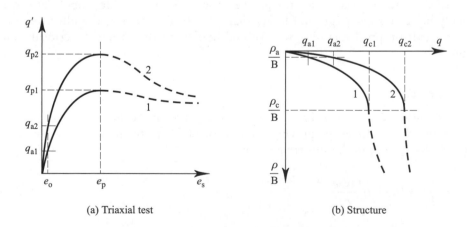

(a) Triaxial test   (b) Structure

Figure 18.5 Behaviour of soils in triaxial tests and behaviour of a structure.

at about the same strain ε_p which, for many cases, would be about 1%. If we wanted to limit the strains in the triaxial test to ε_a by applying a load factor to the peak strength then we would apply allowable stresses given by

$$L_f = \frac{q_{a1}}{q_{p1}} = \frac{q_{a2}}{q_{p2}} \tag{18.8}$$

The load factor is the same in both cases because the stress–strain curves are geometrically similar.

Figure 18.5(b) shows the behaviour of the same structure on the soils whose behaviour in triaxial tests is shown in Fig. 18.5(a). The load is q and the movement ρ has been normalised by dividing it by a characteristic dimension B. Up to the point of failure where the load is the collapse load and the movements are ρ_c the load–movement curves are similar to each other and to the stress–strain curves for the triaxial test. Beyond the point of failure the load–movement curves do not approach each other.

If we wanted to design the structure to have an allowable movement ρ_a with an allowable load q_a we would apply a load factor given by

$$L_f = \frac{q_{a1}}{q_{c1}} = \frac{q_{a2}}{q_{c2}} \tag{18.9}$$

The load factor is the same in both cases because the load–movement curves are geometrically similar.

It is important to understand that the load factor defined above is not a factor of safety: it is a factor to reduce the loads from the collapse load to a point at which movements will be small. Notice that a factor of safety is normally applied to a soil strength while a load factor is normally applied to a load. You may want to apply additional factors, particularly to the soil strength, to take account of uncertainties, such as whether mean or worst credible values have been used as discussed later.

18.7   Serviceability limit state: soil stiffness and design loads

From your courses in structures you will have learned how to calculate the movements of beams, frames, cylinders and plates made of elastic material from the applied loads and elastic parameters. Similar methods are used to calculate ground movements. Often the calculations are complicated and there are a number of standard solutions, especially for foundations: these are discussed in Sec. 22.8.

The parameters in these calculations are normally Young's modulus E and Poisson's ratio ν and these are defined in Sec. 3.8. Other stiffness parameters are the shear modulus G, the bulk modulus K and the one-dimensional modulus M. Relationships between these are given in Sec. 3.8. In selecting values for design it is necessary, as always, to distinguish between drained and undrained loading. For drained loading E' and ν' are measured in drained triaxial tests. For undrained loading E_u is measured in undrained triaxial tests and $\nu_u = 0.5$ because volumetric strains are zero. An important result for an elastic material is $G' = G_u$ and this leads to relationships between the other drained and undrained moduli given in Sec. 22.9.

It is also necessary to recognize that soil stiffness is highly non-linear – it changes with stress and with strain – and it is necessary to select values appropriate to the strains in the soil in the ground. Non-linear soil stiffness was discussed in Chapter 13. The characteristic variation of stiffness with strain is illustrated in Fig. 13.8. At very small strain the value of Young's modulus is E_0. This is found from G_0 which can be measured in dynamic laboratory or in situ tests and it varies with stress and state, as given by Eq. (13.8) and shown in Fig. 13.9. Analyses can be done in one step using secant values or in several steps using tangent moduli. Figure 18.6(a) shows a non linear stress–strain curve for a triaxial test on a soil sample: this is similar to Fig. 3.2 in Chapter 3. The diagram has axes $q = (\sigma_a - \sigma_r)$ and axial strain ε_a and the gradient is Young's Modulus: if the test is drained the gradient is E' and if it is undrained it is E_u. At the point A at some stage of the test the secant modulus is

$$E_{\text{sec}} = \frac{\Delta q}{\Delta \varepsilon_a} \tag{18.10}$$

and the tangent modulus is

$$E_{\text{tan}} = \frac{dq}{d\varepsilon_a} \tag{18.11}$$

where Δ represents the change of stress and strain from the start of the test. For simple analyses the secant modulus method would normally be used and the step taken as the whole foundation loading. Figure 18.6(b) shows the variations of tangent and secant modulus with strain corresponding to the stress–strain curve in Fig. 18.6(a). The stiffnesses have been normalized by dividing by E_0. At the critical state at F the strain is about 10% and $E_{\text{tan}} = 0$. At the peak state at P the strain is about 1% and, again, $E_{\text{tan}} = 0$. As discussed in Sec. 13.4 the average strains in the ground near a typical foundation at working load are about 0.1% but locally they are often in the range 0.01% to 1%. This range is shown in Fig. 18.6(b) and this demonstrates that

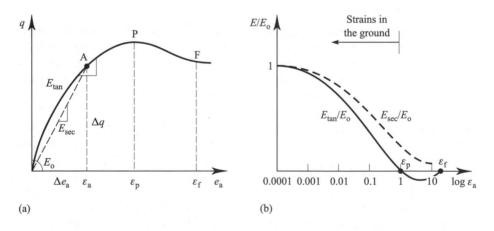

(a)   (b)

Figure 18.6 Tangent and secant moduli.

the probable variation of stiffness in the ground can be very large. For design it is necessary to choose a value of stiffness which corresponds to the mean strains in the ground, as discussed by Atkinson (2000).

18.8   Parameters related to description and classification

Some parameters such as the critical friction angle ϕ'_c are material parameters and they depend only on the nature of the grains. Other parameters, such as undrained strength s_u, are state dependent parameters and they depend both on the nature of the grains and on the current state. There may even be parameters which have the same value for all soils. In the following sections I will give some correlations between values for some soil parameters and the soil descriptions and classifications described in Chapter 5. Some of these are empirical but some arise from the definitions of the parameters themselves.

(a)   Undrained strength and liquidity index

Undrained strength is a state dependent parameter and it depends on voids ratio or water content, as shown in Fig. 9.5. The critical state undrained strength of soil at its liquid limit is about 1.7 kPa and at its plastic limit it is about 170 kPa. The liquidity index I_l given by Eq. (5.10) defines the current water content in terms of the Atterberg limits described in Sec. 5.6. There is a linear relationship between liquidity index and the logarithm of undrained strength, as shown in Fig. 18.7.

(b)   Critical state friction angle

The critical state friction angle ϕ'_c is a material parameter and it depends on the nature of the grains. For fine grained soils it varies with plasticity index I_p, ranging from less than 20° for high plasticity soils to about 28° for low plasticity soils. For coarse grained soils ϕ'_c depends on the shape and roughness of the grains and it ranges from about 30° for soils with smooth rounded grains to more than 40° for soils with rough angular grains. Typical values for ϕ'_c are given by Muir Wood (1991).

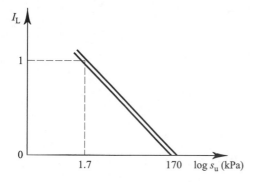

Figure 18.7 Variation of undrained strength with liquidity index.

(c) Compressibility

From Fig. 9.4 the compressibility of a soil is given by the gradient C_c of the normal compression and critical state lines and this is a material parameter. From Fig. 9.5 s_u is proportional to σ' so the critical state line can be drawn as Fig. 18.8 and

$$e_{LL} - e_{PL} = C_c \log 100 = 2C_c \tag{18.12}$$

Since $e = wG_s$ and noting that e is a number while w is a percentage, Eq. (18.12) becomes

$$C_c = \frac{I_p G_s}{100} \tag{18.13}$$

Hence C_c is related to the Atterberg limits and this is a consequence of the 100-fold difference between the undrained strengths at the liquid and plastic limits.

(d) Critical state line

It turns out that the critical state lines with axes e and $\log \sigma'$ for many fine grained soils pass through the same point called the Ω (omega) point. The approximate coordinates of the Ω point given by Schofield and Wroth (1968) are $e_\Omega = 0.25$ and $\sigma'_\Omega = 15$ MPa and these are fundamental constants. From Eq. (9.3)

$$e_\Gamma = 0.25 + C_c \log 15{,}000 \tag{18.14}$$

From Eqs. (18.13) and (18.14) the parameters C_c and e_Γ, which define the critical state line for fine grained soils, can be obtained from the Atterberg limits. There are simple relationships between C_c and λ and between e_Γ and Γ.

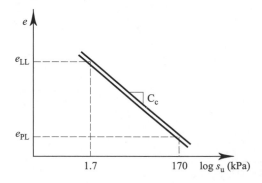

Figure 18.8 Determination of compressibility from the Atterberg limits.

(e)   Peak strength

The peak strength of soil discussed in Chapter 10 can be represented by a linear Mohr–Coulomb criterion given by Eq. (10.6) which contains the parameter ϕ'_p. This includes the critical stress σ'_c which contains the water content, so ϕ'_p is a material parameter. Alternatively the peak strength can be represented by a simple power law criterion given by Eq. (10.14). This contains a parameter B which is also a material parameter.

(f) G_0

The shear modulus for very small strain G_0 was described in Sec. 13.6. It is related to the current stress and overconsolidation ratio by Eq. (13.8) and so it is a state dependent parameter. Equation 13.8 contains parameters A, n and m which are related to the plasticity index I_p (Viggiani and Atkinson, 1995) and so they are material parameters.

From these simple analyses there are a number of important parameters which are material parameters and which are related to simple descriptions and classification parameters, such as the Atterberg limits. Some state dependent parameters such as G_0 are related to the current state through material parameters.

These relationships between soil parameters and soil descriptions and classifications are helpful for preliminary design before a ground investigation has been completed. They are also useful for validation of results obtained from laboratory and in situ tests. If parameter values measured in tests differ from those derived from descriptions and classifications you should find out why. It may be that the test results were wrong or it may be that the soil has some special properties or the effects of its structure are important.

18.9   States of soils in the ground

Most soils were deposited through natural geological processes then compressed and swelled during subsequent deposition and erosion. For example London clay was deposited in a shallow sea and the surface level was once of the order of 200 m above present ground level; it is overconsolidated. In contrast the soils below the nearby Thames marshes have the same nature as the London clay because that is from where they were eroded, but the ground level has never been higher than the present: they are normally consolidated. The ground conditions in the Thames estuary are like those shown in Fig. 4.4 and the states of the clay soils vary with depth in predictable ways.

Changes of stress and water content in soils during deposition and erosion were described in Chapter 16. Figure 16.2(d) shows qualitatively the variations of water content with depth in normally consolidated and overconsolidated soils. When soil is freshly deposited its water content is close to its liquid limit. When soil is at its plastic limit its undrained strength s_u is about 170 kPa, it has been compressed under an effective stress of about 800 kPa which would correspond to a depth of about 80 m. Figure 18.9(a) shows the variation of water content with depth for normally consolidated and overconsolidated soils. Near the surface the water contents

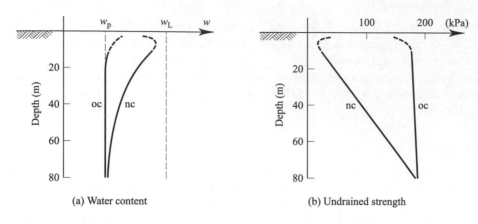

Figure 18.9 Variation of water content and undrained strength with depth in the ground for normally consolidated and overconsolidated soils.

have been modified by the ground conditions as indicated by the broken lines: the normally consolidated soil has dried due to evaporation and vegetation while the overconsolidated soil has wetted and swelled due to rainwater penetrating cracks.

Figure 18.9(b) shows the variations of undrained strength s_u with depth corresponding to the water contents in Fig. 18.9(a). The undrained strength in the overconsolidated soil is about 170 kPa corresponding to the water content close to the plastic limit except near the surface. The undrained strength in the normally consolidated soil increases linearly with depth and is about 170 kPa at a depth of about 80 m, with an increase near the surface.

18.10   Accounting for variability

If you make a number of separate measurements of the same parameter, for example water content of a lorry-load of soil, you will obtain a range of results. Your results will be different because of errors in measurement of small weights of wet and dry soil and because of the variation of true water content throughout the lorry-load. There will be variations in any soil parameter you measure due to experimental errors and due to the natural variation of soil in the ground. In soils the parameter may be a state dependent parameter in which case its true value will vary with state, but we have seen how to normalize test data to take account of state.

There are essentially three main ways in which engineers select design values from a set of test results all measuring the same material parameter. Figure 18.10 shows a typical distribution of results as the number of observations plotted against the observed value and this is a common plot in statistical analysis. (For simplicity I have shown a symmetric distribution but for soil test data it could well be skewed.)

There is a mean value for which there are approximately as many larger values as there are smaller values. There is a worst credible value and if you have measured any values smaller than this you have discarded the results for one reason or another. There is a value called the moderately conservative value which is somewhere

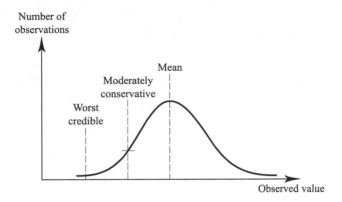

Figure 18.10 Variability of observed values.

between the two. There is no standard definition for moderately conservative but reasonable ones might be the value for which only 25% of the observations are smaller or the value which is one standard deviation away from the mean.

The value chosen for design should be considered together with the value of the factor of safety or load factor which will be used with it. Again there are no hard and fast rules and these choices are a matter of engineering judgment. If you choose a worst credible value of ϕ'_c then you could choose a partial factor of 1.0 to go with it but you would probably choose a mean value to determine the water content of your lorry load of soil.

Remember that, if you are designing for something for which soil must fail, like a tunnelling machine or an excavator, the worst credible strength is the *largest* value you have measured reliably.

18.11   Summary

1. Analyses of slopes, foundations and walls require values for soil parameters and factors of safety or load factors. These depend on the type of structure and on the ground conditions.
2. A factor of safety is applied to soil strength and its purpose is to ensure that the structure does not approach its ultimate limit state. A load factor is applied to a load and its purpose is to ensure that movements are small.
3. A fundamental choice must be made between drained construction requiring effective stress parameters and undrained construction requiring total stress parameters.
4. The critical state strength with a factor of safety should be used to calculate an ultimate limit state unless there is evidence of previous large movements which have already reduced the strength to the residual. The peak strength with a load factor should be used to calculate serviceability limit states.
5. Movements can be calculated using soil stiffness. This is non-linear and a value should be chosen which corresponds to the mean of the strains in the ground.

6. Some parameters are material parameters. They depend on the nature of the grains and are related to soil classification parameters. Other parameters are state dependent parameters; they are often related to the state by parameters which are themselves material parameters.

7. In choosing design parameters allowance should be made for variations in test results due to natural variations of soil in the ground and due to experimental variations.

References

Atkinson, J. H. (2000) 40th Rankine Lecture: Non-linear stiffness in routine design, *Geotechnique*, Vol. 50, No, 5, pp. 487–507.

Viggiani, G. and J. H. Atkinson (1995) Stiffness of fine-grained soil at very small strains, *Geotechnique*, Vol. 45, No. 2, pp. 249–266.

Chapter 19

Ultimate stability of soil structures using bound methods

19.1 Introduction

In previous chapters I considered the behaviour of single elements of soil, either in the ground or in laboratory tests, and I developed simple theories for strength of soil and simple constitutive equations relating increments of stress and strain. What we have to do now is to apply these theories to the behaviour of geotechnical structures such as foundations, slopes and retaining walls. As discussed earlier, solutions for problems in mechanics must satisfy the three conditions of equilibrium, compatibility and material properties. It is fairly obvious that complete solutions, satisfying these conditions with the material properties for soil, will be very difficult to obtain, even for very simple foundations and slopes.

First, I will consider the conditions of ultimate collapse where the important material property is the soil strength. Remember that, as always, it is necessary to distinguish between cases of undrained and drained loading. For undrained loading the strength of soil is given by

$$\tau = s_u \tag{19.1}$$

where s_u is the undrained strength. For drained loading where pore pressures can be determined from hydrostatic groundwater conditions or from a steady state seepage flownet the strength is given by

$$\tau' = \sigma' \tan \phi' = (\sigma - u) \tan \phi' \tag{19.2}$$

where ϕ' is a friction angle. As discussed in Sec. 9.2 soil has a peak strength, a critical state strength and a residual strength which are mobilized at different strains or displacements. The factors which determine which strength should be used in stability calculations are discussed in Chapter 18.

Even with these relatively simple expressions for soil strength it is still quite difficult to obtain complete solutions and the standard methods used in geotechnical engineering involve simplifications. There are two basic methods: the bound methods described in this chapter and the limit equilibrium method described in the next chapter. Both methods require approximations and simplifications which will be discussed in due course.

19.2   Factors

The analyses in this chapter and the next calculate ultimate limit states where the slope, wall or foundation can be said to have collapsed. In practice engineers are required to design safe and serviceable structures and to do this they apply factors to their calculations. These factors are intended to move the design away from a collapse state into a state in which there is no danger of collapse or where movements are acceptably small. Similar procedures are employed throughout engineering design.

Some factors are factors of safety and they are intended to ensure that the structure is not near its ultimate limit state. Other factors are load factors and they are intended to ensure that the deformations remain within a small strain range to limit movements. Factors of safety are not the same as load factors: each has a different purpose. Sometimes, instead of a single factor, partial factors are used. Each partial factor is applied to a separate component of the calculation. Use of factors to ensure safety and to limit movements were discussed in Chapter 18 and will be discussed further in later chapters dealing with different structures.

19.3   Theorems of plastic collapse

In order to simplify stability calculations it is possible to ignore some of the conditions of equilibrium and compatibility and to make use of important theorems of plastic collapse. It turns out that, by ignoring the equilibrium condition, you can calculate an upper bound to the collapse load so that if the structure is loaded to this value it must collapse; similarly, by ignoring the compatibility condition you can calculate a lower bound to the collapse load so that if the structure is loaded to this value it cannot collapse. Obviously the true collapse load must lie between these bounds.

The essential feature of the upper and lower bound calculations is that rigorous proofs exist which show that they will bracket the true collapse load. Thus, although the two methods of calculation have been simplified by ignoring, for the first, equilibrium and, for the second, compatibility, no major assumptions are needed (other than those required to prove the bound theorems in the first place). What has been lost by making the calculations simple is certainty; all you have are upper and lower bounds and you do not know the true collapse load (unless you can obtain equal upper and lower bounds). Usually you can obtain upper and lower bounds that are fairly close to one another so the degree of uncertainty is quite small.

I am not going to prove the plastic collapse theorems here and I will simply quote the results. A condition required to prove the theorems is that the material must be perfectly plastic. This means that, at failure, the soil must be straining at a constant state with an associated flow rule so that the vector of plastic strain increment is normal to the failure envelope (see Chapter 3). The first condition, straining at a constant state, is met by soils at their ultimate or critical states, given by Eqs. (19.1) and (19.2). The second condition is illustrated in Fig. 19.1(a) for undrained loading and in Fig. 19.1(b) for drained loading.

In both cases elastic strains must be zero since the stresses remain constant; thus total and plastic strains are the same. For undrained loading the failure envelope given by Eq. (19.1) is horizontal and the volumetric strains are zero (because undrained means constant volume) and so the vector of plastic strain $\delta \varepsilon^{p}$ is normal to the failure

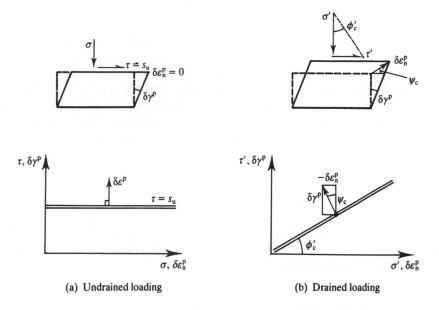

Figure 19.1 Straining of perfectly plastic soil with an associated flow rule.

envelope as shown. For drained loading the failure envelope is given by Eq. (19.2) and if the flow rule is associated the angle of dilation at the critical state ψ_c is

$$-\frac{\delta\varepsilon_n^p}{\delta\gamma^p} = \tan\psi_c = \tan\phi_c' \qquad (19.3)$$

At the critical state, however, soil strains at a constant state (i.e. at a constant volume) and so $\psi_c = 0$, which means that, at failure at the critical state, the flow rule is not associated and soil in drained loading is not perfectly plastic. This does not actually matter very much as you can prove that an upper bound for a material with $\psi_c = \phi_c'$ is still an upper bound, even if ψ_c is less than ϕ_c', but you can not do the same for the lower bound. In practice upper and lower bounds for soil structures calculated with $\psi_c = \phi_c'$ give good agreement with experimental observations and, although the lower bound solution is not absolutely rigorous, the errors seem to be small.

The statements of the bound theorems are simple and straightforward:

1. Upper bound. If you take any compatible mechanism of slip surfaces and consider an increment of movement and if you show that the work done by the stresses in the soil equals the work done by the external loads, the structure must collapse (i.e. the external loads are an upper bound to the true collapse loads).
2. Lower bound. If you can determine a set of stresses in the ground that are in equilibrium with the external loads and do not exceed the strength of the soil,

the structure cannot collapse (i.e. the external loads are a lower bound to the true collapse loads).

To calculate an upper bound you must satisfy the conditions of compatibility and the material properties (which govern the work done by the stresses in the soil), but nothing is said about equilibrium. To calculate a lower bound you must satisfy the conditions of equilibrium and the material properties (which determine the strength), but nothing is said about displacements or compatibility. Because a structure with an upper bound load must collapse this is often known as the unsafe load and because a structure with a lower bound load cannot collapse this is known as the safe load. The basic principles of these upper and lower bound methods are also used to calculate stability of framed structures by using plastic hinges to create mechanisms or by using elastic analysis to calculate yield stresses at critical sections.

In the present context the terms upper and lower bounds have the very specific meanings associated with the bound theorems. Engineers also investigate bounds to structural behaviour by investigating the consequences of optimistic and pessimistic values for material properties, but bounds calculated in this way are obviously quite different from the present meaning.

19.4   Compatible mechanisms of slip surfaces

To calculate an upper bound a mechanism of slip surfaces must meet the requirements of compatibility. These requirements determine both the allowable shape of individual slip surfaces and their general arrangement.

Figure 19.2(b) shows a segment of a curved slip surface represented by a double line and Fig. 19.2(a) shows an enlarged small element. On one side the material is stationary and on the other side there is an increment of displacement δw at an angle ψ. The length along the slip surface is constant so it is a zero extension line (see Sec. 2.6).

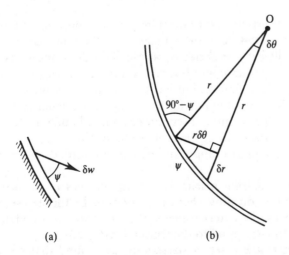

Figure 19.2 Geometry of a slip plane.

From Eq. (2.11) and from the geometry of Fig. 2.9, slip surfaces makes angles α and β to the major principle planes where

$$\alpha = \beta = 45° + \tfrac{1}{2}\psi \tag{19.4}$$

From the geometry of Fig. 19.2(b),

$$\frac{dr}{r\,d\theta} = \tan\psi \tag{19.5}$$

and therefore

$$\frac{r_{\mathrm{B}}}{r_{\mathrm{A}}} = \exp(\Delta\theta \tan\psi) \tag{19.6}$$

where $\Delta\theta$ is the angle between the radii r_{A} and r_{B}. This is the equation of a logarithmic spiral for $\psi > 0$ but, for undrained loading when $\psi = 0$,

$$\frac{r_{\mathrm{B}}}{r_{\mathrm{A}}} = \exp(0) = 1 \tag{19.7}$$

This is the equation of a circular arc. Also, as $r_{\mathrm{A}} \to \infty$, Eqs. (19.6) and (19.7) tend to the equation for a straight line. Thus, for drained loading where $\psi = \phi_{\mathrm{c}}'$, slip surfaces may be straight lines or logarithmic spirals while, for undrained loading where $\psi = 0$, slip surfaces may be straight lines or circular arcs. In Fig. 19.2(b) the radii intersect the curved slip surface at a constant angle $(90° - \psi)$ and hence radii may also be slip surfaces.

Slip surfaces can be assembled to form a compatible mechanism of plastic collapse; a number of simple mechanisms are illustrated in Fig. 19.3. These may consist of straight lines or curves (circular arcs for undrained loading with $\psi_{\mathrm{c}} = 0$ or logarithmic spirals for drained loading with $\psi_{\mathrm{c}} = \phi_{\mathrm{c}}'$) or combinations of straight lines and curves. Notice that in Fig. 19.3(f) the curved section is in fact a fan with radial slip surfaces and these are required to make the mechanism compatible by constructing a displacement diagram, as described in Sec. 2.3.

19.5   Work done by internal stresses and external loads

To determine an upper bound it is necessary to calculate the work done by the internal stresses and by the external loads during an increment of movement of a compatible mechanism. The work done by a force is simply the product of the force and the increment of displacement resolved into the direction of the force. We can always determine the increments of displacements, resolved in any direction, from a displacement diagram.

External loads arise from concentrated forces from small foundations, from distributed stresses below embankments and wide foundations and from the self-weight of the soil. External loads from concentrated forces are easy to determine and are the same for drained and for undrained loading, but for distributed stresses and self-weight drained and undrained loading must be considered separately. Figure 19.4 shows an

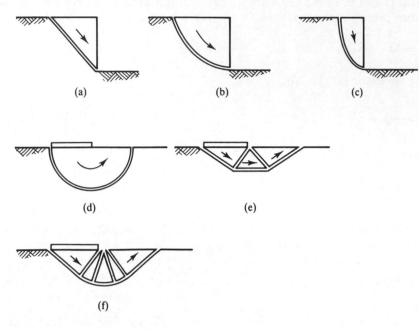

(a)   (b)   (c)

(d)   (e)

(f)

Figure 19.3 Compatible mechanisms.

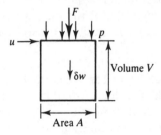

Figure 19.4 Work done by external loads.

element of soil with unit weight γ and with a total stress p and a concentrated load F at the top surface where the pore pressure is u. There is an increment of displacement δw in the direction of the surface stress, the concentrated load and self-weight forces. For undrained loading the increment of work δE is

$$\delta E = F\delta w + pA\delta w + \gamma V \delta w \tag{19.8}$$

For drained loading the water remains stationary so the work is done by the effective stresses only; hence

$$\delta E = F\delta w + (p - u)A\delta w + (\gamma - \gamma_{\mathrm{w}})V\,\delta w \tag{19.9}$$

For dry soil simply put $u = \gamma_{\mathrm{w}} = 0$ in Eq. (19.9).

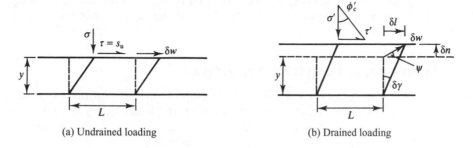

(a) Undrained loading   (b) Drained loading

Figure 19.5 Work done by internal stresses on slip planes.

The work done by the internal stresses is the work dissipated by plastic straining in the material in the thin slip surfaces that make up the compatible mechanism and, again, undrained and drained loading must be considered separately. Figure 18.5 shows short lengths of slip surfaces that have increments of displacement δw as shown. Since the soil is at the critical state in each case the stresses are given by Eqs. (19.1) and (19.2) and, for drained loading, the shear and normal strains are related by Eq. (19.3).

In Fig. 19.5(b) for drained loading the water remains stationary, the work is done by the effective stresses and hence

$$\delta W = \tau' L \delta l - \sigma_{\mathrm{n}}' L \delta n \tag{19.10}$$

Note that for dilation the work done by the normal stress is negative since σ_{n}' and δn are in opposite directions. From Eq. (19.10), with the volume of the slip plane $V = Ly$,

$$\delta W = V(\tau'\delta\gamma + \sigma_{\mathrm{n}}'\delta\varepsilon_{\mathrm{n}}) = V\tau'\delta\gamma \left(1 - \frac{\tan\psi_{\mathrm{c}}}{\tan\phi_{\mathrm{c}}'}\right) \tag{19.11}$$

However, for a perfectly plastic material $\psi_{\mathrm{c}} = \phi_{\mathrm{c}}'$ and so the work dissipated by the internal stresses for drained loading is

$$\delta W = 0 \tag{19.12}$$

This is a very surprising result and presents difficulties which I will not explore here. The implication is that a perfectly plastic factional material is both dissipative and conservative, which is nonsense. The conclusion must be that the flow rule for a frictional material cannot be associated. Nevertheless, the result given by Eq. (19.12) is very convenient and it may be used to calculate upper bounds for frictional materials like soil.

In Fig. 19.5(a) for undrained loading the increment of work done by the total stresses τ and σ is

$$\delta W = \tau L \delta w = s_{\mathrm{u}} L \delta w \tag{19.13}$$

Note that for undrained or constant volume straining no work is done by the normal stress σ_n because there is no displacement normal to the slip surface. For an upper bound calculation you must evaluate Eq. (19.13) for all the slip planes in the compatible mechanism.

19.6   Simple upper bounds for a foundation

In order to illustrate the use of the bound theorems I shall obtain solutions for the bearing capacity of a foundation subject to undrained loading. Figure 19.6 shows a foundation with unit length out of the page so that the width B is equal to the area A. The foundation itself is weightless so the bearing pressure $q = V/B$. As the foundation load V and bearing pressure q are raised the settlement ρ will increase until the foundation can be said to have failed at the collapse load V_c or the bearing capacity q_c. The foundation is smooth so there are no shear stresses between the soil and the foundation. I will obtain solutions using, firstly, a simple mechanism and, secondly, two stress discontinuities, and later I will obtain more complex solutions using a slip fan and a stress fan. The purpose here is to illustrate the principles of the bound solutions; I will consider the bearing capacity of foundations in more detail in Chapter 22.

Figure 19.7(a) shows a simple mechanism consisting of three triangular wedges and Fig. 19.7(b) is the corresponding displacement diagram. The increments of work done by the self-weight forces sum to zero since block B moves horizontally while the vertical components of the displacements of blocks A and C are equal and opposite. Hence, from Eq. (19.8), we have

$$\delta E = V_u \delta w_f \tag{19.14}$$

In order to calculate the work done by the internal stresses on the slip planes, from Eq. (19.13) it is easiest to tabulate s_u, L and δw for each slip plane. Hence, from Table 19.1,

$$\delta W = 6 s_u B \delta w_f \tag{19.15}$$

and, equating δE and δW, an upper bound for the collapse load is

$$V_u = 6 B s_u \tag{19.16}$$

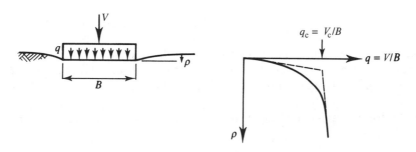

Figure 19.6 Bearing capacity of a simple foundation.

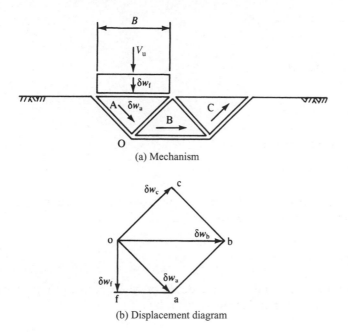

Figure 19.7 Mechanism of collapse for a foundation.

Table 19.1 Work done by internal stresses for mechanism in Figure 19.7

Slip plane	Shear stress	Length	Displacement	$\delta W = s_u L \delta w$
oa	s_u	$\frac{1}{\sqrt{2}}B$	$\sqrt{2}\delta w_f$	$s_u B \delta w_f$
ob	s_u	B	$2\delta w_f$	$2s_u B \delta w_f$
oc	s_u	$\frac{1}{\sqrt{2}}B$	$\sqrt{2}\delta w_f$	$s_u B \delta w_f$
ab	s_u	$\frac{1}{\sqrt{2}}B$	$\sqrt{2}\delta w_f$	$s_u B \delta w_f$
bc	s_u	$\frac{1}{\sqrt{2}}B$	$\sqrt{2}\delta w_f$	$s_u B \delta w_f$
fa	0	B	δw_f	0
			Total	$6s_u B \delta w_f$

19.7   Discontinuous equilibrium stress states

To calculate a lower bound it is necessary to analyse an equilibrium state of stress and to show that it does not exceed one of the failure criteria given by Eqs. (19.1) and (19.2). The equilibrium states of stress may vary smoothly from place to place or there can be sudden changes of stress across stress discontinuities, provided, of course, that the conditions of equilibrium are met across the discontinuities.

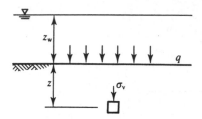

Figure 19.8 Vertical stress in the ground.

The variation of vertical total stress with depth in the ground was given in Sec. 6.2. From Fig. 19.8 the vertical stress on an element at a depth z is

$$\sigma_v = \gamma z + q + \gamma_w z_w \tag{19.17}$$

where q is a uniform surface stress and z_w is the depth of water above ground level. For drained loading the effective vertical stress is given by

$$\sigma'_v = \sigma_v - u \tag{19.18}$$

where u is the (steady state) pore pressure.

In Fig. 19.9(a) there are two regions A and B separated by a discontinuity represented by a single bold line; the stresses in each region are uniform and are characterized by the magnitudes and directions of the major principal stresses σ_{1a} and σ_{1b} as shown. The rotation in the direction of the major principal stress across the discontinuity is $\delta\theta = \theta_b - \theta_a$. The Mohr circles of total stress are shown in Fig. 19.9(b). The point C represents the normal and shear stresses on the discontinuity and the poles of the circles are found by drawing $P_a - C - P_b$ parallel to the discontinuity in Fig. 19.9(a). Hence the directions of the major principal planes are given by the broken lines in Fig. 19.9(b) and, from the properties of the Mohr circle construction given in Sec. 2.5, we can mark $2\theta_a$ and $2\theta_b$, the angles subtended by σ_{1a} and σ_{1b}, and the normal stress on the discontinuity.

As usual it is necessary to consider undrained and drained loading separately. Figure 19.10 shows the analysis for undrained loading. Both Mohr circles of total stress touch the failure line given by Eq. (19.1). From the geometry of Fig. 19.10(b), noting that $AC = s_u$,

$$\delta s = 2s_u \sin \delta\theta \tag{19.19}$$

Hence the change of total stress across a discontinuity is simply related to the rotation $\delta\theta$ of the direction of the major principal stress.

Figure 19.11 shows the analysis for drained loading. Both Mohr circles of effective stress touch the failure line given by Eq. (19.2) and the angle ρ' defines the ratio τ'_n/σ'_n

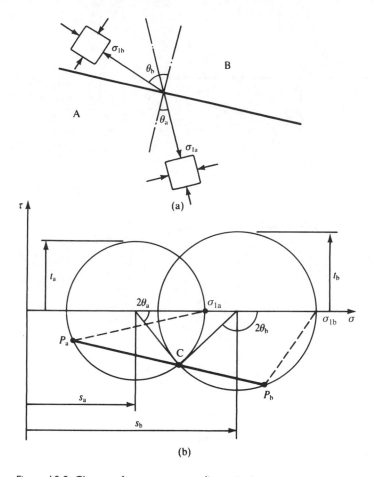

Figure 19.9 Change of stress across a discontinuity.

on the discontinuity. It is convenient to define an angle P as shown in Fig. 19.12, where

$$P = 90° - \delta\theta \tag{19.20}$$

From the geometry of Fig. 19.12, noting that $A'C' = t'_a$,

$$\sin P = \frac{A'D'}{t'_a} \qquad \sin \rho' = \frac{A'D'}{s'_a} \tag{19.21}$$

Hence, making use of Eq. (19.20),

$$\sin \rho' = \sin P \sin \phi'_c = \cos \delta\theta \sin \phi'_c \tag{19.22}$$

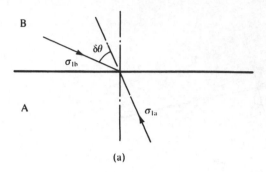

(a)

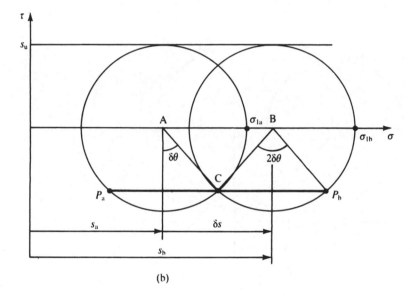

(b)

Figure 19.10 Change of stress across a discontinuity for undrained loading.

With the aid of the constructions in Fig. 19.12 and noting that $O'E' = O'F'$,

$$\sin(P + \rho') = \frac{O'E'}{s'_a} \quad \sin(P - \rho') = \frac{O'F'}{s'_b} \tag{19.23}$$

and hence, making use of Eq. (19.20),

$$\frac{s'_b}{s'_a} = \frac{\cos(\delta\theta - \rho')}{\cos(\delta\theta + \rho')} \tag{19.24}$$

where ρ' is given by Eq. (19.22).

From Eqs. (19.24) and (19.22) the change of effective stress across a discontinuity is simply related to the rotation $\delta\theta$ of the direction of the major principal stress.

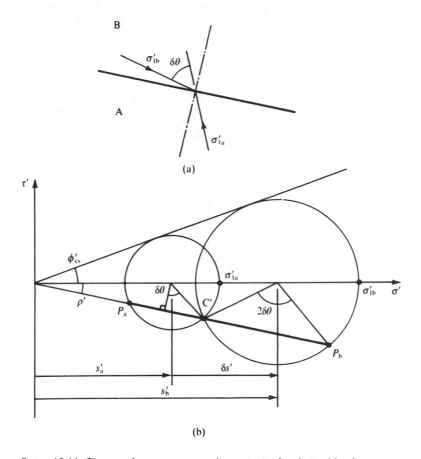

Figure 19.11 Change of stress across a discontinuity for drained loading.

19.8 Simple lower bounds for a foundation

We can now obtain a simple lower bound solution for the foundation shown in Fig. 19.6. Figure 19.13(a) shows a state of stress with two vertical stress discontinuities where the state of stress is symmetric about the centre-line. Shear stresses on horizontal and vertical planes are zero and hence, from Eq. (19.17), the vertical stresses in elements A and C in regions I and III are

$$\sigma_z = \gamma z \tag{19.25}$$

and the vertical stresses in elements B and D in regions II and IV are

$$\sigma_z = q_l + \gamma z \tag{19.26}$$

Figure 19.13(b) shows the Mohr circles of total stress for the elements A and B and Fig. 19.13(c) shows the circles for the elements C and D; the points a and b represent

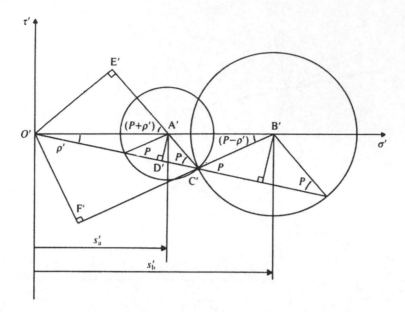

Figure 19.12 Geometrical analysis of Figure 19.11(b).

the stresses on the discontinuities marked α and β in Fig. 19.13(a). From the geometry of Figs 19.13(b) and (c),

$$q_l + \gamma z = 4s_u + \gamma z \tag{19.27}$$

and hence a lower bound for the collapse load is

$$V_l = 4s_u B \tag{19.28}$$

Alternatively, we could consider the rotations of the directions of the major principle stresses across the discontinuities, making use of Eq. (19.19). For each discontinuity $\delta\theta = 90°$ and $\delta s = 2s_u$; hence, from the geometry of Fig. 19.13(b) and (c) we obtain Eqs. (19.27) and (19.28). The mean of the upper and lower bound solutions gives $V_c = 5s_u$ and the bounds differ by about ± 20 per cent from this mean. Bearing in mind the problems in determining true values of s_u for natural soils, which may not be either isotropic or homogeneous, these simple bounds may be adequate for simple routine designs. However, in order to illustrate the use of slip fans and stress fans we will examine some alternative solutions.

19.9   Upper and lower bound solutions using fans

In Fig. 19.3(f) there is a combination of straight and curved slip surfaces and in order to have a compatible mechanism it is necessary to have slip fan surfaces as illustrated.

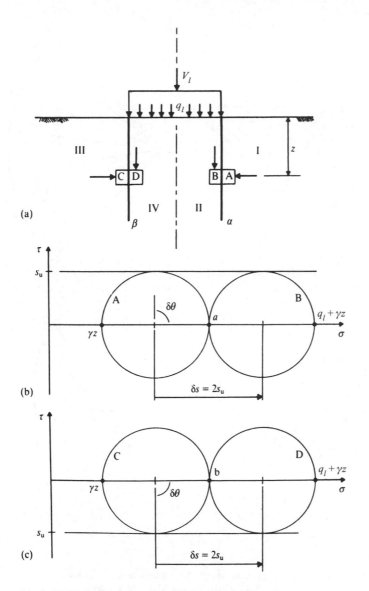

Figure 19.13 Equilibrium stress field for a foundation.

Figure 19.14 shows mechanisms and displacement diagrams for slip fans: Fig. 19.14(a) is for undrained loading and Fig. 19.14(b) is for drained loading. You should work your way through these using the description of the construction of displacement diagrams given in Sec. 2.4. From the geometry of Fig. 19.14(a),

$$r_b = r_a \quad \text{and} \quad \delta w_b = \delta w_a \tag{19.29}$$

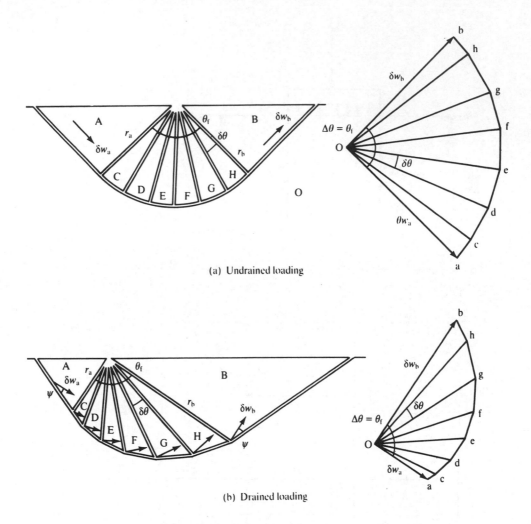

(a) Undrained loading

(b) Drained loading

Figure 19.14 Slips fans and corresponding displacement diagrams.

and so the radius of the fan and the increment of displacement remain constant through a slip fan for undrained loading. From the geometry of Fig. 19.14(b),

$$r_b = r_a \exp(\theta_f \tan \psi) \tag{19.30}$$

$$\delta w_b = \delta w_a \exp(\theta_f \tan \psi) \tag{19.31}$$

where θ_f is the fan angle; thus the outer arcs of the slip fan and the displacement diagram are both logarithmic spirals.

For a slip fan like that shown in Fig. 19.14(a), it is necessary to evaluate the work done on the circular slip surface and on all the radial slip surfaces. From Fig. 19.15, summing for the elements of the circular arc and for the radial slip surfaces,

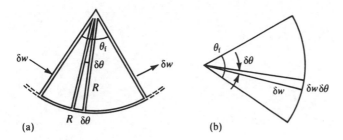

Figure 19.15 Work done in a slip fan.

the increment of work done by the internal stresses through the fan is

$$\delta W = \sum s_u R(\delta w \delta\theta) + \sum s_u (R\delta\theta)\delta w \qquad (19.32)$$

Hence, in the limit,

$$\delta W = 2 s_u R \delta w \int_0^{\theta_f} d\theta \qquad (19.33)$$

and

$$\delta W = 2 s_u R \delta w \theta_f = 2 s_u R \delta w \Delta\theta \qquad (19.34)$$

where θ_f is the fan angle which is equal to the change $\Delta\theta$ in the direction of the vector of displacement δw across slip fan.

We can also consider the change of stress from one region to another across a fan of discontinuities, as shown in Fig. 19.16. (The fan of stress discontinuities in Fig. 19.16 is not necessarily the same as the fan of slip surfaces in Fig. 19.14.) The fan angle

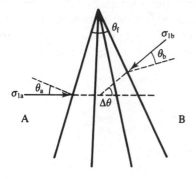

Figure 19.16 Rotation of the direction of the major principal stress across a stress fan.

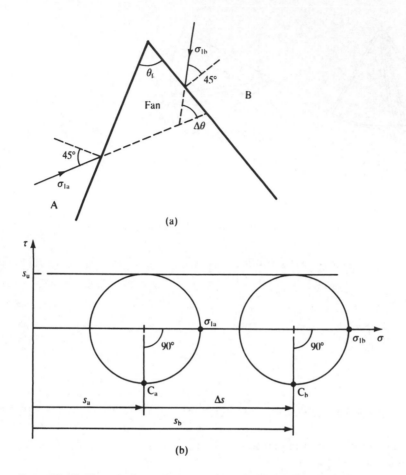

Figure 19.17 Change of stress across a stress fan for undrained loading.

θ_f is equal to the rotation $\Delta\theta$ of the direction of the major principal stress across the fan. Figure 19.17(a) shows a stress fan for undrained loading and Fig. 19.17(b) shows the Mohr circles of total stress for the outermost discontinuities; within the fan there are a great many radial discontinuities and there are equally a great many Mohr circles between those shown. Note that the outermost limits of the fan are defined by $\theta_a = \theta_b = 45°$. From Eq. (19.19), as $\delta\theta \to 0$,

$$\frac{ds}{d\theta} = 2s_u \tag{19.35}$$

and integrating through the fan from region A to region B,

$$\Delta s = 2s_u \Delta\theta = 2s_u\theta_f \tag{19.36}$$

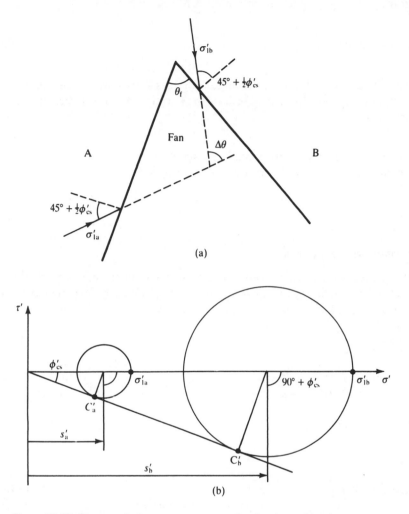

Figure 19.18 Change of stress across a stress fan for drained loading.

Figure 19.18 shows a stress fan and the corresponding Mohr circles for drained loading. As before there will be a great number of additional radial discontinuities and Mohr circles between the outermost ones. Note that the limits of the fan zone are defined by $\theta_a = \theta_b = 45° + \frac{1}{2}\phi_c'$. From Eq. (19.24), the change of stress across a discontinuity can be written as

$$\frac{ds'}{s'} = \frac{2 \sin \delta\theta \sin \rho'}{\cos(\delta\theta + \rho')} \tag{19.37}$$

As $\delta\theta \to 0$, from Eq. (19.22) we have $\rho' = \phi_c'$ and from Eq. (19.37),

$$\frac{ds'}{d\theta} = 2s' \tan \phi_c' \tag{19.38}$$

Hence, integrating through the fan from region A to region B,

$$\frac{s_b'}{s_a'} = \exp(2\tan\phi_c'\Delta\theta) = \exp(2\tan\phi_c'\theta_f) \qquad (19.39)$$

Equations (19.36) and (19.39) give the changes of stress across stress fans in terms of the soil strength s_u or ϕ_c' and the fan angle θ_f or the rotation $\Delta\theta$ of the direction of the major principal stress.

19.10 Bound solutions for the bearing capacity of a foundation for undrained loading using fans

The simple upper and lower bound solutions obtained earlier can now be modified by adding slip fans or stress fans.

(a) Upper bound with a slip fan

Figure 19.19(a) shows a mechanism consisting of two triangular wedges and a slip fan and Fig. 19.19(b) is the corresponding displacement diagram. As before the work done by the self-weight forces sums to zero and, from Eq. (19.8).

$$\delta E = V_u \delta w_f \qquad (19.40)$$

The radius of the fan is $R = B/\sqrt{2}$, the fan angle is $\theta_f = \frac{1}{2}\pi$ and $\delta w_a = \sqrt{2}\delta w_f$. Hence, from Eq. (19.34) the work done by the internal stresses in the slip fan is

$$\delta W = 2s_u R\delta w\theta_f = \pi s_u B\delta w_f \qquad (19.41)$$

and, from Table 19.2, for the whole mechanism

$$\delta W = (2+\pi)s_u B\delta w_f \qquad (19.42)$$

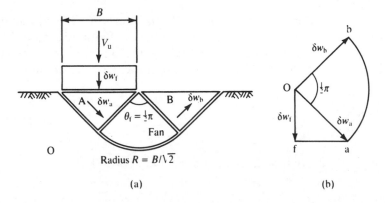

Figure 19.19 Mechanism of collapse for a foundation for undrained loading.

Table 19.2 Work done by internal stresses for mechanism in Figure 19.19

Slip plane	Shear stress	Length	Displacement	$\delta W = s_u L \delta w$
oa	s_u	$\frac{1}{\sqrt{2}} B$	$\sqrt{2} w_f$	$s_u B \delta w_f$
ob	s_u	$\frac{1}{\sqrt{2}} B$	$\sqrt{2} w_f$	$s_u B \delta w_f$
Fan	s_u	–	–	$\pi s_u B \delta w_f$
fa		B	δw_f	0
			Total	$(2 + \pi) s_u B \delta w_f$

Equating δE and δW, the upper bound for the collapse load is

$$V_u = (2 + \pi) B s_u \tag{19.43}$$

(b)   Lower bound with stress fans

Figure 19.20(a) shows a state of stress with two stress fans in regions II and IV. As before, the state of stress is symmetric about the centre-line and Eqs. (19.25) and (19.26) apply in regions I and III respectively. Figure 19.20(b) shows Mohr circles of total stress for elements at A and C and the points a and c represent the stresses on the outermost discontinuities in the fan in region II. From the geometry of Fig. 19.20, the fan angle is $\theta_f = 90° = \pi/2$ and from Eq. (19.36) the change of stress through the fan is

$$\Delta s = s_u \Delta \theta_f = \pi s_u \tag{19.44}$$

From the geometry of Fig. 19.20(b),

$$q_l + \gamma z = (2 + \pi) s_u + \gamma z \tag{19.45}$$

and hence a lower bound for the collapse load is

$$V_l = (2 + \pi) B s_u \tag{19.46}$$

Strictly, we should examine the state of stress in region VI where the stress fans overlap. It is intuitively fairly clear that the stresses in region VI will be less critical than those near the edges of the foundation and that the conditions in the overlapping stress fans will tend to cancel each other out.

Notice that the upper and lower bounds given by Eqs. (19.43) and (19.46) are equal and so they must be an exact solution. We have been very fortunate to obtain an exact solution with such simple upper and lower bound solutions; normally you would only be able to obtain unequal bounds.

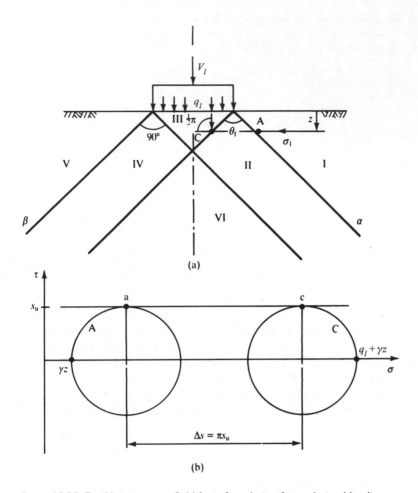

Figure 19.20 Equilibrium stress field for a foundation for undrained loading.

19.11   Bound solutions for the bearing capacity of a foundation for drained loading using fans

Analyses for the bearing capacity of a simple foundation for drained loading are similar to those for undrained loading and, as before, the best solutions are those with a slip fan or a stress fan.

(a)   Upper bound with a slip fan

Figure 19.21(a) shows a mechanism consisting of two triangular wedges and a slip fan and Fig. 19.21(b) is the corresponding displacement diagram: these are essentially the same as those in Fig. 19.14 and, as before, the foundation is assumed to be weightless. Because the mechanism is not symmetric the work done by the self-weight forces do not

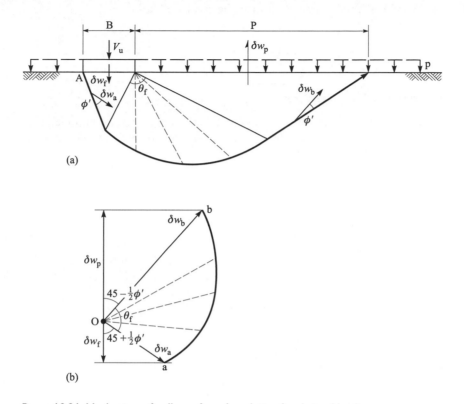

Figure 19.21 Mechanism of collapse for a foundation for drained loading.

sum to zero. With self-weight forces the mathematics becomes quite complicated and, for this book, I will consider the case of a weightless soil. The analyses for the case where the soil has self weight is given by Atkinson (1981).

For drained loading the work done by the internal stresses is zero so the work done by the upper bound foundation load V_u as it moves δw_f is equal to the work done by the surfaces stresses p acting over the length P as they move upwards δw_p: since the stresses p act in the opposite direction to δw_p the work done by the stresses p is negative. Hence:

$$V_u \delta w_f - pP \delta w_p = 0 \tag{19.47}$$

From the geometry of Fig. 19.21, noting that the fan angle $\theta_f = 90°$ and making use of Eq. (19.31)

$$\frac{P}{B} = \frac{\delta w_p}{\delta w_f} = \tan\left(\frac{\pi}{4} + \frac{\phi'}{2}\right) \exp\left(\frac{\pi}{2} \tan\phi'\right) \tag{19.48}$$

And, from Eq. (19.47) the upper bound for the collapse load is

$$V_u = pB \tan^2 \left(\frac{\pi}{4} + \frac{\phi'}{2} \right) \exp \left(\pi \tan \phi' \right) \qquad (19.49)$$

(b) Lower bound with a stress fan

Figure 19.22(a) shows an equilibrium stress field for a weightless soil and Fig. 19.22(b) shows the corresponding Mohr circles. These are similar to those in Fig. 19.19.

From the geometry of the Mohr circles

$$s'_a = p' \left(\frac{1}{1 - \sin \phi'} \right) \qquad (19.50)$$

$$s'_b = q'_l \left(\frac{1}{1 + \sin \phi'} \right) \qquad (19.51)$$

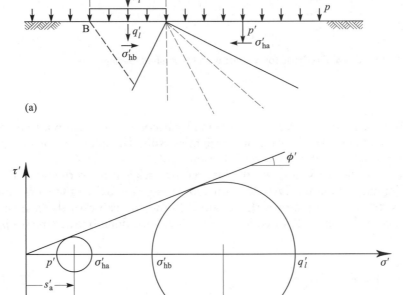

(a)

(b)

Figure 19.22 Equilibrium stress field for a foundation for drained loading.

from Eq. (19.39) noting that the fan angle $\theta_f = \pi/2$

$$\frac{s'_b}{s'_a} = \exp\left(\pi \tan \phi'\right) \tag{19.52}$$

and the lower bound collapse load is:

$$V_l = pB \tan^2\left(\frac{\pi}{4} + \frac{\phi'}{2}\right) \exp\left(\pi \tan \phi'\right) \tag{19.53}$$

The upper bound given by Eq. (19.49) is the same as the lower bound given by Eq. (19.53) so we have an exact solution. This solution is however for the artificial case of a foundation on a weightless soil. The bearing capacity arises from the stresses p on the surface outside the foundation. Bearing capacity of foundations for both the drained and the undrained cases will be considered further in Chapter 22.

19.12   Summary

1. Estimates of the collapse of structures can be found from relatively simple upper and lower bound calculations. An upper bound solution gives an unsafe load and if this load is applied the structure must collapse; a lower bound gives a safe load and with this load the structure cannot collapse.
2. To calculate an upper bound you have to choose a compatible mechanism of collapse and equate the work done by the external loads with the work done by the internal stresses. Mechanisms consist of slip surfaces that have circular arcs, logarithmic spirals or straight lines and may be arranged as fan zones.
3. To calculate a lower bound you need to find a distribution of stress that is in equilibrium with the external loads and does not exceed the appropriate failure criterion. An equilibrium state of stress may have strong discontinuities or stress fans.

The cases discussed in this chapter have been relatively simple and were intended simply to illustrate the basic principles of the upper and lower bound calculations. Other, more complicated, cases are given by Atkinson (1981).

Worked examples

Example 19.1: Loads on trench struts for undrained soil The   trench   shown   in Fig. 19.23 is supported by smooth sheet piles held apart by struts, 1 m apart out of the page, placed so that the piles do not rotate.

(a) Upper bound. Figure 19.24(a) shows a collapse mechanism and Fig. 19.24(b) is the corresponding displacement diagram. The forces acting on the moving block

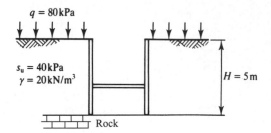

Figure 19.23 Trench supported by propped sheet piles.

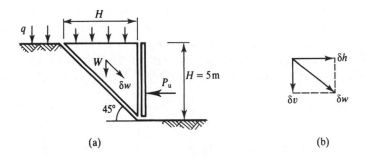

(a)   (b)

Figure 19.24 Mechanism of collapse for trench in Figure 19.23.

(for a slice 1 m thick out of the page) are

$$Q = qH = 80 \times 5 = 400\,\text{kN}$$

$$W = \tfrac{1}{2}\gamma H^2 = \tfrac{1}{2} \times 20 \times 5^2 = 250\,\text{kN}$$

From the displacement diagram, for $\delta v = 1$,

$$\delta v = \delta h = 1 \quad \text{and} \quad \delta w = \sqrt{2}$$

Hence, from Eq. (19.8), the work done by the external forces is

$$\delta E = Q\,\delta v + W\delta v - P_\text{u}\delta h = 400 + 250 - P_\text{u}$$

From Eq. (19.13), the work dissipated in the slip plane with length $5\sqrt{2}$ m is

$$\delta W = s_\text{u}L\delta w = 40 \times 5\sqrt{2} \times \sqrt{2} = 400$$

Hence, equation $\delta E = \delta W$,

$$P_\text{u} = 400 + 250 - 400 = 250\,\text{kN}$$

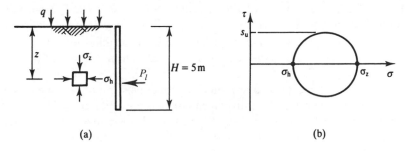

(a)   (b)

Figure 19.25 Equilibrium stress field for the trench in Figure 19.23.

(b)   Lower bound. Figure 19.25(a) shows a typical element in an equilibrium stress field and Fig. 19.25(b) is the corresponding Mohr circle of total stress. From these,

$$\sigma_z = q + \gamma z = 80 + 20z$$

$$\sigma_h = \sigma_z - 2s_u = (80 + 20z) - (2 \times 40) = 20z$$

Hence, integrating over the height of the trench,

$$P_l = \int_0^H 20z\,dz = \tfrac{1}{2} \times 20 \times 5^2 = 250 \text{ kN}$$

Example 19.2: Drained bearing capacity of a foundation   Figure 19.26(a) shows a long foundation, 3 m wide and carrying a load V per metre out of the page, buried 1 m below

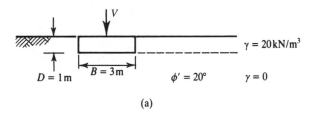

(a)

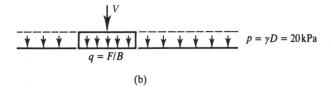

(b)

Figure 19.26 Loads on a foundation.

the ground surface in dry soil, which has a friction angle $\phi' = 20°$. The bearing pressure (i.e. the total stress on the underside of the foundation) is $q = V/B$. For simplicity the soil is assumed to be weightless ($\gamma = 0$) except above foundation level, where $\gamma = 20$ kN/m^3, so that the 1 m deep layer applies a uniform surcharge $\gamma D = 20$ kPa at foundation level. The idealized loads and stresses are shown in Fig. 19.26(b).

(a) Upper bound. Figure 19.27(a) shows a mechanism consisting of two wedges and Fig. 19.27(b) is the corresponding displacement diagram. For $\phi' = 20°$ suitable angles for the slip surfaces are $45° \pm 10°$ and all displacements are at angles $\psi = \phi' = 20°$ to the slip surfaces. (Notice that if $\phi' > 30°$ the directions ob and ab diverge and the mechanism is not compatible.)

From the geometry of Fig. 19.27(a),

$$L = B\tan^2(45° + \tfrac{1}{2}\phi') = 3\tan^2 55° = 6.1 \text{ m}$$

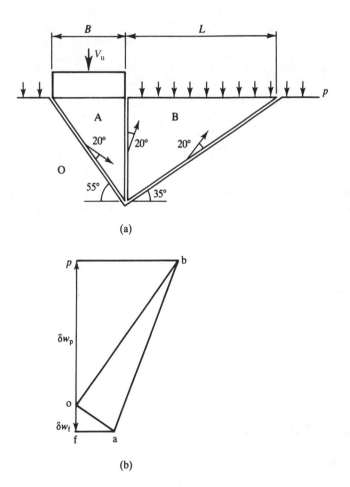

(a)

(b)

Figure 19.27 Mechanism of collapse for the foundation in Figure 19.26.

and so the force applied by the stress p is

$$P = pL = 20 \times 6.1 = 122 \text{ kN}$$

From the geometry of Fig. 19.27(b), taking $\delta w_f = 1$,

$$\delta w_p = \tan(45° + \tfrac{1}{2}\phi')\tan(45° + \tfrac{3}{2}\phi') = \tan 55° \tan 75° = 5.3$$

For drained loading, from Eq. (19.12), $\delta W = 0$. The work done by the external loads is given by Eq. (19.9) with $\gamma = \gamma_w = 0$ for weightless and dry soil:

$$\delta E = V_u \delta w_f - P \delta w_p = (V_u \times 1) - (122 \times 5.3)$$

Hence, equating $\delta E = \delta W$,

$$V_u = 122 \times 5.3 = 647 \text{ kN}$$

(b) Lower bound. Figure 19.28(a) shows a simple equilibrium stress field, symmetric about the centre-line, with two discontinuities. Figure 19.28(b) shows two Mohr circles of effective stress for the two regions of uniform stress below and to the side of the foundation. From the geometry of Fig. 19.28(b),

$$\frac{q_l}{\sigma_h'} = \frac{\sigma_h'}{p'} = \tan^2(45° + \tfrac{1}{2}\phi')$$

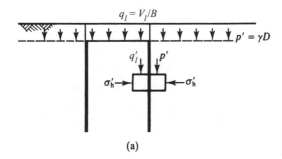

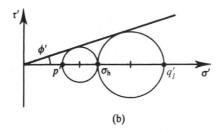

Figure 19.28 Equilibrium stress field for the foundation in Figure 19.26.

Hence,

$$V_l = Bp' \tan^4(45° + \tfrac{1}{2}\phi') = 3 \times 20 \times \tan^4 55° = 250 \text{ kN}$$

Example 19.3: Drained bearing capacity of a foundation   Better bound solutions for the foundation in Fig. 19.26 can be found using a mechanism which includes a slip fan and a stress field which includes a stress fan.

(a)   Upper bound. Figure 19.29(a) shows a mechanism consisting of two wedges and a logarithmic spiral slip fan and Fig. 19.27(b) is the corresponding displacement diagram. For $\phi' = 20°$ the angles in the mechanism and in the displacement diagram are $45 \pm \tfrac{1}{2}\phi' = 55°$ or $35°$.
From the geometry of Fig. 19.29(a), and making use of Eq. (19.6) with $\psi = \phi'$,

$$L = B \tan(45° + \tfrac{1}{2}\phi') \exp(\Delta\theta \tan \phi')$$

$$L = 3 \tan 55° \exp(\pi/2 \tan 20°) = 7.6 \text{ m}$$

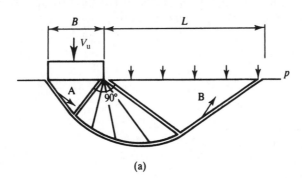

(a)

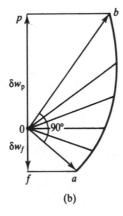

(b)

Figure 19.29 Mechanism of collapse for the foundation in Figure 19.26.

and

$$P = pL = 20 \times 7.6 = 152 \text{ kN}$$

From the geometry of Fig. 19.29(b), taking $\delta w_f = 1$,

$$\delta w_p = \tan(45° + \tfrac{1}{2}\phi') \exp(\Delta\theta \tan\phi')$$

$$\delta w_p = \tan 55° \exp(\pi/2 \tan 20°) = 2.53$$

For drained loading, from Eq. (19.12), $\delta W = 0$. The work done by the external loads is given by Eq. (19.9) with $\gamma = \gamma_w = 0$ for weightless and dry soil:

$$\delta E = V_u \delta w_f - P\delta W_p$$

Equating $\delta E = \delta W$,

$$V_u = 152 \times 2.53 = 385 \text{ kN}$$

(b)   Lower bound. Figure 19.30(a) shows an equilibrium stress field consisting of a region B where σ_1' $(= q_l)$ is vertical and a region A where σ_3' $(= p' = 20 \text{ kPa})$ is vertical. These are separated by a fan zone with a fan angle of 90°. There could be a similar stress field at the left-hand edge of the foundation. Figure 19.30(b)

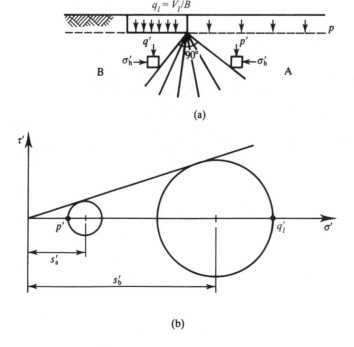

Figure 19.30 Equilibrium stress field for the foundation in Figure 19.26.

shows the two Mohr circles of effective stress for the two regions of uniform stress.

From Eq. (18.39),

$$s_b' = s_a' \exp(2\theta_f \tan \phi') = s_a' \exp(\pi \tan 20°) = 3.14 s_a'$$

From the geometry of the Mohr circles,

$$s_a' = p'\left(\frac{1}{1 - \sin \phi'}\right) = 20\left(\frac{1}{1 - \sin 20°}\right) = \frac{20}{0.66}$$

$$s_b' = q_l'\left(\frac{1}{1 + \sin \phi'}\right) = q_l'\left(\frac{1}{1 + \sin 20°}\right) = q_l'\frac{1}{1.34}$$

Hence,

$$q_l' = 1.34 \times 3.14 s_a' = 1.34 \times 3.14 \times \frac{20}{0.66} = 128 \text{ kPa}$$

and

$$V_l = q_l' A = 128 \times 3 = 384 \text{ kN}$$

Notice that these last upper and lower bound solutions are the same; this is because the mechanism in Fig. 19.29 corresponds to the stress field in Fig. 19.30.

Further reading

Atkinson, J. H. (1981) *Foundations and Slopes*, McGraw-Hill, London.
Calladine, C. R. (1969) *Engineering Plasticity*, Pergamon Press, London.
Chen, W. F. (1975) *Limit Analysis and Soil Plasticity*, Elsevier, New York.
Sokolovskii, V. V. (1965) *Statics of Granular Media*, Pergamon Press, Oxford.

Chapter 20

Limit equilibrium method

20.1 Theory of the limit equilibrium method

The limit equilibrium method is by far the most commonly used analysis for the stability of geotechnical structures. The steps in calculating a limit equilibrium solution are as follows:

1. Draw an arbitrary collapse mechanism of slip surfaces; this may consist of any combination of straight lines or curves arranged to give a mechanism.
2. Calculate the statical equilibrium of the components of the mechanism by resolving forces or moments and hence calculate the strength mobilized in the soil or the external forces (whichever is unknown).
3. Examine the statical equilibrium of other mechanisms and so find the critical mechanism for which the loading is the limit equilibrium load.

Remember that, as always, we must distinguish between cases of undrained and drained loading. For undrained loading the ultimate strength of the soil is given by

$$\tau = s_u \tag{20.1}$$

where s_u is the undrained shear strength. For drained loading where pore pressures can be determined from hydrostatic groundwater conditions or from a steady state seepage flownet, the strength is given by

$$\tau' = \sigma' \tan \phi' = (\sigma - u) \tan \phi' \tag{20.2}$$

where ϕ' is a friction angle.

As discussed in Sec. 9.2 soil has a peak strength, a critical state strength and a residual strength which are mobilized at different strains or displacements. The factors which determine which strength should be used in stability calculations are discussed in Chapter 18.

As in the previous chapter on the upper and lower bound methods, the analyses in this chapter calculate ultimate limit states where the slope, wall or foundation can be said to have collapsed. For design of safe and serviceable structures factors have to be applied. The total or partial factors of safety or load factors applied in limit equilibrium analyses are the same as those applied in bound analyses. They are discussed in Chapter 18 and will be discussed in later chapters dealing with different structures.

The limit equilibrium method combines features of the upper and lower bound methods. The geometry of the slip surfaces must form a mechanism that will allow collapse to occur, but they may be any shape so they need not meet all the requirements of compatibility (see Sec. 19.3). The overall conditions of equilibrium of forces on blocks within the mechanism must be satisfied, but the local states of stress within the blocks are not investigated. Although there is no formal proof that the limit equilibrium method leads to correct solutions, experience has shown that the method usually gives solutions that agree quite well with observations of the collapse of real structures and the method is firmly established among the techniques of geotechnical engineering.

20.2   Simple limit equilibrium solutions

Two simple problems, one for drained loading and one for undrained loading, are shown in Figs. 20.1 and 20.2. These illustrate the general principles of the limit equilibrium method. Figure 20.1(a) shows part of a very long slope in soil where the pore pressures are zero. The problem is to determine the critical slope angle i_c when the

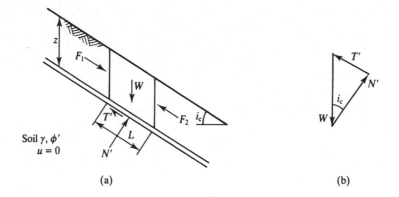

(a)   (b)

Figure 20.1 Limit equilibrium solution for stability of an infinite slope for drained loading.

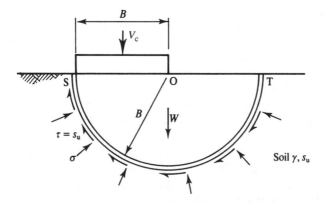

Figure 20.2 Limit equilibrium solution for the bearing capacity of a foundation for undrained loading.

slope fails. A mechanism could be a straight slip surface at a depth z as shown, and the forces on the block with length L down the surface are marked on the diagram. If the slope is very long, F_1 and F_2 are equal and opposite. The normal and shear forces on the slip surface are $T' = \tau' L$ and $N' = \sigma' L$ and the weight is $W = \gamma L z \cos i_c$. Figure 20.1(b) is a polygon of these forces which closes (i.e. the forces are in equilibrium) when

$$\frac{T'}{N'} = \frac{\tau'}{\sigma'} = \tan i_c \tag{20.3}$$

Hence, from Eq. (20.2), the limit equilibrium solution is

$$i_c = \phi' \tag{20.4}$$

Strictly we should consider other possible mechanisms with combinations of curved and straight slip surfaces, but it is fairly obvious that the mechanism illustrated in Fig. 20.1 is one of the most critical. The solution $i_c = \phi'$ can also be obtained as an upper bound and as a lower bound so it is an exact solution.

Figure 20.2(a) shows a section of a foundation with width B and unit length out of the page so that the width B is equal to the foundation area A. The foundation is loaded undrained and the undrained strength of the soil is s_u. The problem is to determine the collapse load V_c or the ultimate bearing capacity $q_c = V_c/A$. A mechanism could be a circular slip surface with centre O at the edge of the foundation. The rotating block of soil is in equilibrium when the moments about O balance and

$$V_c \times \tfrac{1}{2} B = s_u B \widehat{ST} \tag{20.5}$$

where $\widehat{ST} = \pi B$ is the length of the arc ST. Notice that the lines of action of the weight W of the soil block and the normal stresses on the circular slip surfaces act through O and so their moments about O are zero. From Eq. (20.5) we have

$$V_c = 2\pi B s_u \tag{20.6}$$

As before we should now consider other possible mechanisms with combinations of straight and curved slip surfaces to seek the minimum value of F_c, which will be the limit equilibrium solution. Figure 20.3 shows a circular slip surface with its centre at a height h above the ground surface. Readers should show that the minimum value for this mechanism is $V_c = 5.5 B s_u$ when $h/B = 0.58$; one way to do this is to take trial values of h and plot V_c against h to determine the minimum value of V_c.

Remember that in Chapter 19 we obtained equal upper and lower bound solutions (i.e. an exact solution) for a foundation on undrained soil as $V_c = (2 + \pi)B s_u$ (see Eq. 19.46) and so, in this case, the best limit equilibrium solution with a circular arc slip surface overestimates the true solution by less than 10%.

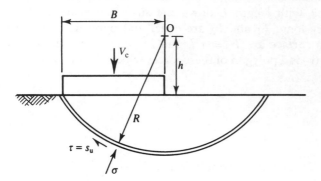

Figure 20.3 Limit equilibrium solution for the bearing capacity of a foundation for undrained loading.

20.3   Coulomb wedge analyses

Calculation of the loads required to maintain the stability of a retaining wall provides a convenient example to illustrate both the basic features of the limit equilibrium method and a number of special features of the method. Solutions are particularly simple as a mechanism can be constructed from a single straight slip surface. This calculation was first developed by Coulomb in about 1770 and is one of the earliest engineering calculations still in current use, although with a number of modifications.

Figure 20.4(a) shows a section of a smooth wall with unit length out of the page supporting soil that is undrained. The horizontal force on the wall necessary to prevent the

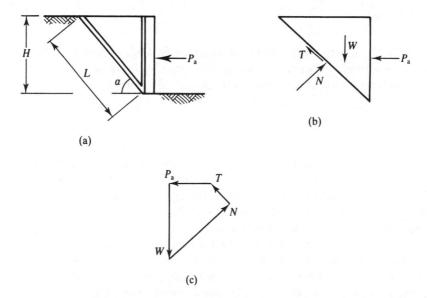

Figure 20.4 Coulomb wedge analysis for a smooth wall for undrained loading.

soil collapsing into the excavation is P_a and this is called the active force (see Sec. 24.1). (In practice, vertical cracks may form in the ground near the top of the wall; I will consider the influence of tension cracks later, but for the present I will assume that they do not occur.) A mechanism can be constructed from a single straight slip surface at an angle α and there must be slip surfaces between the soil and the wall as shown.

The forces acting on the triangular wedge are shown in Fig. 20.4(b). There is no shear force between the soil and the smooth wall. The directions of all the forces are known and the magnitudes of P_a and N are unknown; the magnitude of the shear force T is given by

$$T = s_u L \tag{20.7}$$

where s_u is the undrained strength and L is the length of the slip surface; T acts up the surface as the wedge moves down into the excavation. With two unknowns the problem is statically determinate and a solution can be found by resolution of the forces; notice that if you resolve in the direction of the slip surface N does not appear and P_a can be found directly. Alternatively, the solution can be found graphically by constructing the closed polygon of forces in Fig. 20.4(c).

To obtain the limit equilibrium solution you must vary the angle α to find the maximum, or critical, value for P_a. If you do this you will find that the critical angle is $\alpha = 45°$ and the limit equilibrium solution is

$$P_a = \tfrac{1}{2}\gamma H^2 - 2s_u H \tag{20.8}$$

Notice that if we put $P_a = 0$ we obtain

$$H_c = \frac{4s_u}{\gamma} \tag{20.9}$$

which is a limit equilibrium solution for the undrained stability of an unsupported trench.

This analysis can be extended quite simply to include the effects of foundation loads, water in the excavation and shear stresses between the soil and a rough wall. The additional forces are shown in Fig. 20.5(a) and the corresponding polygon of forces is shown

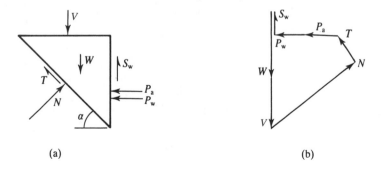

(a)   (b)

Figure 20.5 Coulomb wedge analysis for a rough wall for undrained loading.

in Fig. 20.5(b). The shear force on the wall S_w is given by

$$S_w = s_w H \qquad (20.10)$$

where s_w is the shear stress between the soil and the wall; obviously s_w must be in the range $0 \leq s_w \leq s_u$ depending on the roughness of the wall. Free water in the excavation applies a total force P_w to the wall, given by

$$P_w = \tfrac{1}{2} \gamma_w H_w^2 \qquad (20.11)$$

where H_w is the depth of water in the excavation. For the undrained case the pore pressures in the soil do not come into the calculation and will not be in equilibrium with the water pressures in the excavation. Again the only unknowns are the magnitudes of the forces N and P_a, so the problem is statically determinate. The limit equilibrium solution is the maximum value of P_a and coincides with the critical slip surface.

The case shown in Fig. 20.6 is similar to that in Fig. 20.4 except that the soil is drained and dry so pore pressures are zero. The forces on the triangular wedge are shown in Fig. 20.6(a). There are now three unknown forces, T', N' and P'_a, but the forces T' and N' are related by Eq. (20.2) so the resultant of T' and N', shown by the broken line, is at an angle ϕ to the direction of N'. (The primes are added to these forces because they are associated with the effective stresses in the dry soil.) This now provides sufficient information to construct the force polygon shown in Fig. 20.6(b) to calculate the magnitude of P_a. To obtain the limit equilibrium solution you must vary the angle α to find the critical value for P_a. This occurs when $\alpha = 45° + \tfrac{1}{2}\phi'$, and the limit equilibrium solution is

$$P_a = \tfrac{1}{2} \gamma H^2 \tan^2 \left(45 - \tfrac{1}{2}\phi'\right) \qquad (20.12)$$

This solution was developed by Rankine in about 1850 (but in a different way) and is really a case of the Coulomb wedge analysis.

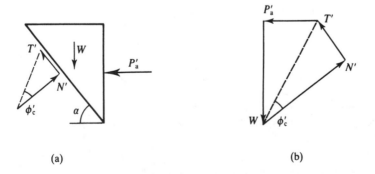

(a)   (b)

Figure 20.6 Coulomb wedge analysis for a smooth wall for drained loading.

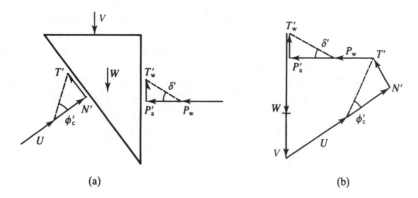

(a)   (b)

Figure 20.7 Coulomb wedge analysis for a rough wall for drained loading.

Again the analysis can be extended to include external loads, water in the excavation, pore pressures and shear stresses between the soil and a rough wall. The additional forces are shown in Fig. 20.7(a) and the corresponding polygon of forces is shown in Fig. 20.7(b). For simplicity the water table is assumed to be the same in the soil and in the excavation, so there is no seepage; later I will examine the case where the excavation is dewatered and there is a steady state seepage flownet in the soil. The force U is the sum (or integral) of the pore pressures over the slip surface and is found by summing Eq. (6.4) over the length L of the slip surface. The shear force T' is given by

$$T' = N' \tan \phi' = (N - U) \tan \phi' \tag{20.13}$$

Similarly, the shear force between the soil and the wall is given by

$$T'_w = p'_a \tan \delta' \tag{20.14}$$

where δ' is the friction angle between the soil and the wall; obviously δ' must be in the range $0 \leq \delta' \leq \phi'$ depending on the roughness of the wall. Notice that the total normal force on the vertical face of the soil is $P'_a + P_w$ (i.e. the sum of the force from the support prop and the force from the free water).

In Figs. 20.4 and 20.6 the major principal planes are horizontal because the shear stress on the wall is zero and $\sigma_z > \sigma_h$. In Sec. 2.6 we found that zero extension lines (i.e. lines of zero strain) were at angles $\alpha = 45° + \frac{1}{2}\psi$ to the major principal plane and planes where the stress ratio was $\tau'/\sigma' = \tan \rho'$ were at angles $\alpha = 45° + \frac{1}{2}\rho'$ to the major principal plane. For undrained loading $\psi = 0$ and for drained loading, at the critical state $\rho' = \phi'$. Hence the critical surfaces in these limit equilibrium solutions coincide with the critical planes and zero extension lines obtained from the Mohr circle constructions discussed in Chapter 2. In Figs. 20.5 and 20.7 there are shear stresses between the wall and the soil, so horizontal and vertical planes are not principal planes and the critical surfaces are not necessarily at angles $\alpha = 45°$ or $45° + \frac{1}{2}\phi'$ to the horizontal.

20.4   Simple slip circle analyses for undrained loading

A mechanism in which the slip surface is a circular arc – or a slip circle – as shown in Fig. 20.2, is very commonly used in routine limit equilibrium analyses in geotechnical engineering. The methods of solution are different for drained and for undrained loading and we will consider each separately.

Figure 20.8 shows a section of a slope with a foundation at the top and water in a river or lake at the toe. There is a mechanism consisting of a single circular arc with centre at O. The forces on the mechanism are due to the foundation load V, the weight of the soil W, the free water P_w and the shear stresses in the soil $T = s_u\widehat{AB}$ where $\widehat{AB}$ is the length of the arc AB; these forces have lever arms x and R as shown. Taking moments about O, the foundation and slope are just stable when

$$Wx_w + Vx_f - P_wx_u = s_u\widehat{AB}R \tag{20.15}$$

The limit equilibrium solution must be found by searching for the critical slip circle by varying the radius and the position of the centre. Notice in Fig. 20.8 that the normal stresses on the slip circle are radial and pass through the origin, so they have no moment about O. Calculation of values for Wx_w and $s_u\widehat{AB}R$ can be simplified by dividing the mechanism into a number of vertical slices and tabulating the results as in Fig. 20.18 in Example 20.3.

20.5   Slip circle method for drained loading – the method of slices

Figure 20.9(a) shows a slope with part of a steady state seepage flownet to a drain at the toe of the slope. The broken line in Fig. 20.9(a) is the slip circle shown in Fig. 20.9(b) and there is a standpipe with its tip on the slip circle and on an equipotential. The height

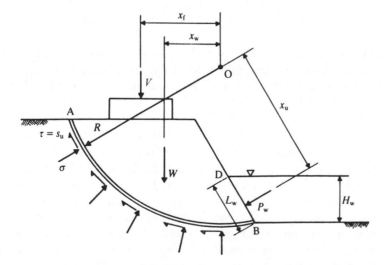

Figure 20.8 Slip circle method for undrained loading.

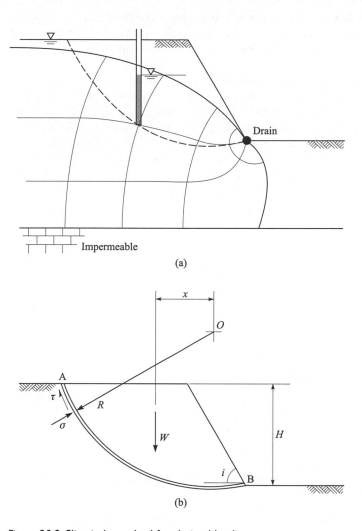

Figure 20.9 Slip circle method for drained loading.

of water in the standpipe corresponds to the level at which the equipotential meets the top flowline where the pore pressure is zero. From the heights of water in similar standpipes the pore pressures anywhere round the slip circle can be found as described in Chapter 14.

Taking moments about the centre of the slip circle O in Fig. 20.9(b) the slope is just stable when

$$Wx = R \int_{AB} \tau' dl \tag{20.16}$$

where the shear stresses are given by

$$\tau' = (\sigma - u) \tan \phi' \tag{20.17}$$

Although we can calculate values for the pore pressure u around the slip circles we cannot, at present, calculate the normal stresses σ. Thus the simple analysis which served for undrained loading for which $\tau = s_u$ cannot be used for drained loading.

The approach adopted for the method of slices is to subdivide the mechanism into a number of approximately equal vertical slices and examine the statical equilibrium of the slices and, by summation, of the whole mechanism. Figure 20.10(a) shows the mechanism of Fig. 20.9 divided into four slices, of which a typical slice FGHJ is shown in Fig. 20.10(b). The total forces on the slice shown in Fig. 20.10(b) are its weight W, and total normal and shear forces N and T on the base FJ, and forces F_1 and F_2 from adjacent slices. The interslice forces F_1 and F_2 are not necessarily equal and opposite, and their resultant F acts at a height a above the centre of the base of the slice and at an angle θ to the horizontal. The total normal and shear forces on the base of the slice are related by

$$T = (N - U)\tan\phi' \tag{20.18}$$

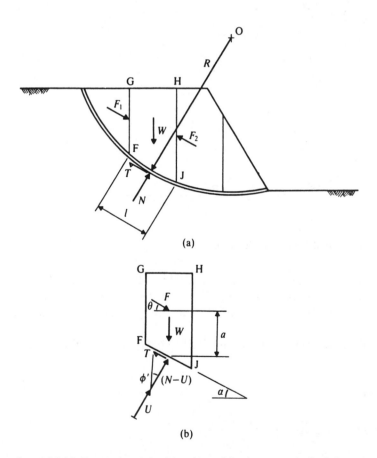

Figure 20.10 Slip circle method for drained loading – method of slices.

where the forces $T = \tau l$, $N = \sigma l$ and $U = u l$, where l is the length of the base FJ. Summing for all the slices gives

$$\sum T = \sum (N - U) \tan \phi' \qquad (20.19)$$

The interslice forces such as F may be decomposed into horizontal and vertical components E and X. In the slip circle method the boundaries between adjacent slices are not slip surfaces and so nothing can be said at present about the magnitude, direction or point of application of the force F in Fig. 20.10. Considering the forces on the block FGHJ in Fig. 20.10(b), the magnitudes, direction and points of application are known for W and U, the directions and points of application are known for T and N, but nothing is known about the force F. Thus there are five unknowns: T, N, F, a and θ. We can obtain three equations by resolution of forces and by taking moments following the usual rules of statics. These, together with Eq. (20.18), lead to a possible total of four equations and each slice is statically indeterminate. To obtain a solution for the method of slices for drained loading we are obliged to make at least one simplifying assumption in order to make the problem statically determinate. There are a number of such solutions, each based on a different simplifying assumption. For the present I will consider the two commonest of these solutions.

(a)   The Swedish method of slices (Fellenius, 1927)

Here it is assumed that the resultant F of the interslice forces is zero for each slice and thus F, a and θ vanish. Each slice is then statically determinate, and from Fig. 20.11

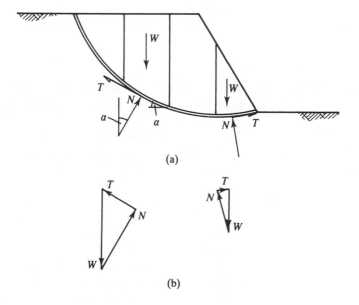

(a)

(b)

Figure 20.11 Slip circle method for drained loading – Swedish method.

we have

$$T = W \sin \alpha \quad N = W \cos \alpha \tag{20.20}$$

where α is the average inclination of the slip surface at the base of the slice. Hence we may calculate T and N for each slice and, for equilibrium, making use of Eq. (20.19),

$$\sum W \sin \alpha = \sum (W \cos \alpha - ul) \tan \phi' \tag{20.21}$$

where u is the average pore pressure over the length l of the base of each slice. Instead of making use of Eq. (20.20) we may calculate T and N for each slice from force polygons like those shown in Fig. 20.10(b). The calculations are assisted by the use of a table such as that shown in Fig. 20.19(c) in Example 20.4. As before, it is necessary to examine a number of different mechanisms to locate the critical slip circle; the slope is taken to be in a state of collapse if Eq. (20.21) is satisfied for any mechanism.

(b)   The Bishop routine method (Bishop, 1955)

Here it is assumed that the resultant of the interslice forces is horizontal. Hence $\theta = 0$ as shown in Fig. 20.12 and each slice is statically determinate. After resolving, taking moments and summing over the whole mechanism, the solution comes out in the form

$$\sum W \sin \alpha = \sum \frac{(W - ub)\sec \alpha \tan \phi'}{1 + \tan \alpha \tan \phi'} \tag{20.22}$$

where b is the width of each slice. In practice, evaluation of Eq. (20.22) is simplified if use is made of a table similar to that in Fig. 20.19(c). As before, it is necessary to examine a number of different mechanisms to locate the critical slip circle; the slope is then taken to be in a state of collapse if Eq. (20.22) is satisfied for any mechanism.

20.6   Other limit equilibrium methods

So far we have considered mechanisms consisting either of a single straight slip surface or a circular arc. The limit equilibrium method is not restricted to these geometries and there are two other commonly used arrangements of slip surfaces.

Figure 20.13 shows a mechanism consisting of several straight slip surfaces forming two triangular wedges and a block; this mechanism is appropriate where a layer of relatively weak soil occurs within the slope as shown. The shear and normal forces across each slip surface are marked. In this case, unlike the method of slices, the soil in the vertical slip surfaces is at failure and so the shear stresses can be determined from either Eq. (20.1) or (20.2) and the lengths of the slip surfaces. Working from the left-hand wedge towards the right, the forces on each block are statically determinate.

Figure 20.14 shows a mechanism in which there is a single continuous slip surface of general shape. The solution is found using the method of slices, as described above, for which at least one simplifying assumption is required. Thus the Swedish method (X and $E = 0$) or the Bishop routine method ($X = 0$) can be applied to general slip surfaces. Other solutions were developed by Janbu (1973) and by Morgenstern and Price (1965).

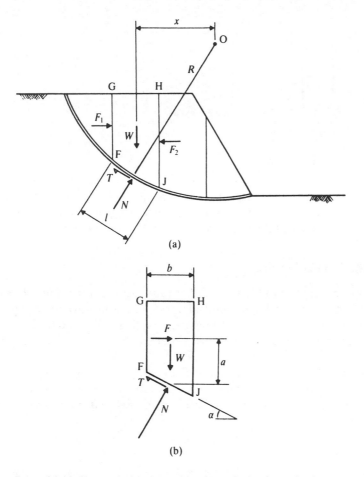

(a)

(b)

Figure 20.12 Slip circle for drained loading – Bishop's method.

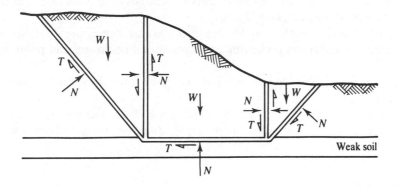

Figure 20.13 Wedge method.

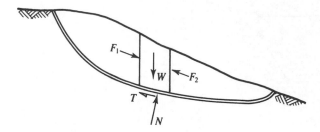

Figure 20.14 General slip surface method.

You can see that all these named methods (Swedish, Bishop, Janbu, Morgenstern and Price, and others) are basically limit equilibrium solutions using the method of slices with different assumptions to avoid the problem of statical indeterminacy.

20.7   Limit equilibrium solutions

Although the limit equilibrium method is approximate and requires a number of basic assumptions it has advantages over other methods. It is quite general and can be applied to walls, slopes or foundations, or to any combination of these. The method can be adapted for cases where the soil has layers with different properties or irregularly shaped boundaries.

The calculations for determining the forces on slices and for varying the geometry of the mechanism of slip surfaces are largely repetitive and there are a number of computer programs for the stability of geotechnical structures that make use of the limit equilibrium method.

20.8   Summary

1. The basic limit equilibrium method requires that blocks of soil inside a mechanism of slip surfaces are in statical equilibrium.
2. Mechanisms consist of slip surfaces which may be straight lines, arcs of circles (in the slip circle method) or any general shape.
3. Coulomb and Rankine analyses apply for mechanisms consisting of a single straight slip surface and the equilibrium calculations can be done using polygons of forces.
4. For undrained analyses with slip circles solutions can be found by taking moments about the centre of the circle.
5. For drained analyses with slip circles or with any general slip surface the problem is statically indeterminate and solutions are found using the method of slices with one of a number of alternative assumptions.

Worked examples

Example 20.1: Coulomb wedge analysis for undrained loading   The trench shown in Fig. 20.15 is supported by rough sheet piles held apart by struts, 1 m apart out of

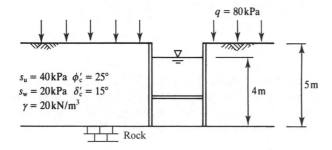

Figure 20.15 Trench supported by propped sheet piles.

the page, placed so that the piles do not rotate. The trench is part filled with water as shown.

For undrained loading a suitable Coulomb wedge is formed by a single slip plane at 45° to the horizontal and Fig. 20.16(a) shows the forces on the wedge. The magnitudes of the known forces are

$$V = qH = 80 \times 5 = 400 \text{ kN}$$

$$W = \tfrac{1}{2}\gamma H^2 = \tfrac{1}{2} \times 20 \times 5^2 = 250 \text{ kN}$$

$$T = \sqrt{2}Hs_u = \sqrt{2} \times 5 \times 40 = 283 \text{ kN}$$

$$S_w = s_u H = 20 \times 5 = 100 \text{ kN}$$

$$P_w = \tfrac{1}{2}\gamma_w H_w^2 = \tfrac{1}{2} \times 10 \times 4^2 = 80 \text{ kN}$$

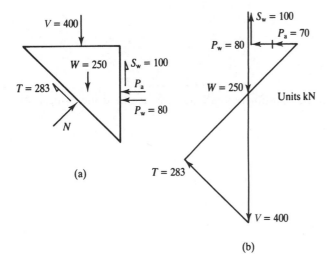

Figure 20.16 Forces on sheet piles in Figure 20.15 for undrained loading.

The force polygon is shown in Fig. 20.16(b). Scaling from the diagram, or by calculation,

$$P_a = 70 \text{ kN}$$

Example 20.2: Coulomb wedge analysis for drained loading   The angle of friction of the soil is $\phi' = 25°$ and for drained loading a suitable Coulomb wedge is formed by a slip plane at $\alpha = 45° + \frac{1}{2}\phi' = 57\frac{1}{2}°$ to the horizontal. Figure 20.17(a) shows the forces on the wedge. The magnitudes of the known forces are

$$V = qH \tan(90° - \alpha) = 80 \times 5 \times \tan 32.5° = 255 \text{ kN}$$

$$W = \frac{1}{2}\gamma H^2 \tan(90° - \alpha) = \frac{1}{2} \times 20 \times 5^2 \times \tan 32.5° = 159 \text{ kN}$$

$$P_w = \frac{1}{2}\gamma_w H_w^2 = \frac{1}{2} \times 10 \times 4^2 = 80 \text{ kN}$$

$$U = \frac{1}{2}\gamma_w H_w^2 \times \frac{1}{\sin \alpha} = \frac{1}{2} \times 10 \times 4^2 \times \frac{1}{\sin 57.5°} = 95 \text{ kN}$$

and the other information is

$$T' = (N - U) \tan 25°$$

$$T'_w = P' \tan 15°$$

The force polygon is shown in Fig. 20.17(b). Scaling from the diagram or by calculation,

$$P'_a = 245 \text{ kN}$$

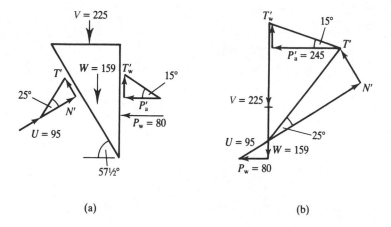

(a)   (b)

Figure 20.17 Forces on sheet piles in Figure 20.15 for drained loading.

Example 20.3: Undrained slope stability   Figure 20.18(a) shows a slope and a slip circle divided into slices. For the case where the soil is undrained, replacing s_u with s_u/F_s and making use of Eq. (20.15),

$$F_s = \frac{\sum s_u R l}{\sum W x}$$

The table in Fig. 20.18(b) gives the calculations for each slice and, summing over the whole mechanism,

$$F_s = \frac{9648}{3238} = 2.98$$

You should now repeat the calculations with different values of the radius R and different positions for the centre O to find the lowest value of F_s.

Example 20.4: Drained slope stability   Figure 20.19(a) shows a slope and a slip circle divided into slices and Fig. 20.19(b) shows part of a flownet sketched for steady state seepage towards a drain at the toe of the slope. The pore pressure at any point on

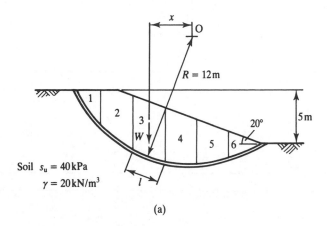

(a)

Slice	Area (m^2)	W (kN)	x (m)	Wx (kNm)	l (m)	$s_u R l$ (kNm)
1	3.8	76	9.2	699	4.0	1920
2	14.4	288	6.8	1958	3.7	1776
3	15.3	306	3.8	1163	3.2	1536
4	13.8	276	0.8	221	3.0	1440
5	10.2	204	−2.2	−449	3.0	1440
6	3.8	75	−4.7	−353	3.2	1536
Totals				3239		9648

(b)

Figure 20.18 Slope stability analysis – undrained loading.

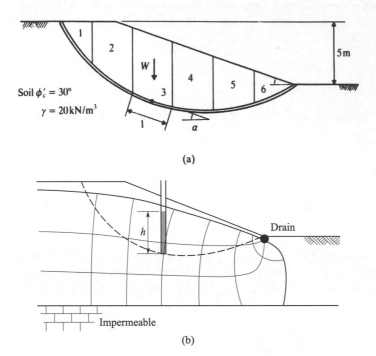

(a)

(b)

Slice	Area (m²)	W (kN)	α	W sin α (kN)	W cos α (kN)	h (m)	u (kN/m²)	l (m)	ul (kN)	W cos α − ul (kN)
1	3.8	76	54°	61	45	1.4	13.7	4.0	55	−10
2	14.4	288	36°	169	233	3.0	29.4	3.7	109	124
3	15.3	306	20°	105	288	4.0	39.2	3.2	126	162
4	13.8	276	4°	19	275	3.8	37.3	3.0	112	163
5	10.2	204	−10°	−35	201	2.8	27.5	3.0	82	119
6	3.8	76	−25°	−32	69	1.4	13.7	3.2	44	25
Totals				287						583

(c)

Figure 20.19 Slope stability analysis – drained loading.

the slip circle can be estimated from the height of the water in a standpipe on an equipotential as shown.

Replacing $\tan \phi'$ with $\tan \phi'/F_s$ and making use of Eq. (20.21) for the Swedish method of slices,

$$F_s = \frac{\sum (W \cos \alpha - ul)}{\sum W \sin \alpha} \tan \phi'$$

The table in Fig. 20.19(c) gives the calculations for each slice and, summing over the whole mechanism,

$$F_s = \frac{583}{287} \tan 30° = 1.17$$

You should now repeat the calculations with different circles to find the lowest value of F_s. Notice that near the toe of the slope the seepage becomes approximately parallel to the surface and there is the possibility of local instability, which should be investigated.

References

Bishop, A. W. (1955) 'The use of the slip circle in the stability analysis of earth slopes', *Geotechnique*, 5, 7–17.

Fellenius, W. (1927) *Erdstatische Berechnungen*, W. Ernst und Sohn, Berlin.

Janbu, N. (1973) 'Slope stability computations', in *Embankment Dam Engineering, Casagrande Memorial Volume*. Hirschfield and Poulos (eds), Wiley, New York.

Morgenstern, N. R. and V. E. Price (1965) 'The analysis of the stability of general slip surfaces', *Geotechnique*, 15, 79–93.

Further reading

Atkinson, J. H. (1981) *Foundations and Slopes*, McGraw-Hill, London.

Bromhead, E. N. (1992) *The Stability of Slopes*, Taylor and Francis, London.

Heyman, J. (1972) *Coulomb's Memoir on Statics*, Cambridge University Press, Cambridge.

Chapter 21

Stability of slopes

21.1   Introduction

The surface of the earth is very rarely flat and so there are slopes nearly everywhere. Even relatively flat ground often has rivers and drainage channels with side slopes. Slopes may be natural, due to erosion by rivers or the sea, or man-made by excavation or fill. Man-made slopes for roads and dams are permanent, but temporary slopes are required during construction of foundations and underground structures.

The geometry of a slope may be characterized by its angle i and height H, as shown in Fig. 21.1. The loads on the slope are due to the self-weight of the soil and to external loads, which may come from foundations at the top or water in the excavation. A special case of a slope is a vertical cut, such as the sides of a trench, where $i = 90°$. In the soil behind any slope there will be shear stresses and these are required to maintain the slope. Materials that cannot sustain shear stresses cannot have slopes, so water in a glass has a level surface.

During excavation of a slope the mean normal total stresses will be decreased due to removal of soil from the excavation, while during construction of an embankment the mean normal total stresses will increase as more fill is placed. In both cases, however, the shear stresses increase as the height and/or slope angle increase. I will call any kind of slope construction *loading* because the shear stresses increase irrespective of what happens to the mean normal total stress.

If a slope is too steep or too high it will fail and there will be a slip or landslide, as illustrated in Fig. 21.2. The landslide will stop when the height and angle are critical (H_c and i_c) and the slope has a factor of safety of unity. Rock slopes can be very steep, but soil slopes are much more modest, with angles from $10°$ to $30°$ and, for steeper angles, heights up to 20 m. The best laboratory to study slope stability is at the seaside where you should dig a hole in the beach and construct a sandcastle.

21.2   Types of instability

Slope instabilities involve large ground movements and usually require a mechanism of slip surfaces. Mechanisms can have a number of different configurations and some typical ones are illustrated in Fig. 21.3. In Fig. 21.3(a) and (b) the soil is homogeneous and the position of the slip surface (deep or shallow) is governed largely by the pore pressures. In Fig. 21.3(c) the geometry of the slip surface is controlled by a weak layer. Figure 21.3(d) illustrates a mud flow where there are very large homogeneous strains.

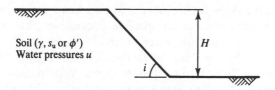

Figure 21.1 Geometry of a simple slope.

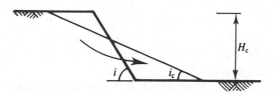

Figure 21.2 Simple slope failure.

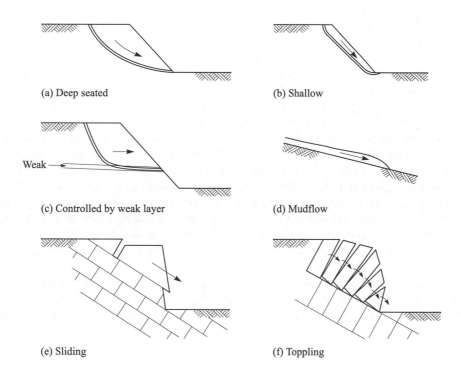

(a) Deep seated

(b) Shallow

(c) Controlled by weak layer

(d) Mudflow

(e) Sliding

(f) Toppling

Figure 21.3 Types of slope failure.

Figures 21.3(e) and (f) illustrate mechanisms of failure of steep slopes in jointed rock. The rock itself is very strong and slope failure occurs as blocks of rock move on pre-existing joints. These mechanisms are not compatible because cracks have opened. The relative spacing of the joints controls whether the failure is predominantly by sliding, as in Fig. 21.3(e), or by toppling, as in Fig. 21.3(f). Figure 21.3 illustrates only a few mechanisms, primarily related to sliding. Often combinations of mechanisms occur with one mechanism changing into another or both taking place simultaneously.

21.3 Parameters and factors for design of slopes

The limiting height and angle of a slope depend on the strength of the soil and there are a number of different soil strengths. The most important distinctions are between undrained strength and drained or effective stress strength and between peak, critical state and residual strengths. Choices of soil parameters and factors for design of slopes and other structures were discussed in Chapter 18.

For slopes the choice between undrained strength and drained strength is relatively straightforward. For temporary slopes and cuttings in fine grained soils with low permeability you should choose the undrained strength s_u and do analyses using total stresses. If you do this, remember that the analysis is valid only so long as the soil remains undrained and the stability will deteriorate with time as pore pressures rise and the soil swells and softens, as shown in Sec. 21.4. For a permanent slope the critical conditions are at the end of swelling when pore pressures have reached equilibrium with a steady state seepage flownet or with hydrostatic conditions. In this case you can calculate the pore pressures, choose the drained strength and do analyses using effective stresses. Analyses for slopes where the excess pore pressures have only partly dissipated are beyond the scope of this book.

As discussed in Chapter 18 slopes should be designed for an ultimate limit state with a single factor of safety or with partial factors applied to each uncertainty. The designs should use either the critical state strength (s_u or ϕ_c') or, if there has been previous landsliding, the residual strength (ϕ_r'). The first thing to investigate is whether a landslide or slope failure has occurred in a clay slope in the past, in which case the soil in a slip plane may have already reached its residual state. New construction, either excavation or loading, may reactivate the old movements and the appropriate strength is the residual strength. Detection of old landslides, some of which may be geologically old, requires very detailed and careful ground investigation. In the absence of preexisting failures you should design safe slopes using the critical state strength.

For design of safe slopes you can apply a single factor of safety or apply partial factors to the parameters which control the stability. Usually a single factor of safety is applied to the soil strength so the safe shear stresses in the soil τ_s or τ_s' are given by

$$\tau_s = \frac{s_u}{F_s} = s_{us} \tag{21.1}$$

$$\tau_s' = \sigma' \frac{\tan \phi_c'}{F_s} = \sigma' \tan \phi_s' \tag{21.2}$$

where s_{us} and ϕ'_s are strength parameters required for a safe design, as discussed in Sec. 18.5. The idea is to reduce the soil strength and then ensure that the slope is in equilibrium with the lower strength. Values for F_s depend on a number of things, including the consequences of failure and the reliability of the measurements of soil strength. Typically engineers use values for F_s in the range 1.25 to 1.5. This is usually enough for a safe design but may not be enough to prevent ground movements which are sufficiently large to damage nearby structures.

Alternatively, engineers apply partial factors to each variable in the analysis. For undrained, total stress analyses the variables are the slope height and angle, the unit weight of the soil and the undrained strength and partial factors can be applied to each. For drained effective stress analyses the variables are the slope height and angle, the unit weight of the soil, the critical state friction angle and the pore pressure and, again, partial factors can be applied to each. The values of the partial factors should reflect the uncertainty with which each variable can be determined.

Usually the slope height and angle and the unit weight of the soil can be determined reliably and the partial factors for these can be close to 1.0. The critical state friction angle can also be measured reliably and it is the worst credible value: the strength will not be less than the critical state strength unless movements are sufficiently large to mobilize the residual strength. In practice measurements of undrained strength are variable and tend to increase with depth so the partial factor will depend on whether the values for s_u are worst credible, moderately conservative or some other value. For effective stress analyses the pore pressure must be determined from a flownet or from considerations of the long term groundwater conditions. It is usually quite easy to determine the long term pore pressures in which case the partial factor can be 1.0. For effective stress stability analyses it is logical to take the critical state friction angle, the worst credible pore pressures and apply partial factors of 1.0: it cannot get worse than the worst credible. In the end, however, it is for the engineer to choose the design parameters and their associated partial factors.

21.4 Stress changes in slopes

Natural slopes are usually eroded very slowly and the soil is essentially drained so that pore pressures are governed by steady state seepage from the ground towards the excavation. Man-made slopes are often constructed quite quickly and in clays the soil will be essentially undrained during the excavation.

The changes of total and effective stress during undrained slope excavation are illustrated in Fig. 21.4. In Fig. 21.4(a) the total stresses on a slip surface are τ and σ and the pore pressure is illustrated by the rise of water in a standpipe. (For simplicity the excavation is kept full of water so that the phreatic surface is level and the initial and final pore pressures are the same.)

In Fig. 21.4(b) the total stress path is A → B; this corresponds to a reduction in σ due to the excavation and an increase in τ because the slope height and/or angle are increased. The effective stress path is A′ → B′, which corresponds to undrained loading at constant water content, as shown in Fig. 21.4(c). The exact effective stress path A′ → B′ in Fig. 21.4(b) will depend on the characteristics of the soil and its initial state or overconsolidation ratio, as discussed in Chapter 11.

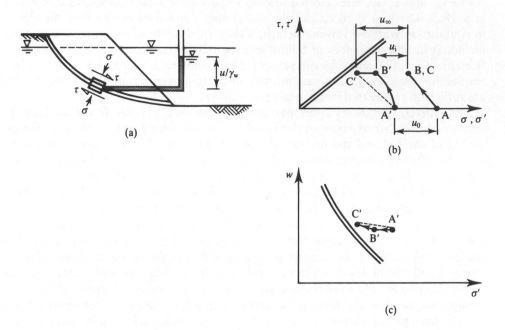

Figure 21.4 Stress and pore pressure changes in a stable slope.

As shown in Fig. 21.4(b), the pore pressure immediately after construction u_i is less than the steady pore pressure u_∞ and so the initial excess pore pressure $\bar{u}_i$ is negative (i.e. the level of water in the standpipe is below the phreatic surface, as shown in Fig. 21.4(a). As time passes the total stresses remain unchanged at B (because the geometry of the slope remains the same) but the negative excess pore pressures dissipate and the pore pressure rises. The effective stress path is $B' \rightarrow C'$ and this corresponds to swelling and a reduction in mean normal effective stress, as shown in Fig. 21.4(b) and (c). The final state at C' corresponds to a steady state pore pressure after swelling u_∞; in the example shown $u_\infty = u_0$ but the arguments would be the same if u_∞ was different from u_0, which would be the case if the excavation was drained of water.

The slope will fail if the states of all elements along the slip surface reach the critical state line: if B' reaches the critical state line the slope fails during undrained excavation and if C' reaches the critical state line the slope fails some time after construction. The distance of the effective stress points B' or C' from the critical state line is a measure of the factor of safety of the slope and Fig. 21.4 demonstrates that the factor of safety of a slope decreases with time.

This means that the critical time in the life of a slope is in the long term when the pore pressures have come into equilibrium with the steady state seepage flownet. Consequently, a permanent slope should be designed for the long-term, fully drained, condition. Temporary slopes that are required to stand for very short periods are often designed as undrained, but remember that just because a slope or a trench is standing now does not mean that it will still be stable in 10 minutes time. Slopes and excavations are very dangerous; many people are killed by trench failures which occur as the effective stresses move from B′ towards C′ in Fig. 21.4. In the design of temporary excavations the important question is not so much the undrained stability but how quickly the pore pressures will increase. In Fig. 21.4(b) and (c) the broken lines A′ → C′ represent the drained case in which pore pressures remain constant.

If a slope fails the total stresses change as the angle and height reduce as shown in Fig. 21.5(a). Figure 21.5(b) shows stress paths for a steep slope failing during undrained excavation. The effective stress path is A′ → B′ and this ends on the critical state line where the undrained strength is s_u. The total stress path would like to continue to X, corresponding to the initial slope angle i_x, but cannot; therefore the slope geometry changes and the mean slope angle i_c and height H_c correspond to total stresses at B. Figure 21.5(c) shows stress paths for a slope that fails some time after excavation. The state immediately after excavation is B and B′ and failure occurs at C and C′ when the pore pressure is u_f. Subsequently, as the pore pressures continue to rise, the effective stresses move along C′ → D′ down the critical state line and the total stresses move more or less along C → D due to unloading (i.e. reduction) of the shear stress as the slope angle decreases. The slope will reach a stable state when the pore pressure is the final steady state pore pressure u_∞.

These analyses and the stress paths shown in Figs. 21.4 and 21.5 are simplified and idealized but they illustrate the essential features of the behaviour of slopes during and

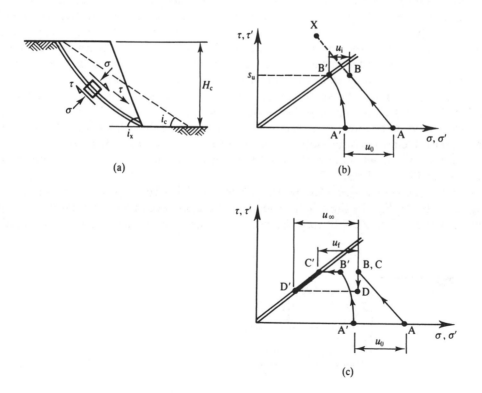

Figure 21.5 Stress and pore pressure changes in failing slopes.

", "", "", "", "", ""]

after construction. Notice the critical importance of changing pore pressures with time and their influence on stability. The examples were for excavated slopes where pore pressures decreased during undrained excavation. In man-made compacted soils the initial pore pressures are negative because the fill is unsaturated and so the initial states at B and B' are more or less the same for cut and fill slopes.

21.5   Influence of water on stability of slopes

Water influences slope stability in several fundamentally different ways and these are illustrated by commonly observed failures. Firstly, slopes may fail well after completion of excavation due to dissipation of negative excess pore pressures and swelling and softening of the soil, as discussed in Sec. 21.4. Secondly, slopes in river banks, lakes and trenches may fail if the external water level is quickly lowered. Thirdly, slopes often fail after periods of heavy rainfall.

Free water in a river or lake, or in a water-filled trench, applies total stresses σ_w to a soil surface, as shown in Fig. 21.6. These total stresses help to support the slope which may fail if the support is removed. (In practice temporary excavations for piles and retaining walls are supported by a slurry of bentonite clay, or some other natural or artificial mud, with unit weight greater than that of water.) Notice that after undrained excavation the pore pressures in the soil may not be in equilibrium with the free water in the excavation.

Slope failures after rainfall, or after changes in the groundwater conditions, are due to increases in the pore pressures which lead to reductions in effective stress and strength. (Notice that the soil remains saturated while pore pressures change and there is no question of the rainwater lubricating the soil – this is an entirely false interpretation.) In order to calculate the pore pressures in a slope under steady state conditions it is necessary to draw a flownet as described in Chapter 14.

Figure 21.7 shows flownets for steady state seepage towards an excavation. In Fig. 21.7(a) there is a drain at the toe of a slope, while in Fig. 21.7(b) the excavation is partly filled with water. In both cases there is impermeable rock below the soil and the water table far from the slope is near the ground surface as shown. The pore pressure in the soil anywhere in each flownet can be determined from the equipotentials, as described in Sec. 14.3.

Often slopes fail with a shallow slip plane parallel with the surface, as shown in Fig. 21.3(b). In Fig. 21.7(a) the flowlines near the toe become nearly parallel to the slope, while in Fig. 21.7(b) they are normal to the slope which is an equipotential.

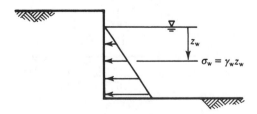

Figure 21.6 Loads on slopes from water in the excavation.

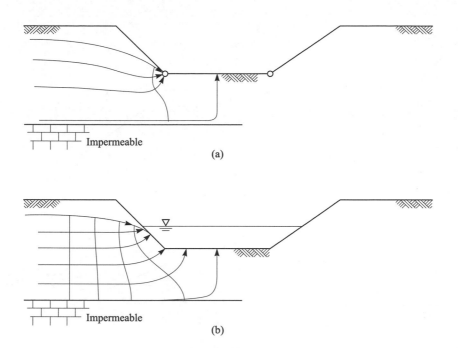

Figure 21.7 Steady state seepage towards excavations.

We can consider the general case where flow lines meet the surface from different directions. If the slope is submerged as in Fig. 21.7(b) the pore pressures are governed by the whole flownet, but if the slope is not submerged as in Fig. 21.7(a) the pore pressures are governed by the local conditions.

The point to remember is that water will rise to the same height in standpipes whose tips are on the same equipotential and where the equipotential meets the phreatic surface the pore pressure is zero. Figure 21.8 shows h_w for a standpipe whose tip is at a depth z for various directions of seepage flow near a slope. In Fig. 21.8(a) the flow is vertically downwards. This would correspond to heavy rain on soil which was drained from below. The equipotentials are horizontal so $u = 0$ everywhere. In Fig. 21.8(b) the seepage is down the slope, the flowlines are parallel with the slope and the equipotentials are normal to the slope. From the geometry of the figure

$$u = \gamma_w h_w = \gamma_w z \cos^2 i \tag{21.3}$$

where i is the slope angle. In Fig. 21.8(c) the seepage is horizontal and the equipotentials are vertical. From the geometry of the figure

$$u = \gamma_w h_w = \gamma_w z \tag{21.4}$$

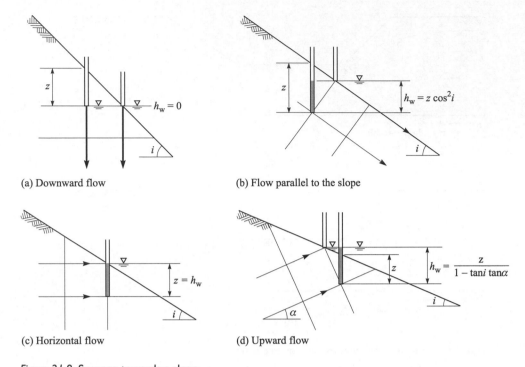

(a) Downward flow

(b) Flow parallel to the slope

(c) Horizontal flow

(d) Upward flow

Figure 21.8 Seepage towards a slope.

In Fig. 21.8(d) the seepage is inclined upwards at an angle α. From the geometry of the figure

$$u = \gamma_w h_w = \gamma_w z \left(\frac{1}{1 - \tan\alpha \tan i} \right) \tag{21.5}$$

Equation 21.5 gives $u = 0$ for $\alpha = -90°$ and the expressions in Eqs. (21.3) and (21.4) with the appropriate values for α. Notice that as the direction of the seepage rotates from downwards to inclined upwards the pore pressure at a depth z increases. This will mean that the limiting slope angle i_c will decrease as the direction of seepage rotates and this will be discussed later in Sec. 21.6.

21.6   Stability of infinite slopes

From now on I will examine the limiting stability of slopes with the critical state strengths s_u or ϕ'_c; to apply a factor of safety you can do the same calculations using s_{us} or ϕ'_s obtained from Eqs. (21.1) and (21.2). For slope stability calculations you can use the upper and lower bound method described in Chapter 19 or the limit equilibrium method described in Chapter 20. A simple but very useful case is for shallow sliding on a slip surface parallel to the slope, as illustrated in Fig. 21.3(b). The depth to the slip surface will be controlled by geological or groundwater conditions; a common case is where there is a mantle of soil over rock in a hillside and the slip surface is close to the interface between the soil and the rock.

(a) Undrained loading

Figure 21.9(a) shows an infinite slope where the angle is an upper bound i_u with a mechanism of plastic collapse consisting of a slip surface through the soil at the rock level; there is a block of soil length l measured down the slope. The corresponding displacement diagram for an increment of displacement δw is shown in Fig. 21.9(b). For an infinitely long slope, the forces on any such block are the same as those on any other similar block and so the forces F_1 and F_2 are equal and opposite. From the geometry of Fig. 21.9(a) the weight of the block (for unit thickness normal to the page) is

$$W = \gamma H l \cos i_u \tag{21.6}$$

and from Fig. 21.9(b) the vertical component of displacement is

$$\delta v = \delta w \sin i_u \tag{21.7}$$

Hence, noting that the increments of work done by the equal and opposite forces F_1 and F_2 sum to zero, we have

$$\delta W = s_u l \, \delta w \tag{21.8}$$

$$\delta E = \gamma H l \cos i_u \, \delta w \sin i_u \tag{21.9}$$

and, equating $\delta W = \delta E$, an upper bound for the critical slope angle is given by

$$\sin i_u \cos i_u = \frac{s_u}{\gamma H} \tag{21.10}$$

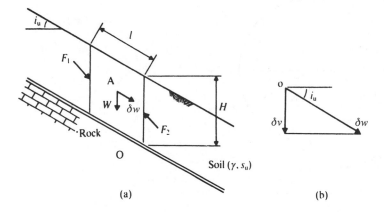

(a)   (b)

Figure 21.9 Mechanism of plastic collapse for an infinitely long slope for undrained loading.

and

$$i_u = \frac{1}{2} \sin^{-1} \frac{2s_u}{\gamma H} \tag{21.11}$$

Figure 21.10(a) shows forces and stresses on an element in an infinite slope where the angle is a lower bound i_l. The state of stress increases linearly with depth from zero at the surface and the maximum shear stress $\tau = s_u$ occurs on a surface parallel with the slope. For an infinite slope, as before, the forces F_1 and F_2 are equal and opposite and the weight of a block of soil of length l is $W = \gamma H l \cos i_l$. Hence, resolving normal to and along the slope, we have

$$\sigma_s = \gamma H \cos^2 i_l \quad \tau_s = \gamma H \sin i_l \cos i_l \tag{21.12}$$

where σ_s and τ_s are the normal and shear stresses in the soil on the surface parallel to the slope at a depth H. The Mohr circle of total stress for an element of soil just above the rock is shown in Fig. 21.10(b). The pole is at P and points a and b represent the states of stress on a horizontal plane and on a plane parallel

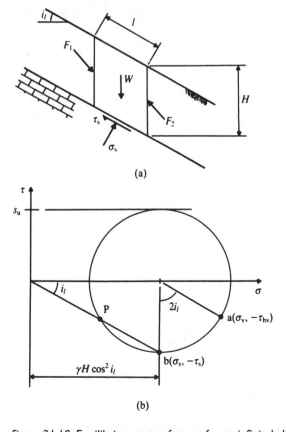

(a)

(b)

Figure 21.10 Equilibrium state of stress for an infinitely long slope for undrained loading.

with the slope respectively; the angle subtended at the centre of the circle is $2i_l$. The Mohr circle just touches the undrained failure envelope and so the state of stress in the slope does not exceed the undrained failure criterion. From the geometry of Fig. 21.10(b), making use of Eq. 21.12, a lower bound for the critical slope angle is given by

$$\tan i_l = \frac{\tau_s}{\sigma_s} = \frac{s_u}{\gamma H \cos^2 i_l} \tag{21.13}$$

and hence

$$i_l = \frac{1}{2} \sin^{-1} \frac{2 s_u}{\gamma H} \tag{21.14}$$

Comparing Eqs. (21.11) and (21.14), the upper bound solution exactly equals the lower bound solution and so both must equal the exact solution. Hence the critical slope angle i_c for undrained loading of an infinite slope is given by

$$i_c = \frac{1}{2} \sin^{-1} \frac{2 s_u}{\gamma H} \tag{21.15}$$

(b) Drained loading – no seepage

Figure 21.11(a) shows a mechanism of plastic collapse for an infinitely long slope whose angle to the horizontal is an upper bound i_u. The mechanism is a single slip surface at a depth z and there is a block of soil length l measured down the slope; as before, the forces F_1 and F_2 that act on the vertical sides are equal and opposite. The displacement diagram for an increment of displacement δw is shown in Fig. 21.11(b), where the direction of the increment of displacement makes an angle $\psi = \phi'_c$ to the slip surface.

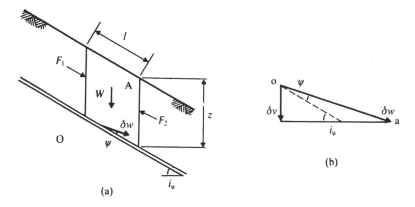

Figure 21.11 Mechanism of plastic collapse for an infinitely long slope in dry soil.

For drained loading the increment of work done by the internal stresses for an increment of plastic collapse is $\delta W = 0$ and, noting that $F_1 = F_2$, the increment of work done by the external loads for dry soil is

$$\delta E = \delta v\, \gamma V \tag{21.16}$$

where $V = zl \cos i_\mathrm{u}$ is the volume of the block. Hence, equating $\delta E = \delta W$, an upper bound is given by

$$\delta v\, \gamma V = 0 \tag{21.17}$$

Since the volume V is non-zero, the upper bound is given by $\delta v = 0$ and hence, from the geometry of Fig. 21.11(b), an upper bound for the critical slope angle is given by

$$i_\mathrm{u} = \phi'_\mathrm{c} \tag{21.18}$$

Figure 21.12 shows an infinite slope whose angle with the horizontal is a lower bound i_l, and a block of soil of length l measured down the slope and depth z measured vertically; the forces on the faces of the block are shown and, as before, the forces F_1

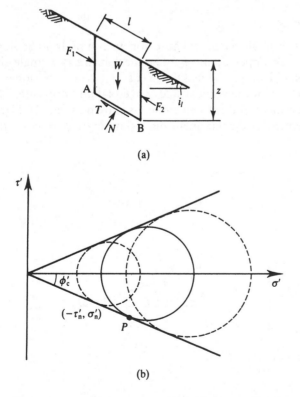

(a)

(b)

Figure 21.12 Equilibrium states of stress for an infinitely long slope in dry soil.

and F_2 are equal and opposite. Resolving normal to and parallel with the base AB the normal and shear forces N and T are

$$N = W \cos i_l = \gamma zl \cos^2 i_l \tag{21.19}$$

$$T = W \sin i_l = \gamma zl \sin i_l \cos i_l \tag{21.20}$$

For dry soil, where pore pressures are zero and total and effective stresses are equal, the effective normal and shear stresses on the plane AB are given by

$$\sigma'_n = \gamma z \cos^2 i_l \tag{21.21}$$

$$\tau'_n = \gamma z \sin i_l \cos i_l \tag{21.22}$$

and hence

$$\tau'_n = \sigma'_n \tan i_l \tag{21.23}$$

which is valid for all planes such as AB at any depth. The limiting values of τ'_n and σ'_n are given by

$$\tau'_n = \sigma'_n \tan \phi'_c \tag{21.24}$$

and hence a lower bound for the limiting slope angle is given by

$$i_l = \phi'_c \tag{21.25}$$

The Mohr circle of effective stress for the state of stress in an element on AB is shown in Fig. 21.12(b); the circles shown with broken lines correspond to the states of stress in elements above and below AB. All the Mohr circles just touch the drained failure envelope. The pole of the Mohr circle is at P and hence we may calculate the stresses on any other plane in the slope; the normal and shear stresses on vertical planes are equal in magnitude to those on planes parallel to the slope.

From Eqs. (21.18) and (21.25) the upper and lower bounds are equal and hence the critical slope angle for dry soil is

$$i_c = \phi'_c \tag{21.26}$$

Figure 21.13(a) shows a partly submerged slope and Fig. 21.13(b) shows a fully submerged slope; in both cases there are no hydraulic gradients and no seepage. Since Eq. (21.26) does not contain either the unit weight or the pore pressure it applies to the submerged slopes in Fig. 21.13.

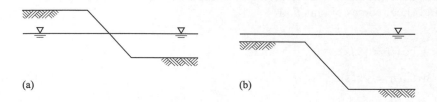

Figure 21.13 Submerged slopes with no seepage.

21.7   Influence of seepage on the stability of infinite slopes

In Sec. 21.5 I examined the pore pressures in infinite slopes with different seepage conditions near the slope and we can now investigate how these different pore pressures influence the critical slope angle. In the previous section I used upper and lower bound analyses and showed that the critical angle of a slope in dry soil is $i_c = \phi'_c$. In this section I will use the limit equilibrium method.

Figure 21.14(a) shows a mechanism consisting of a slip surface parallel to the slope at a depth z and the forces acting on a block length l down the slope and Fig. 21.14(b) is the polygon of forces acting on the block. These are the weight W, the shear force $T' = \tau' l$ and the total normal force $N = \sigma_n l$ which is made up of an effective normal force $N' = \sigma'_n l$ and a pore pressure force $U = ul$.

From the force polygon

$$T' = N \tan i_c = (N - U) \tan \phi'_c \qquad (21.27)$$

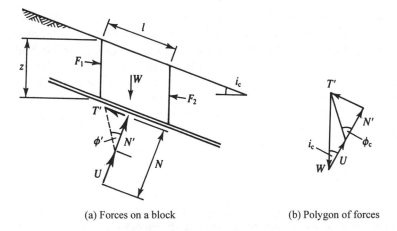

(a) Forces on a block   (b) Polygon of forces

Figure 21.14 Limit equilibrium solution for an infinitely long slope with steady state seepage parallel with the slope.

And the critical slope angle is given by

$$\tan i_c = \tan \phi'_c \left(1 - \frac{U}{N}\right) = \tan \phi'_c \left(1 - \frac{u}{\sigma_n}\right) \tag{21.28}$$

and, from Eq. (21.21)

$$\tan i_c = \tan \phi'_c \left(1 - \frac{u}{\gamma z \cos^2 i_c}\right) \tag{21.29}$$

The value of the pore pressure u depends on the direction of the seepage near the slope, as discussed in Sec. 21.5 and as shown in Fig. 21.8. For the case of seepage vertically downwards, as shown in Fig. 21.8(a) $u = 0$ and $i_c = \phi'_c$ and the critical slope angle is the same as that in dry soil or in a submerged slope. For the case where the seepage is parallel with the slope $u = \gamma_w z \cos^2 i_c$ and

$$\tan i_c = \tan \phi'_c \left(1 - \frac{\gamma_w}{\gamma}\right) \tag{21.30}$$

In many cases $\gamma \approx 2\gamma_w$ and

$$i_c \approx \tfrac{1}{2}\phi'_c \tag{21.31}$$

so, with seepage parallel to the slope, the critical slope angle is about half the critical angle of a submerged or dry slope.

From Sec. 21.5 and Fig. 21.8 the pore pressure increases as the seepage firstly becomes horizontal and then becomes inclined upwards and the critical slope angle decreases as the pore pressure increases. From Fig. 21.8(d) the critical angle of the slope below the water where the flowlines are inclined upwards will be less than $\tfrac{1}{2}\phi'_c$. This is one of the reasons why it is difficult to dig a hole in the beach below the water table.

The solutions for the stability of infinite slopes given by Eqs. (21.28), (21.29) and (21.30) are relatively simple. Notice that for the undrained slope the critical angle i_c is governed by the depth H of the slip surface; if this depth is relatively large the mechanism cannot be approximated to sliding parallel to the surface and the solution is no longer valid. For the drained case the critical angles for dry and submerged slopes are the same, $i_c = \phi'_c$ (because neither the unit weight nor the pore pressure appear in the final solution), but if there is steady state seepage towards the slope the critical slope angle is reduced. These results demonstrate the very significant influence of pore pressures on slope stability.

21.8 Stability of vertical cuts

A simple experiment with dry sand or sugar demonstrates that you cannot make a vertical cut in a drained soil. We can, however, make vertical cuts in soils that are undrained where the negative pore pressures generate positive effective stresses.

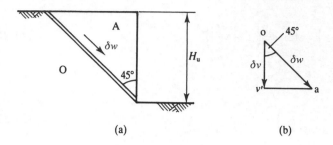

Figure 21.15 Mechanism of plastic collapse for a vertical cut slope for undrained loading.

A simple collapse mechanism consisting of a single straight slip surface at an angle of 45° to the vertical is shown in Fig. 21.15(a) and Fig. 21.15(b) is the corresponding displacement diagram for an increment of displacement δw down the slip surface. From the geometry of Fig. 20.14(a), the length L of the slip surface and the volume V of the wedge (for unit thickness normal to the page) are given by

$$L = \sqrt{2}H_u \quad V = \tfrac{1}{2}H_u^2 \tag{21.32}$$

where H_u is an upper bound for the height of the slope at collapse. From the geometry of Fig. 21.15(b) we have

$$\delta v = \frac{1}{\sqrt{2}}\,\delta w \tag{21.33}$$

The only external forces are those due to the self-weight of the sliding soil and

$$\delta W = s_u\sqrt{2}H_u\,\delta w \tag{21.34}$$

$$\delta E = \frac{1}{\sqrt{2}}\,\delta w\,\gamma\,\frac{1}{2}H_u^2 \tag{21.35}$$

Hence, equating $\delta W = \delta E$, an upper bound for the height of the cut slope at collapse is given by

$$H_u = \frac{4s_u}{\gamma} \tag{21.36}$$

For a lower bound Fig. 21.16(a) shows a state of stress in which shear stresses on vertical and horizontal planes are zero. The vertical and horizontal stresses are $\sigma_z = \gamma z$ and $\sigma_h = 0$, and these are principal stresses. Mohr circles of stress for the elements A

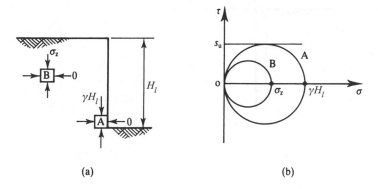

Figure 21.16 Equilibrium state of stress for a vertical cut slope for undrained loading.

and B in Fig. 21.16(a) are shown in Fig. 21.16(b). The Mohr circle A does not cross the undrained failure envelope when

$$\gamma H_l = 2s_u \tag{21.37}$$

and hence a lower bound for the height of the cut is given by

$$H_l = \frac{2s_u}{\gamma} \tag{21.38}$$

These upper and lower bound solutions are not really very close to one another and it is very difficult to obtain better solutions. The best solution, and the one that is commonly used in design, is

$$H_c = \frac{3.8s_u}{\gamma} \tag{21.39}$$

which is close to the upper bound given by Eq. (21.36). If the excavation is filled with water the critical height is given by

$$H_c = \frac{3.8s_u}{\gamma - \gamma_w} \tag{21.40}$$

Comparing Eqs. (21.39) and (21.40), the critical height of a dry excavation is only about one-half that of an excavation filled with water.

You have probably noticed that the ground surface is often cracked and fissured, particularly near the top of a slope or excavation. Each vertical crack is like a small trench, as in Fig. 21.17, and the maximum depth of the crack is given by Eq. (21.39) or (21.40), depending on whether it is empty or filled with water. Notice that as pore pressures rise, the soil softens and weakens and the depth of the crack decreases; in the end, when the pore pressures are hydrostatic with a phreatic surface at ground level the cracks will have closed.

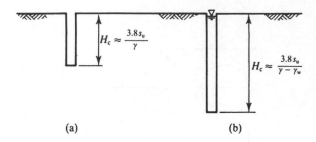

Figure 21.17 Stability of vertical cut slopes and vertical cracks filled with water.

21.9   Routine slope stability analyses

The most common procedure for slope stability analysis is to use the limit equilibrium method with a slip circle or a general curved slip surface. These methods were described in Chapter 20. For undrained loading the problem is statically determinate and the solution is relatively simple. For drained loading the problem is statically indeterminate and solutions using the method of slices require assumptions; there are a number of different solutions (e.g. Bishop, Janbu, Morgenstern and Price), each developed from different assumptions. In these solutions the calculations are largely repetitive and standard computer programs are available for slope stability analysis.

For slopes with relatively simple geometries, standard solutions are available in the form of non-dimensional tables and charts. These are very useful for preliminary design studies.

(a)   Stability numbers for undrained loading

The solution for an infinite slope for undrained loading was given by Eq. (21.15), which can be rewritten as

$$H_c = \frac{2}{\sin 2i} \frac{s_u}{\gamma} \tag{21.41}$$

or

$$H_c = N_s \frac{s_u}{\gamma} \tag{21.42}$$

where N_s is a stability number that depends principally on the geometry of the slope.

Figure 21.18(b) shows a more general case where strong rock occurs at a depth $n_d H$ below the top ground level and Fig. 21.18(a) shows values of the stability number N_s in terms of the slope angle i and the depth factor n_d. The data in Fig. 21.18 are taken from those given by Taylor (1948, p. 459) and were obtained from the limit equilibrium slip circle method.

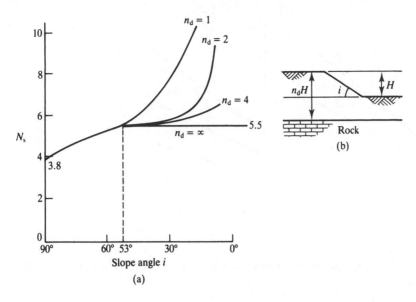

Figure 21.18 Stability numbers for undrained loading. (After Taylor, 1948.)

(b) Stability numbers for drained loading

The safe slope angle for drained loading with steady state seepage is obtained from Eq. (21.28) substituting the safe friction angle ϕ'_s for ϕ'_c and is given by

$$\tan i = \tan \phi'_s \left(1 - \frac{u}{\sigma_n}\right) \tag{21.43}$$

and with Eq. 21.2

$$F_s = \frac{\tan \phi'_c}{\tan i} \left(1 - \frac{u}{\sigma_n}\right) \tag{21.44}$$

Equation (21.44) can be written as

$$F_s = m - nr_u \tag{21.45}$$

where m and n are stability numbers that depend on the geometry of the slope and on the friction angle ϕ' and $r_u = u/\sigma_z$ is a pore pressure coefficient. Figure 21.19 shows values for the stability numbers m and n for simple slopes calculated by Bishop and Morgenstern (1960) from slip circle analysis using the method of slices. For a particular slope an average value of r_u must be estimated from a steady state seepage flownet and from the position of the critical slip circle: in many practical cases r_u is taken as about $\frac{1}{3}$.

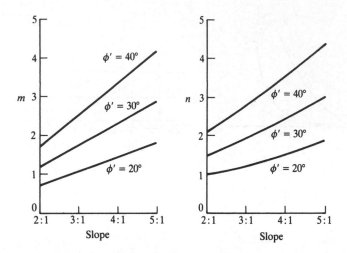

Figure 21.19 Stability numbers for drained loading. (After Bishop and Morgenstern, 1960.)

21.10   Behaviour of simple excavations

All the features of slope stability described in the previous section can be observed by digging a hole in the beach. What you will see is illustrated in Fig. 21.20. In the dry sand at the surface the slope angle is ϕ_c'. In the unsaturated sand above the water table the pore pressures are negative and it is possible to excavate a vertical cut BC. The cut

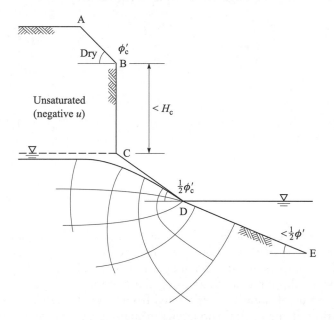

Figure 21.20 Stability of a simple excavation.

will fail if the depth exceeds the critical height H_c; this is given by Eq. (21.15) where s_u can be found from an unconfined compression test carried out on a sand-castle at the same density and water content. The vertical cut cannot be continued below the water table C where the pore pressures are zero. (The cut often fails just above the water table where the sand is saturated and the negative pore pressures are small.) Notice that pore pressures behind the cut BC are negative so the face should look dry.

You know that it is very difficult to dig the hole below the water table. If you excavate slowly there will be steady state seepage so the angle of the slope CD will be about $\frac{1}{2}\phi'_c$ but, if you can excavate below water the angle of the slope DE will be smaller. In practice seepage into the excavation along CD usually causes erosion and you cannot dig much below the water table.

When you do this experiment remember that the factor of safety of the vertical cut BC is probably reducing with time and you must be very careful that it does not collapse on you. You should also observe what happens to your hole as the tide comes in or as the sun shines on to the face BC.

21.11 Summary

1. Slopes fail as soil moves on slip surfaces and there are several possible mechanisms depending on the ground and groundwater conditions.
2. Immediately after excavation or filling pore pressures are reduced and, as time passes, pore pressures rise, effective stresses reduce and the safety of a slope deteriorates.
3. For slope stability calculations the factor of safety accounts for uncertainties in the determination of the soil parameters and the analyses. For routine analyses the critical state strength will give safe designs with factors of safety accounting for uncertainties in the pore pressures. If previous landsliding has occurred the strength may have reduced to the residual before construction starts.
4. Slope stability calculations can be done using the upper and lower bound methods or the limit equilibrium method; preliminary designs can be carried out making use of routine stability numbers.

Worked examples

Example 21.1: Undrained slope stability   Figure 21.21 shows the geometry of a simple slope. From Eq. (21.42) and replacing s_u with s_u/F_s,

$$F_s = \frac{N_s s_u}{\gamma H}$$

From Fig. 21.18 for $i = 20°$ and $n_d = \infty$ we have $N_s = 5.5$ and

$$F_s = \frac{5.5 \times 40}{20 \times 5} = 2.2$$

Notice that this is rather less than the result $F_s = 2.98$ obtained for Example 20.3, indicating that the slip circle in Fig. 20.18 was not the critical one.

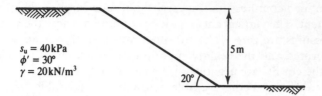

Figure 21.21 Stability of a simple slope – Examples 21.1 and 21.2.

Example 21.2: Drained slope stability   For drained loading of the slope in Fig. 21.21, from Eq. (21.45),

$$F_s = m - nr_u$$

For $i = 20°$ the gradient is 2.75:1 and, from Fig. 21.19, for $\phi' = 30°$ we have $m \approx 1.6$ and $n \approx 1.8$. Taking a characteristic value for $r_u = 0.3$,

$$F_s = 1.6 - (0.3 \times 1.8) = 1.06$$

Near the toe of the slope the flowlines will be approximately parallel to the slope and the phreatic surface is close to ground level. From Eq. (21.30), replacing $\tan \phi'_c$ with $\tan \phi'_c/F_s$,

$$F_s = \left(1 - \frac{\gamma_w}{\gamma}\right) \frac{\tan \phi'_c}{\tan i}$$

The calculated factor of safety is

$$F_s = \left(1 - \frac{10}{20}\right) \frac{\tan 30°}{\tan 20°} = 0.80$$

and local instability will occur near the toe. In order to stabilize the slope the drain in Fig. 20.19 should be lowered to reduce the pore pressures.

References

Bishop, A. W. and N. R. Morgenstern (1960) 'Stability coefficients for earth slopes', *Geotechnique*, **10**, 129–150.
Taylor, D. W. (1948) *Fundamentals of Soil Mechanics*, Wiley, New York.

Further reading

Atkinson, J. H. (1981) *Foundations and Slopes*, McGraw-Hill, London.
Bromhead, E. N. (1992) *The Stability of Slopes*, Taylor and Francis, London.
Chandler, R. J. (ed) (1991) Slope stability engineering, *Proc. Int. Conf, Slope Stability*, I. C. E., London.

Chapter 22

Bearing capacity and settlement of shallow foundations

22.1 Types of foundations

Any structure that is not flying or floating rests on or in the ground and the base of the structure and the soil together make up the foundation. Buildings and embankments must have foundations and so must vehicles and people. The criteria for the design of a foundation are that the settlements should be limited so that the building does not become damaged, vehicles can still move about and you do not lose your boots in mud. All foundations settle because nothing (not even tarmac or rock) is absolutely rigid, but obviously some settle more than others; look at the Tower of Pisa for instance. When you walk across the beach and leave a footprint it is simply a mark of the settlement of a foundation and so too is a tyre track.

In civil engineering foundations are shallow, deep or piled, as illustrated in Fig. 22.1. (The distinction $D/B = 1$ to 3 for a deep foundation is made for convenience.) We know that, in general, the strength and stiffness of soil increases with depth (because effective stresses increase with depth) and so one advantage of a deep foundation and a pile is that they are founded in stronger and stiffer soil; often the tip of a pile rests on very stiff and strong soil or rock. Another advantage is that shear stresses between the soil and the sides of a deep foundation or a pile contribute to the load capacity; in a shallow foundation the contribution of the side shear stresses is negligible.

The characteristics of a typical foundation are illustrated in Fig. 22.2. The weight of the foundation is W and it supports a vertical load V, a horizontal load H and a moment M. The force V is usually just the weight of the structure while the horizontal force H and moment M arise from wind and wave loads and unexpected impacts. Normally both W and V are known with some certainty but the loads H and M can only be estimated. For most of this chapter I will consider simple foundations which have only a vertical load V on them.

The base width is B; for unit length out of the page this is the base area, so the gross bearing pressure q is

$$q = \frac{V + W}{A} \tag{22.1}$$

Many simple foundations, including piles, are constructed from solid concrete which has unit weight γ_c only a little larger than that of soil, so W ($\approx \gamma_c AD$) depends on the size of the foundation. Some foundations are hollow, particularly where they are used

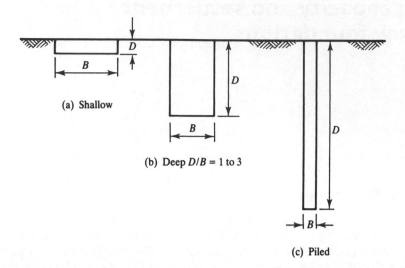

Figure 22.1 Types of foundation.

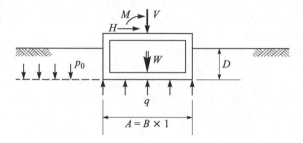

Figure 22.2 Loads and stresses on a foundation.

for parking cars, in which case the weight W is relatively small. It is a good idea to consider the weight W of the foundation separately from the loads from the structure above the ground.

Outside the foundation the total vertical stress at depth D is $\sigma_z = p_0$, where

$$p_0 = \gamma D \qquad\qquad (22.2)$$

and the net bearing pressure q_n is given by

$$q_n = q - p_0 \qquad\qquad (22.3)$$

The net bearing pressure is the change of total stress at the base of the foundation and it is this stress which causes ground movements. Notice that q_n could be either positive or negative depending on the magnitudes of V and W, both of which would be very small for an underground car park or a submerged tank. If q_n is positive the foundation will settle, but if it is negative (i.e. the total stress at foundation level reduces) the foundation will rise. By careful design of a compensated foundation it is possible to have $q_n \approx 0$ so that settlements are negligible.

22.2 Foundation behaviour

Figure 22.3(a) shows a simple shallow foundation with a gross bearing pressure q, a net bearing pressure q_n and a settlement ρ. If the foundation is rigid (e.g. concrete) the settlement ρ will be uniform and the bearing pressure will vary across the foundation. If, on the other hand, the foundation is flexible (e.g. an earth embankment) the bearing pressure will be uniform but the settlements will vary. Figure 22.3 illustrates mean values of q and ρ for each case. Figure 22.3(b) shows the relationship between net bearing pressure q_n and settlement ρ. The general form of Fig. 22.3(b) is the same for drained and undrained loadings but the magnitudes of the stresses and settlements will be different for each case. As the bearing pressure increases the settlements start to accelerate and at some point the foundation can be said to have failed because the settlements have become large. Foundations do not fail in the sense that they can no longer support a load or the load on them has reached a maximum or starts to decrease. Instead they continue to settle and the bearing pressure continues to increase slowly as the depth of the foundation increases with further settlement. Sometimes a foundation under eccentric loading starts to rotate like the leaning Tower of Pisa and then it can reach a state where the mean bearing capacity starts to decrease.

Notice that I have defined the bearing capacity q_c as a net bearing pressure at which settlements accelerate. Some other texts and codes define the bearing capacity as a gross bearing pressure. You should watch out for this.

Obviously you cannot load a building foundation close to its bearing capacity q_c as the settlements would then be too large and the building would probably be damaged (although it may not fall down). To limit the settlements to some allowable value ρ_a it is necessary to reduce the bearing pressure to some allowable bearing pressure q_a, as shown in Fig. 22.3(b). In practice this is usually achieved by applying a load factor to the bearing capacity.

Figure 22.4(a) shows the net bearing pressure of a foundation increased to q_a slowly so that the loading is drained. The foundation settlements increase in parallel with

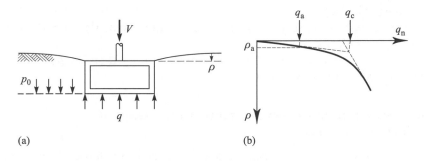

Figure 22.3 Loading and settlement of a foundation.

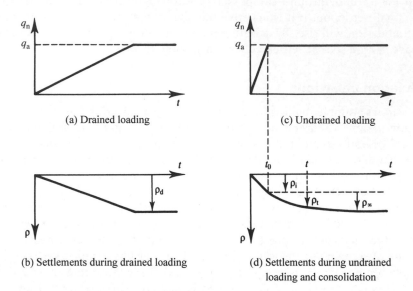

(a) Drained loading

(c) Undrained loading

(b) Settlements during drained loading

(d) Settlements during undrained loading and consolidation

Figure 22.4 Loading and settlement of foundations.

the loading and terminate as ρ_d as shown in Fig. 22.4(b). Figure 22.4(c) shows the same loading increased quickly so the loading is undrained and there is an immediate, undrained settlement ρ_i as shown in Fig. 22.4(d). The undrained loading raises the pore pressure in the soil below the foundation and dissipation of the excess pore pressures causes consolidation settlements to occur. The settlement at some time t after the start of consolidation is ρ_t and the final consolidation settlement which occurs after a relatively long time is ρ_∞. (Notice that the loadings and settlements shown in Fig. 22.4 are similar to those shown in Figs. 6.9 and 6.10 which describe the fundamental differences between drained and undrained loading and consolidation.)

Generally, engineers designing foundations will need to calculate all, or some, of the following:

1. The bearing capacity q_c (to ensure that the foundation has an adequate margin of safety against collapse).
2. The allowable bearing pressure q_a and either the drained settlements ρ_d or the (undrained) immediate settlement ρ_i.
3. For consolidation after loading, the final consolidation settlement ρ_∞, and the variation of settlement ρ_t, with time.

22.3   Soil strength parameters and factors for design of shallow foundations

The main consideration for design of safe slopes described in Chapter 21 is to prevent the slope reaching an ultimate limit state in which it fails. The appropriate soil strength for this is the critical state strength and the factors applied are factors of safety. Normally engineers do not worry about relatively small ground movements near safe slopes. Foundations, however, are designed to a serviceability limit state in which settlements are smaller than those which might lead to damage of the supported structure. It is still necessary of course to check that the foundation loads do not come close to the ultimate bearing capacity but normally settlement criteria control the design of a foundation. The loads are limited to the allowable bearing capacity shown in Fig. 22.3(b) which cause allowable settlements ρ_a.

Two methods for design for serviceability limit states were described in Chapter 18. In one method a load factor is applied to a the bearing capacity to bring the design to a point where settlements are small. In the other method settlements are related to bearing pressures through soil stiffness and this will be covered later in this chapter.

The allowable net bearing pressure q_a is related to the net bearing capacity q_c by

$$q_a = L_f q_c \tag{22.4}$$

where L_f is a load factor. (Notice that the range of a load factor is from 0 to 1.0 while a factor of safety is ≥ 1.0.) The question is which strength, drained or undrained, peak, critical state or residual should be used to calculate the net bearing capacity.

Changes of stress and pore pressure in soil below a foundation as it is loaded drained or undrained are considered in Sec. 22.4. These analyses show that the state in the soil approaches failure more rapidly during undrained loading than during drained loading and, after undrained loading pore pressures during subsequent consolidation fall, effective stresses rise so the soil becomes stronger. So, for foundations on fine grained soils the bearing capacity should be found from the undrained strength. In the case of a foundation on a coarse grained soil the soil would be drained throughout loading and the bearing capacity should be found from the effective stress strength.

As discussed in Sec. 18.6 movements in soil below foundations are too small to develop residual strengths and use of the critical state strength is illogical because then you would design the same foundation on dense and loose sand. The peak strength should be used to calculate the value of the net bearing capacity for Eq. (22.4). This is because, as discussed in Sec. 18.6, all samples of the same soil will reach their peak states at about the same strain so stiffness is related to peak strength.

If the width of a typical foundation is 10 m and the allowable settlement is 10 mm the value of ρ_a/B is 0.1% and this is representative of the strains in the ground beneath the foundation. If $\varepsilon_p = 1\%$, then $\varepsilon_p \approx 10\varepsilon_a$ and if the curves in Fig. 18.5(a) can be approximated by parabolas then

$$L_f = \frac{q_a}{q_c} \approx \frac{1}{3} \tag{22.5}$$

In practice many shallow foundations are designed with load factors of 1/4 to 1/3 and they usually have small settlements.

It is important to understand that the load factor defined above is not a factor of safety: it is a factor to reduce the bearing pressure from the bearing capacity to a point at which settlements will be small. You may want to apply additional factors, particularly to the soil strength, to take account of uncertainties in your values of applied load and soil strength, such as whether mean or worst credible values have been adopted.

22.4   Stress changes in foundations

The changes of stress and water content during undrained loading and subsequent consolidation of a foundation are illustrated in Fig. 22.5. In Fig. 22.5(a) the total stresses on a typical element below the foundation are τ and σ and the pore pressure is illustrated by the rise of water in a standpipe. In Fig. 22.5(b) the total stress path $A \to B$ corresponds to increases of σ and τ due to the loading of the foundation. The effective stress path is $A' \to B'$ which corresponds to undrained loading with constant water content, as shown in Fig. 22.5(c). The exact effective stress path $A' \to B'$ will depend on the characteristics of the soil and its initial overconsolidation ratio, as discussed in Chapter 11. Normally the long term steady state pore pressure u_∞ is the same as the intial pore pressure u_0.

As shown in Fig. 22.5(b), the pore pressure immediately after construction u_i is greater than the final steady state pore pressure u_∞ and so the initial excess pore pressure $\bar{u}_i$ is positive. As time passes the total stresses remain essentially unchanged at B, since the foundation loading does not change, but the pore pressures drop. The effective stress path is $B' \to C'$, which corresponds to compression and an increase in the mean normal effective stress, as shown in Figs. 22.5(b) and (c).

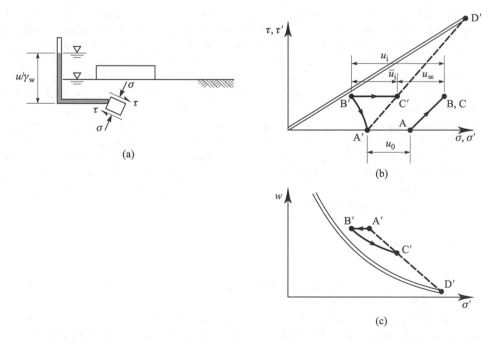

Figure 22.5 Changes of total and effective stress during loading and consolidation of a foundation.

The foundation can be said to fail if all the elements along a critical slip surface such as that in Fig. 19.3 reach the critical state line. The distance of B' from the critical state line is a measure of the factor of safety of the foundation and Fig. 22.5 demonstrates that the factor of safety of a foundation initially loaded undrained generally increases with time but there will be continuing settlements due to consolidation.

The stress path for drained loading of a foundation is the broken line A' → C' → D' in Fig. 22.5(b) and (c). From the geometry of the diagram the stress path approaches the critical state line slowly and you can continue adding load to a foundation on sand although settlements will become large.

22.5   Bearing capacity of shallow foundations

The bearing capacity of a foundation can be calculated using the upper and lower bound methods (Chapter 19) or the limit equilibrium method (Chapter 20). There are standard solutions which are routinely used in practice.

(a)   Undrained loading

The gross undrained bearing capacity of the simple shallow foundation shown in Fig. 22.6(a) is

$$q_c = s_u N_c + p_0 \tag{22.6}$$

where N_c is a bearing capacity factor and p_0 is the total stress at the level of the base of the foundation. For a long foundation at ground level the equal upper and lower bounds obtained in Sec. 19.10 given by Eqs. (19.43) and (19.46) are equivalent to

$$N_c = (2 + \pi) \tag{22.7}$$

The value of N_c depends on the shape and depth of the foundation and values given by Skempton (1951) are shown in Fig. 22.6(b). The allowable gross bearing pressure q_a is obtained by applying a load factor to the net bearing pressure and is given by

$$q_a = L_f s_u N_c + p_0 \tag{22.8}$$

Remember that the gross bearing pressure is the total stress at the base of the foundation and it includes the applied load and the weight of the foundation so the allowable applied load V_a is

$$V_a + W = L_f s_u N_c B + \gamma B D \tag{22.9}$$

If the soil is water so $s_u = 0$ and $\gamma = \gamma_w$ Eq. (22.9) states that the weight of the foundation and the applied load is equal to the weight of water displaced, which is Archimedes' principle.

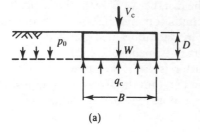

(a)

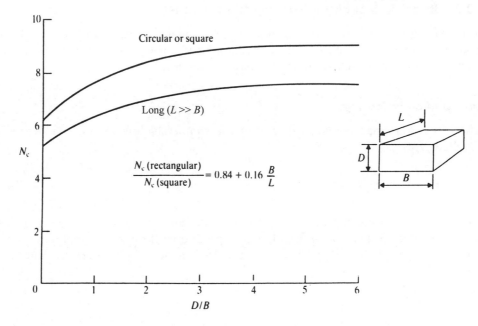

Figure 22.6 Bearing capacity factors for undrained loading of foundations.

(b) Drained loading

The gross bearing capacity of the simple shallow foundation shown in Fig. 22.7(a) for drained loading is

$$q_c = \left[\tfrac{1}{2}(\gamma - \gamma_w)\, BN_\gamma + (\gamma - \gamma_w)\,(N_q - 1)\, D\right] + p_0 \tag{22.10}$$

where N_γ and N_q are bearing capacity factors and p_0 is the total stress at the level of the base of the foundation.

N_q is the contribution to the bearing capacity arising from the surcharge stress acting at the level of the base of the foundation. In Sec. 19.11 we obtained equal upper and lower bounds for a simple long foundation with a surcharge p acting at the surface

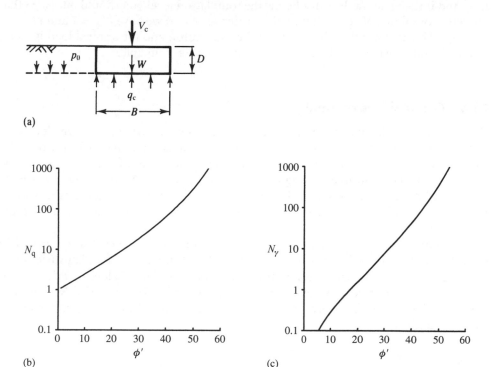

Figure 22.7 Bearing capacity factors for drained loading of foundations.

and these are equivalent to

$$N_q = \tan^2\left(\frac{\pi}{4} + \frac{\phi'}{2}\right)\exp\left(\pi \tan\phi'\right) \qquad (22.11)$$

Values for N_q given by Eq. (22.11) are shown plotted against ϕ' in Fig. 22.7(b). N_γ is the contribution to the bearing capacity created by the self weight of the soil beneath the foundation. There are no simple upper and lower bounds for this case. The values of N_γ shown in Fig. 22.7(c) were found from a numerical method given by Martin (2003).

The allowable bearing pressure is obtained by applying a load factor to the net bearing pressure. Remember that the gross bearing pressure is the total stress at the base of the foundation and it includes the applied load and the weight of the foundation, so the allowable applied load V_a is

$$V_a + W = L_f\left[\frac{1}{2}(\gamma - \gamma_w)B^2 N_\gamma + (\gamma - \gamma_w)(N_q - 1)BD\right] + \gamma BD \qquad (22.12)$$

If the soil is dry you should put $\gamma_w = 0$ in Eq. (22.12). The term containing N_γ is the contribution to the bearing capacity from the unit weight of the soil below foundation level and if the water table is just below the foundation level you should put $\gamma_w = 0$ in the term containing N_q only. If the soil is water so $\phi' = 0$ we have $N_q = 1$ and $N_\gamma = 0$ and Eq. (22.12) states that the weight of the foundation and the applied load is equal to the weight of water displaced, which is Archimedes' principle.

22.6   Foundations on sand

Foundations on sand will be drained and the settlements ρ_d will occur as the loads are applied, as shown in Fig. 22.4(b). Figure 18.5 illustrates the different behaviour of a foundation on a dense sand and a loose sand and shows that for a given allowable settlement ρ_a the allowable bearing pressure q_a depends on the initial state. A simple and logical design procedure would be to relate the allowable bearing pressure directly to the distance of the initial state from the critical state line measured in some suitable *in situ* test.

The routine test to measure state in the ground is the standard penetration test (SPT) described in Sec. 16.5. The result is given as a blowcount value N which varies from small values (1 to 5) when the soil is at its loosest state to large values (over 50) when the soil is at its most dense state. A simple relationship between the SPT-N value and the allowable bearing pressure was given by Terzaghi and Peck (1967) and a simple rule of thumb is

$$q_a = 10N \text{ kPa} \qquad (22.13)$$

This bearing pressure will give settlements of the order of 25 mm (1 inch). Because at relatively small loads the load settlement curve in Fig. 18.5(b) is approximately linear, halving the bearing pressure will give about half the settlement and so on.

22.7   Combined vertical and horizontal loading on shallow foundations

Normally the loading on a foundation is vertical but there are many examples where a foundation is required to support both vertical and horizontal loads. Horizontal loads may be due to wind, waves or earthquakes or from the design of the structure. Figure 22.8 shows a foundation with a horizontal load H and a vertical load V. We have already obtained solutions for the bearing capacity for the case where $H = 0$ and the loading is vertical. If V is small failure will occur when the shear stress on the base of the foundation exceeds the soil strength and the foundation slides sideways. There are other combinations of V and H which cause the foundation to fail.

A simple and effective approach is to construct a failure envelope which separates safe from unsafe states. This can also be considered to be a plastic potential from which movements as the foundation fails can be found. The principles are similar to those shown in Figs. 3.14 and 3.15.

Figure 22.8(b) shows a failure envelope for a simple foundation for undrained loading. (The axes have been plotted in the directions of the loads and they have been normalized by dividing by $s_u B$.) When $H = 0$ the foundation fails when V is given by Eq. (22.6) with $V/S_u B = (2 + \pi)$. When $V = 0$ the foundation slides sideways when

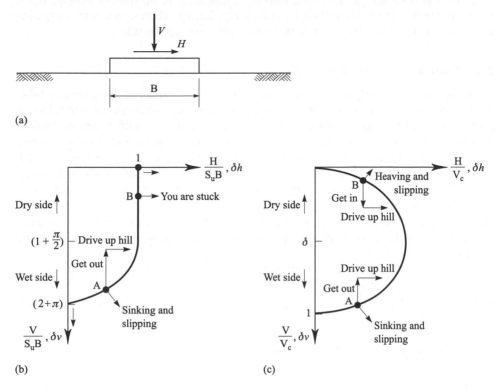

Figure 22.8 Combined loading on a shallow foundation.

$H = s_u B$ and it will continue to do so as V is increased. When $V/s_u B = (1 + \frac{1}{2}\pi)$ the horizontal force to cause failure starts to decrease and when $V/s_u B = (2 + \pi)$ the foundation cannot support any horizontal load. If the envelope in Fig. 22.8(b) is a plastic potential the directions of movement δh and δv are given by the directions of the arrows which are normal to the envelope.

Figure 22.8(c) shows a failure envelope for a simple foundation for drained loading. Most of the features in Fig. 22.8(b) for undrained loading are also in Fig. 22.8(c) for drained loading. If $V = 0$ then $H = 0$ because the shearing resistance is frictional. For small values of V failure by increasing H will cause dilation and the foundation will heave and slide sideways. Because Figs. 22.8(b) and (c) have been normalized with respect to the failure load V_c the upper parts correspond to states on the dry side of critical (dense sand and overconsolidated clay) and the lower parts correspond to states on the wet side (loose sand and normally consolidated clay).

Figures 22.8(b) and (c) illustrate what you should do if you are in a car, off road, and you come to a hill. There will have to be shear stresses between the tyres and the ground to get you up the hill so some horizontal load H must be mobilized. If you are at point A on loose sand or normally consolidated clay the wheels are spinning and the tyres are sinking into the ground. You should unload the car to reduce V and then drive up the hill slowly to keep H small. If you are at point B in dense sand in

Fig. 22.8(c) you should get everyone into the car and drive up the hill slowly. But if you are at point B in overconsolidated clay in Fig. 22.8(b) there is no point in getting into or out of the car: you are stuck and you cannot get up the hill.

22.8   Foundations in elastic soil

An assumption commonly made in practice is that soil is elastic and there are a number of standard solutions for distributions of stresses and ground movements around foundations subjected to a variety of loads. These solutions have generally been obtained by integrating solutions for point loads and so they employ the principle of superposition which is valid only for linear materials. We have seen earlier (Chapters 12 and 13) that soils are usually neither elastic nor linear and so these solutions are not strictly valid, although the errors in calculation of stresses are likely to be considerably less than those in the calculation of ground movement.

The changes of the vertical stress $\delta\sigma_z$ and the settlements $\delta\rho$ at a point in an elastic soil due to a change δQ of a point load at the surface, shown in Fig. 22.9, are given by

$$\delta\sigma_z = \frac{3\delta Q}{2\pi R^2} \left(\frac{z}{R}\right)^3 \tag{22.14}$$

$$\delta\rho = \frac{\delta Q(1+\nu)}{2\pi E R} \left[\left(\frac{z}{R}\right)^2 + 2(1-\nu)\right] \tag{22.15}$$

where E and ν are Young's modulus and Poisson's ratio. Although these expressions lead to infinite stresses and settlements immediately below the point load where $z = R = 0$, they can be used to calculate stresses and settlements some way below small foundations.

For circular or rectangular foundations on elastic soil the changes of vertical stress $\delta\sigma_z$ and settlement $\delta\rho$ at a point below a foundation due to a change of bearing pressure

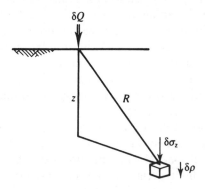

Figure 22.9 Stresses and settlements due to a point load.

δq are given by

$$\delta\sigma_z = \delta q I_\sigma \qquad\qquad (22.16)$$

$$\delta\rho = \delta q B \frac{1 - \nu^2}{E} I_\rho \qquad\qquad (22.17)$$

where I_σ and I_ρ are dimensionless influence factors and B is the width or the diameter of the foundation. The values for the influence factors depend principally on the geometry of the foundation and, to a lesser extent, on the value of Poisson's ratio. Notice Eqs. (22.16) and (22.17) do not contain either E or ν and so the vertical stress in elastic soil depends only on the shape and loading of the foundation. A comprehensive set of tables and charts for influence factors for a wide variety of loading cases are given by Poulos and Davis (1974). Values for the most common simple cases for circular and rectangular loaded areas are shown in Figs. 22.10 and 22.11.

To determine values inside or outside a rectangular or irregularly shaped area you can simply divide the region into a number of rectangles, determine $\delta\sigma_z$ or $\delta\rho$ at the corners of the various rectangles and, making use of the principle of superposition, add or subtract the individual effects. For example, for the L-shaped building in Fig. 22.12 the stresses and settlements at the corner E can be found by adding the effects of the rectangles DABE, BCFE and HGDE; the stresses and settlements at the external

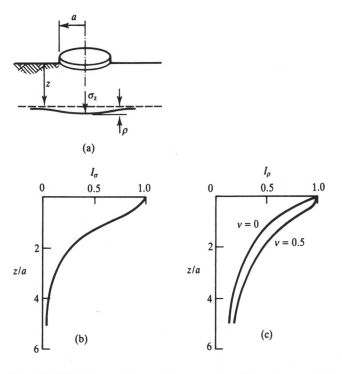

(a)

(b)

(c)

Figure 22.10 Influence factors for stresses and settlements below the centre of a circular foundation.

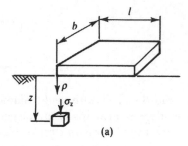

(a)

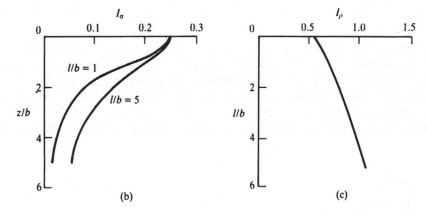

(b)   (c)

Figure 22.11 Influence factor for stresses and settlements below the corner of a rectangular foundation.

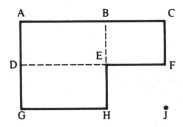

Figure 22.12 Division of rectangular loaded areas.

point J can be found by subtracting the effects of the rectangle HEFJ from those of the rectangle GACJ.

Figures 22.10(b) and 22.11(b) give the depths of influence of surface loads so the increase of stress due to a foundation can be compared with the original stress in the ground. During ground investigations samples should taken and tested at all depths where the increase of stress due to the foundation is more than about 10% of the original *in situ* stress.

22.9   Parameters for elastic analyses

The parameters in these elastic calculations are Young's modulus and Poisson's ratio. In selecting values for design of foundations it is necessary, as always, to distinguish between drained and undrained loading. It is also necessary to recognise that soil stiffness is highly nonlinear and values appropriate to the strains in the soil in the ground should be selected.

Analyses can be done in one step using secant moduli or in several steps using tangent moduli, as described in Sec. 18.7 where the secant modulus is

$$E_{sec} = \frac{\Delta q}{\Delta \varepsilon_a} \tag{22.18}$$

and the tangent modulus is

$$E_{tan} = \frac{dq}{d\varepsilon_a} \tag{22.19}$$

where Δ represents the change of stress and strain from the start of the loading. For simple analyses the secant modulus method would normally be used and the step taken as the whole foundation loading. For drained loading you should choose the parameters E' and ν' corresponding to effective stresses and for undrained loading you should choose E_u and $\nu_u = 0.5$ corresponding to undrained, constant volume loading. The basic relationship between the elastic shear modulus G and the elastic Young's modulus E (see Sec. 3.8) is

$$G = \frac{E}{2(1 + \nu)} \tag{22.20}$$

For an elastic material for which shear and volumetric effects are decoupled we have $G' = G_u$ and hence

$$\frac{E'}{2(1 + \nu')} = \frac{E_u}{2(1 + \nu_u)} \tag{22.21}$$

or, with $\nu_u = 0.5$.

$$E_u = \frac{3E'}{2(1 + \nu')} \tag{22.22}$$

The settlements of a foundation for drained loading ρ_d or for undrained loading ρ_u are given by Eq. (22.17) or with the appropriate values for E and ν. Hence, making use of Eq. (22.22)

$$\frac{\rho_u}{\rho_d} = \frac{3E'}{4\left(1 - \nu'^2\right)E_u} = \frac{1}{2(1 - \nu')} \tag{22.23}$$

and, taking a typical value of $\nu' = 0.25$ we have $\rho_u = 0.67\rho_d$. Thus, for foundations on a deep bed of elastic soil the settlements for undrained loading are of the order of two-thirds those for drained loading of the same foundation; the difference is made up by the additional settlements that occur due to consolidation after undrained loading. If the depth of the soil is relatively small compared to the width of the foundation so that the conditions in the soil are one-dimensional (see Sec. 22.10), $\rho_u = 0$.

Non-linear soil stiffness was discussed in Chapter 13. The characteristic variation of stiffness with strain is illustrated in Fig. 13.8. At very small strain the value of Young's modulus is E_0. This can be determined from values of G_0 measured in dynamic laboratory or *in situ* tests and it varies with stress and state, as given by Eq. (13.8) and shown in Fig. 13.9. Figure 18.6(b) shows the variations of tangent and secant modulus with strain corresponding to the stress–strain curve in Fig. 18.6(a). The stiffnesses have been normalized by dividing by E_0. At the critical state at F the strain is about 10% and $E_{tan} = 0$. At the peak state at P the strain is about 1% and, again, $E_{tan} = 0$.

As discussed in Sec. 13.4 the average strains in the ground near a typical foundation at working load are about 0.1% but locally they cover a very wide range from less than 0.01% to more than 1%, as shown in Fig. 18.6(b). For design it is necessary to choose a value of stiffness which corresponds to the mean strains in the ground, as discussed by Atkinson (2000).

22.10   Consolidation settlements for one-dimensional loading

An assumption commonly made is that the thickness of a compressible soil layer is small compared to the width of the loaded foundation: so that the horizontal strains can be neglected. In this case the conditions of stress strain and consolidation in the ground, shown in Fig. 22.13(a) are the same as those in the one-dimensional oedometer test described in Secs. 7.6 and 8.5 and shown in Fig. 22.13(b).

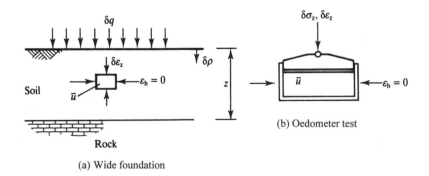

(a) Wide foundation

(b) Oedometer test

Figure 22.13 One-dimensional consolidation in foundations.

In the oedometer test the vertical strains $\delta\varepsilon_z$ are given by Eq. (8.9) as

$$\delta\varepsilon_z = m_v\,\delta\sigma'_z \qquad (22.24)$$

where, for complete consolidation when $\bar{u} = 0$ we have $\delta\sigma'_z = \delta\sigma_z$. Notice that, as discussed in Sec. 8.5 the value on m_v is not a soil constant but it depends on the current stress σ'_{z0} on the change of stress $\delta\sigma'_z$ and is different for loading and unloading. At the ground the surface settlements due to consolidation $\delta\rho_c$ are given by

$$\frac{\delta\rho_c}{z} = \delta\varepsilon_z = m_v\delta\sigma'_z \qquad (22.25)$$

where, for complete consolidation, we have $\delta\sigma'_z = \delta q$, where δq is the net bearing pressure at the surface. Final consolidation settlements for wide foundations can be calculated using Eq. (22.25). However, because the one-dimensional compression and swelling behaviour of soil is non-linear, m_v is not a soil constant and it is necessary to measure m_v in an oedometer test in which the initial stress and the change of stress both correspond to those in the ground.

The rate at which consolidation settlements occur in one-dimensional oedometer tests was considered in Chapter 15. General solutions for rates of consolidation emerge as relationships between the degree of consolidation U_t and the time factor T_v. These are defined as

$$U_t = \frac{\Delta\rho_t}{\Delta\rho_\infty} \qquad (22.26)$$

$$T_v = \frac{c_v t}{H^2} \qquad (22.27)$$

where $\Delta\rho_t$, and $\Delta\rho_\infty$ are the settlements at times t and $t = \infty$, c_v is the coefficient of consolidation and H is the drainage path length.

Relationships between U_t and T_v depend on the geometry of the consolidating layer and its drainage conditions and on the distribution of initial excess pore pressure but not on its absolute value. The most common drainage conditions are shown in Fig. 22.14. For one-dimensional drainage the seepage may be one-way towards a drainage layer at the surface, two-way towards drainage layers at the base and at the surface or many-way towards silt or sand layers distributed through the deposit. For radial drainage seepage is towards vertical drains placed on a regular grid. In each case the drainage path length, H or R, is the maximum distance travelled by a drop of water seeping towards a drain.

For one-dimensional consolidation the relationships between U_t and T_v for different initial excess pore pressure conditions are given in Fig. 15.9 in terms of T_v. These could also be given in terms of T_v plotted to a logarithmic scale, as shown in Fig. 22.15(a), which corresponds to consolidation with the initial excess pore pressure $\bar{u}_i$ uniform

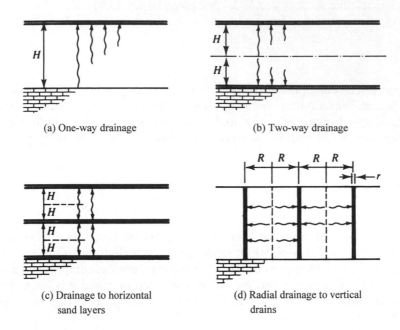

(a) One-way drainage

(b) Two-way drainage

(c) Drainage to horizontal
sand layers

(d) Radial drainage to vertical
drains

Figure 22.14 Drainage conditions in foundations.

with depth. Figure 22.15(b) is for radial consolidation where

$$T_r = \frac{c_r t}{R^2} \tag{22.28}$$

$$n = \frac{R}{r} \tag{22.29}$$

As discussed in Chapter 15, these can be used to calculate either the settlement after a given time or the time for a given settlement. Although, in theory, complete consolidation will require infinite time a reasonable approximation is that T_v or $T_r \approx 1.0$ at $U_t = 1.0$.

22.11   Heave and settlement of foundations due to changes of groundwater

We have seen that in general loading a foundation causes it to settle and unloading will cause it to heave. Foundation movements can be caused by changes other than loading such as underground mining, nearby construction and changes of groundwater. Groundwater can change for a number of reasons, including water extraction, dewatering nearby excavations and changes in vegetation. Foundation movements due to its loading are usually called settlement and those due to other changes are usually called subsidence.

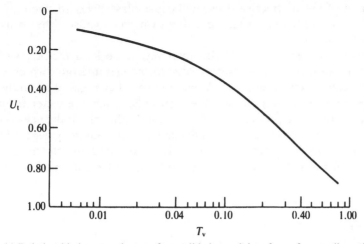

(a) Relationship between degree of consolidation and time factor for one-dimensional consolidation. (*After* Taylor, 1948.)

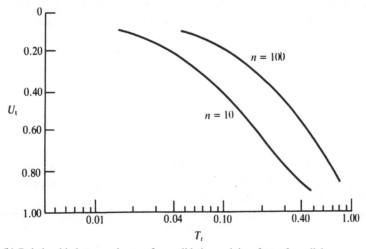

(b) Relationship between degree of consolidation and time factor for radial consolidation. (*After* Barron, 1948.)

Figure 22.15 Solutions for rate of consolidation.

From Eq. (22.8) the drained bearing capacity of a foundation depends on the factors N_γ and N_q and on the effective stresses which arise from the terms $(\gamma - \gamma_w)B$ and $(\gamma - \gamma_w)D$. If the groundwater and pore pressures rise the effective stresses in the ground below the foundation reduce and the bearing capacity decreases. If the foundation has a fixed bearing pressure due to the weight of the structure it is supporting the load factor decreases and the foundation settles due to increased shear straining. On the other hand reduction of mean effective stress will cause plastic clay to swell and the foundation will then heave. Rising or falling groundwater can cause either heave or settlement of foundations.

Foundations which are lightly loaded with relatively large load factors, such as house foundations, are susceptible to subsidence due to changes in groundwater which cause high plasticity clay to shrink or swell. A major cause of changes of groundwater near houses is vegetation or leaking drains. Trees and plants remove water from the ground and so reduce the pore pressures in the ground within the influence of their roots. Allowing vegetation to grow can cause subsidence while removing vegetation or allowing drains to leak can cause heave. If the soil is unsaturated and the foundation has a relatively heavy loading, flooding can cause collapse settlements, as described in Chapter 26.

22.12   Summary

1. Foundations transmit loads to the ground. As the load increases the foundation settles and it fails when the settlements become very large. Foundations may be shallow or they may be deep to take advantage of the general increase of strength and stiffness of soils with depth.
2. The bearing pressure q is the contact stress between the foundation and the soil. The net bearing pressure of a deep foundation is the change of bearing pressure; this may be positive so the foundation settles or it may be negative so it heaves. The bearing pressure q and the net bearing pressure q_n are given by

$$q = \frac{V + W}{A} \tag{22.1}$$

$$q_n = q - p_0 \tag{22.3}$$

3. Under a foundation pore pressures generally increase with undrained loading and, with time, these dissipate as the soil consolidates. As a result further settlements occur but effective stresses and safety factors increase.
4. The bearing pressure when the foundation fails is the bearing capacity q_c given by

$$q_c = s_u N_c + p_0 \tag{22.6}$$

$$q_c = \left[\tfrac{1}{2}(\gamma - \gamma_w)BN_\gamma + (\gamma - \gamma_w)(N_q - 1)D \right] + p_0 \tag{22.10}$$

for undrained and drained loading respectively, where N_c, N_γ and N_q are bearing capacity factors.
5. An important criterion for foundation design is the need to limit the settlements. This may be done by applying a load factor to the net bearing pressure. Alternatively, settlements may be calculated assuming that the soil in the foundation is elastic. For foundations on sand settlements are related to the relative density which may be estimated from the results of SPT tests.

6. For wide foundations on relatively thin beds of soil the strains during consolidation are one-dimensional. The magnitude of the settlement is given by

$$\delta \rho_c = z m_v \, \delta \sigma'_z \qquad\qquad (22.25)$$

The rate of settlement is given by the relationship between the degree of consolidation and the time factor, which are given by

$$U_t = \frac{\Delta \rho_t}{\Delta \rho_\infty} \qquad\qquad (22.26)$$

$$T_v = \frac{c_v t}{H^2} \qquad\qquad (22.27)$$

A reasonable approximation is $T_v = 1$ when $U_t = 1$.

Worked examples

Example 22.1: Undrained bearing capacity of a foundation   For the foundation in Fig. 22.16 the ultimate load for undrained loading is found from Eq. (22.6)

$$V_c + W = s_u N_c B + \gamma D B$$

If the unit weights of soil and concrete are the same, $W = \gamma D B$. From Fig. 22.6(b), for a long foundation with $D/B \approx 1$ we have $N_c = 6$ and

$$V_c = 30 \times 6 \times 2.5 = 450 \, \text{kN/m}$$

If the applied load is $V_a = 300 \, \text{kN/m}$ the load factor is 1.5.

Example 22.2: Drained bearing capacity of a foundation   For the foundation in Fig. 22.16 the ultimate load for drained loading is found from Eq. (22.10)

$$V_c + W = \tfrac{1}{2}(\gamma - \gamma_w) B^2 N_\gamma + (\gamma - \gamma_w)(N_q - 1) BD + \gamma BD$$

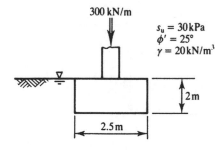

Figure 22.16 Loading of a foundation – Examples 22.1 and 22.2.

As before, $W = \gamma BD$. From Fig. 22.7(b), for $\phi' = 25$, $N_\gamma = 8$ and $N_q = 11$ and

$$V_c = \tfrac{1}{2}(20 - 10)2.5^2 \times 8 + (20 - 10)(11 - 1)2.5 \times 2 = 750\,\text{kN/m}$$

If the applied load is $V_a = 300\,\text{kN/m}$ the load factor is 2.5.

Example 22.3: Settlement of an embankment   The embankment in Fig. 22.17 is sufficiently wide so that the strains and seepage in the soil can be assumed to be one-dimensional. From Eq. (22.25) the magnitude of the final consolidation settlement is

$$\rho_c = m_v z \Delta\sigma_z' = 5 \times 10^{-4} \times 8 \times 100 = 0.40\,\text{m}$$

(a)   From Fig. 22.15(a) the time when the settlement is complete (i.e. when $U_t = 1.0$) corresponds to $T_v = 1.0$. Hence, from Eq. (22.27),

$$t = \frac{T_v H^2}{c_v} = \frac{1.0 \times 8^2}{2} = 32\,\text{years}$$

(b)   After 5 years the time factor is

$$T_v = \frac{c_v t}{H^2} = \frac{2 \times 5}{8^2} = 0.16$$

From Fig. 22.15(a) this corresponds to a degree of consolidation $U_t \approx 0.50$ and the settlement after 5 years is

$$\rho_t = U_t \rho_\infty = 0.50 \times 0.40 = 0.20\,\text{m}$$

Example 22.4: Settlement with drains   In order to speed up the settlement of the embankment in Example 22.3 sand drains are installed in the clay. The drains are 200 mm in diameter ($r = 100\,\text{mm}$) and they are spaced 2 m apart ($R = 1.0\,\text{m}$).

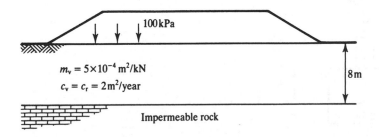

Figure 22.17 Embankment settlement – Examples 22.3 and 22.4.

From Fig. 22.15(b), with $n = R/r = 10$, the time when settlement is complete (i.e. when $U_t = 1.0$) corresponds to $T_v = 1.0$. Hence, from Eq. (22.28),

$$t = \frac{T_r R^2}{c_r} = \frac{1.0 \times 1.0}{2} = 0.5 \, \text{years}$$

Example 22.5: Calculation of stresses and settlements in elastic soil   Figure 22.18 shows a circular water tank at the surface of a deep bed of elastic soil. For $\delta q = 5 \times 10 = 50 \, \text{kPa}$ the changes of vertical stress and the settlements for drained loading are given by Eqs. (22.16) and (22.17).

$$\delta \sigma_z' = \delta q I_\sigma = 50 I_\sigma \, \text{kPa}$$

$$\delta \rho = \delta q B \frac{1 - v'^2}{E'} I_\rho = 50 \times 10 \frac{(1 - 0.25^2) \times 10^3}{10 \times 10^3} I_\rho = 47 I_\rho \, \text{mm}$$

where I_σ and I_ρ are given in Fig. 22.10.

(a)   At point A, $z/a = 0$ so $I_\sigma = 1.0$ and $I_\rho = 1.0$; hence

$$\delta \sigma_z' = 50 \, \text{kPa}$$

$$\delta \rho = 47 \, \text{mm}$$

(b)   At point B, $z/a = 1$ so $I_\sigma = 0.65$ and, for $v' = 0.25$ (interpolating between the data for $v' = 0$ and $v' = 0.5$), $I_\rho = 0.65$; hence

$$\delta \sigma_z' = 33 \, \text{kPa}$$

$$\delta \rho = 31 \, \text{mm}$$

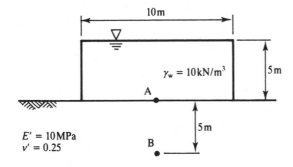

Figure 22.18 Settlement of a water tank on elastic soil – Example 22.5.

References

Atkinson, J. H. (2000) 40th Ranking Lecture. Non-linear soil stiffness in routine design. Geotechnique, Vol. 50, No. 5, pp. 487–508.

Barron, R. A. (1948) 'Consolidation of fine grained soils by drain wells', *Trans. Am. Soc. Civil Engng*, 113, 718–754.

Martin, C. M. (2003) New Software for rigorous bearing capacity calculations, *Proc. Int. Conf. on Foundations*, Dundee, 581–592.

Poulos, H. G. and E. H. Davis (1974) *Elastic Solutions for Soil and Rock Mechanics*, Wiley, New York.

Skempton, A. W. (1951) 'The bearing capacity of clays', *Proceedings of Building Research Congress*, Vol. 1, pp. 180–189, ICE, London.

Taylor, D. W. (1948) *Fundamentals of Soil Mechanics*, Wiley, New York.

Terzaghi, K. and R. B. Peck (1967) *Soil Mechanics in Engineering Practice*, Wiley, New York.

Further reading

Atkinson, J. H. (1981) *Foundations and Slopes*, McGraw-Hill, London.

Burland, J. B. and C. P. Wroth (1975) 'Settlement of buildings and associated damage', *Proceedings of Conference on Settlement of Structures*, Pentech Press, London.

Padfield, C. J. and M. J. Sharrock (1983) *Settlement of Structures on Clay Soils*, CIRIA, Special Publication 27, London.

Peck, R. B., W. E. Hanson and T. H. Thorburn (1974) *Foundation Engineering*, Wiley, New York.

Tomlinson, M. J. (1995) *Foundation Design and Construction*, Longman.

Chapter 23

Piled foundations

23.1 Types of piled foundations

Piles are long slender columns installed into the ground, often in groups. The principal purpose of piling is to transfer loads to stronger and stiffer soil or rock at depth, to increase the effective size of a foundation and to resist horizontal loads. Typically piles are made from steel or reinforced concrete and possibly timber. They may be driven or pushed into the ground or concrete piles may be cast *in situ* by pouring concrete into a drilled hole.

Some typical pile types are illustrated in Fig. 23.1. Figure 23.1(a) shows an end bearing pile where most of resistance is developed at the toe and Fig. 23.1(b) shows a friction pile where a significant contribution to the pile capacity is developed by shear stresses along the sides. Figure 23.1(c) shows raking piles to resist horizontal loads and Fig. 23.1(d) is a pile group joined at the top by a pile cap. Notice that the pile on the left in Fig. 23.1(c) is in tension and so all the resistance comes from shear stress on the sides of the pile.

Figure 23.2 shows the loads on a single pile: the applied load Q is resisted by a force at the base Q_b and a force Q_s due to the shear stresses between the soil and the pile shaft; hence

$$Q = Q_s + Q_b \tag{23.1}$$

In conventonal pile analysis the weight of the pile is taken to be the same as the weight of soil displaced by the pile and both are neglected. In any case these forces are usually small compared with the applied loads, which are typically in the range 500 to 5000 kN and may be considerably larger. Figure 23.2(b) illustrates the increase in base resistance and shaft friction with displacement. The shaft friction increases more quickly than the base resistance and reaches an ultimate state at relatively small strains.

Piles or pile groups may be loaded drained or undrained and the basic total and effective stress paths will be similar to those for shallow foundations, shown in Fig. 22.5. Generally, piles installed in a clay soil will settle with time as the excess pore pressures generated by undrained loading dissipate and the effective stresses and strength of the soil increases. There may, however, be stress changes caused by installation, which would cause swelling and softening of the soil around a bored and cast *in situ* pile or compression and consolidation around a driven pile.

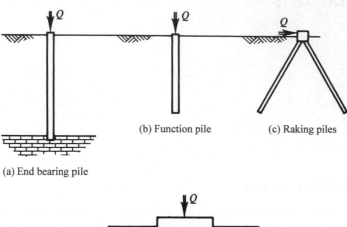

(b) Function pile   (c) Raking piles

(a) End bearing pile

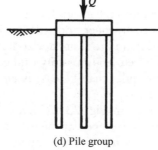

(d) Pile group

Figure 23.1 Types of piled foundations.

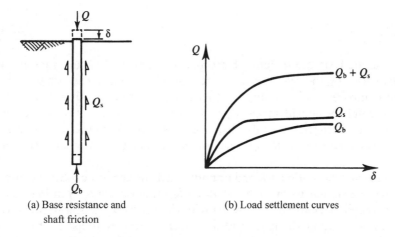

(a) Base resistance and   (b) Load settlement curves
    shaft friction

Figure 23.2 Pile resistance.

23.2 Base resistance of single piles

The base resistance of a single pile is given by

$$Q_b = q_b A_b \qquad (23.2)$$

where q_b is the bearing capacity at the toe and A_b is the area of the pile base. The general principles for calculation of the bearing capacity of piles are similar to those for shallow foundations described in Chapter 22. The mechanism of slip surfaces at the tip of a pile appropriate for an upper bound or limit equilibrium calculation will be similar to that shown in Fig. 23.3 and we would expect the bearing capacity factors for piles to be larger than those for shallow foundations. For undrained loading the bearing capacity is given by

$$q_b = s_u N_c \qquad (23.3)$$

and, for square or circular piles, $N_c \approx 9$ (Skempton, 1951). For drained loading the bearing capacity is given by

$$q_b = \sigma_z' N_q \qquad (23.4)$$

where σ_z' is the vertical effective stress at the level of the toe of the pile. Values for the bearing capacity factor N_q depend principally on ϕ' and there are a number of published relationships based on theory and experiment. The values shown in Fig. 23.3(b) are those given by Berezantzev, Khristoforov and Golubkov (1961).

The choice of the appropriate value of ϕ' is problematical. Soil below the toe of a driven pile will be highly strained during driving while there is the possibility of stress relief and softening at the base of a bored and cast *in situ* pile during construction.

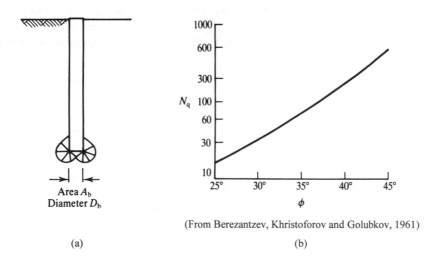

(From Berezantzev, Khristoforov and Golubkov, 1961)

(a)   (b)

Figure 23.3 Base resistance of piles.

Consequently, in both cases a rational design method would take the critical friction angle ϕ_c' to determine a value of N_q for pile design. However, experiments and *in situ* tests indicate that use of ϕ_c' with the values of N_q in Fig. 23.3(b) leads to overconservative designs and often a peak friction angle ϕ_p' is used in practice.

The base resistance of a single pile may also be estimated from the *in situ* probing tests described in Chapter 16. The end bearing capacity of a pile is often equated with the cone resistance measured during a static cone test (sometimes with a correction for the different sizes of the pile and the cone) or derived from the standard penetration test N value.

23.3   Shaft friction on piles

From Fig. 23.4 resistance due to shaft friction on a circular pile, diameter D, is given by

$$Q_s = \pi D \int_0^L \tau_s \, dz \tag{23.5}$$

where τ_s is the shear stress mobilized between the pile and the soil. The value of τ_s is very difficult to determine; it depends on soil, on the pile material and particularly on the method of installation. For undrained loading of piles in clay,

$$\tau_s = \alpha s_u \tag{23.6}$$

where α must be in the range $0 \le \alpha \le 1$. Typically α is taken to be about 0.5 for both driven and cast *in situ* piles. For drained loading,

$$\tau_s' = \sigma_h' \tan \delta' = K\sigma_z' \tan \delta' \tag{23.7}$$

where K is the ratio of the horizontal and vertical effective stresses σ_h'/σ_z' and must be in the range $K_a \le K \le K_p$ (where K_a and K_p are the active and passive earth pressure coefficients discussed in Chapter 21); δ' is the friction angle for shearing between the

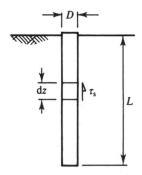

Figure 23.4 Shaft resistance of piles.

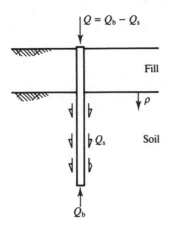

Figure 23.5 Negative shaft friction due to ground settlement.

pile and the soil and for a rough pile this will be in the range $\phi_r' \leq \delta' \leq \phi_p'$. For clays, Eq. (23.7) is often simplified to

$$\tau_s' = \beta \sigma_z' \tag{23.8}$$

where $\beta = K \tan \delta'$ is an empirical parameter that depends on the nature of the soil and on the method of pile installation.

Pile installation influences both δ' and K but differently. When a pile is driven into the ground there will be very large shear displacements between the pile and the soil, and in clays these displacements will probably be enough to reduce the soil strength to its residual value. However, pile driving is likely to increase the horizontal effective stresses which will tend to increase the shaft friction. On the other hand if a pile is driven into cemented soil, the horizontal stress after driving and the available shaft friction could be very small indeed. A cast *in situ* concrete pile is likely to have very rough sides and so the available shearing resistance will lie between the peak and the critical state strength of the soil. However, boring a hole in the ground to construct a cast *in situ* pile will reduce the horizontal stresses which may be reduced still further as the concrete shrinks during setting and curing. For both driven and cast *in situ* piles there are compensating effects on δ' and on K.

Notice that in a soil that is settling, perhaps due to the weight of fill placed at the surface or due to groundwater lowering, the shaft friction will act downwards on the pile as shown in Fig. 23.5, causing negative shaft friction.

23.4   Pile testing and driving formulae

Because of the considerable uncertainties in the analysis of pile load capacity, both in calculation of base resistance and shaft friction, some of the piles on a job will often be subjected to load tests to demonstrate that their capacity is adequate. In typical

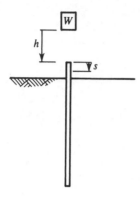

Figure 23.6 Pile driving formulae.

tests, loads will be applied in excess of the design working load and the deflections measured. The loads may be applied in stages and maintained at each increment (like in an oedometer test) or applied at a constant rate of penetration. The latter method is found to give more consistent results and better definition of failure loads.

The capacity of a pile can be inferred from its resistance to driving. The basis of these so-called pile driving formulae is that the work done by the hammer (less any losses) is equal to the work done as the pile penetrates the ground. For the simple drop hammer weight W falling through h shown in Fig. 23.6 the pile capacity Q is related to the set s (i.e. the displacement) for a single blow by

$$Qs = Wh \qquad (23.9)$$

Equation (23.9) is a very simple driving formula, too approximate to be used in practice, but it is the basis of other formulae which include terms to take account of energy losses in the hammer and in the pile.

23.5 Capacity of pile groups

In a group of piles like that shown in Fig. 23.1(d), there will be interactions between neighbouring piles so that the capacity of each pile in the group will be reduced. A group efficiency η is given by

$$V = n\eta Q \qquad (23.10)$$

where V is the total load on the group, n is the number of piles in the group and Q is the capacity of an individual pile on its own. Values for the efficiency η decrease with reduced spacing of the piles, roughly as shown in Fig. 23.7(b).

If the pile spacing is relatively close, as shown in Fig. 23.7(c), it is more appropriate to consider the group as an equivalent foundation of base area A and depth L_g, where $L_g \approx \frac{2}{3}L$. The bearing capacity q_c of the block is calculated using the methods for

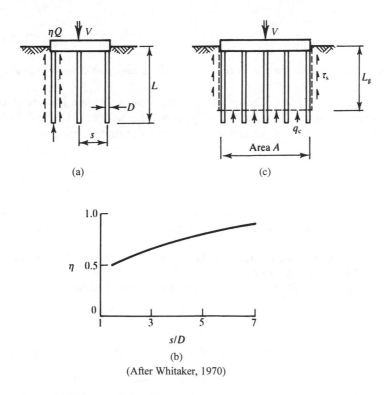

Figure 23.7 Capacity of pile groups.

shallow foundations described in Chapter 22 and the shear stresses on the sides of the block are calculated assuming that the ultimate shear stresses developed correspond to the strength of the soil.

23.6   Summary

1. Piled foundations are used to lower the foundation into soil which is stiffer and stronger. The load capacity of a pile arises from base resistance and shaft friction.
2. Base resistance of a single pile is given by

$$q_b = s_u N_c \tag{23.3}$$

$$q_b = \sigma'_v N_q \tag{23.4}$$

for undrained and drained loading respectively.
3. The shaft friction of a single pile is given by

$$\tau_s = \alpha s_u \tag{23.6}$$

or

$$\tau'_s = \beta \sigma'_z \tag{23.8}$$

where α is a shaft friction factor for undrained loading and, for drained loading, $\beta = K \tan \delta'$.

4.   In practice the capacity of piles is often determined from full-scale load tests or from pile driving formulae. The capacity of groups of piles can be found from the capacity of a single pile with an efficiency factor or from the geometry of an equivalent foundation.

References

Berezantzev, V. G., V. S. Khristoforov and V. N. Golubkov (1961) 'Load bearing capacity and deformation of piled foundations', *Proceedings of 5th International SMFE Conference, Paris*, Vol. 2.

Skempton, A. W. (1951) 'The bearing capacity of clays', *Proceedings of Building Research Congress*, Vol. 1, pp. 180–189, ICE, London.

Whitaker, T. (1976) *The Design of Piled Foundations*, Pergamon Press, London.

Further reading

Fleming, W. G. K., A. J. Weltman, M. F. Randolph and W. K. Elson (1992) *Piling Engineering*, Taylor & Francis.

Tomlinson, M. J. (1994) *Pile Design and Construction Practice*, Spon Press (UK).

Whitaker, T. (1976) *The Design of Piled Foundations*, Pergamon Press, London.

Randolph M. F., M. B. Jamiolkowski and L. Zdravkovic (2004) Load carrying capacity of foundations, in advances in geotechnical engineering, *Proc. The Skempton Conference*, London, 2004. Ed. Jardine, R.J., D.M. Potts and K.G. Higgins, Thomas Telford, London, Vol. l, pp. 207–240.

Chapter 24

Earth pressures and stability of retaining walls

24.1   Introduction

Retaining walls are used to support slopes and vertical cuts that are too steep or too deep to remain stable if unsupported. The principal characteristics of a retaining structure are illustrated in Fig. 24.1. The wall is a structural member that acts as a beam with various loads on either side. Slender walls are embedded into the ground below the excavation level and they may be supported by props or anchors. Thick heavy gravity walls derive their resistance principally from the shear stresses between the soil and the base of the wall. During excavation (or filling on the high side) slender walls will tend to move and bend as indicated as the earth pressures develop. Walls move towards the passive side and away from the active side.

Retaining walls should be designed so they have a margin of safety against failure which might occur in any of the modes discussed in Sec. 24.4. There is often a requirement to limit ground movements so nearby buildings and tunnels are not damaged. Construction of a supported excavation will probably change the groundwater conditions and water pressures contribute to the loads on a wall.

24.2   Earth pressures

Loads on retaining walls arise from the horizontal stresses in the ground (known as earth pressure) and from any props or anchors used to support it. The earth pressure depends principally on whether the wall is moving towards or away from the ground and on the properties of the soil.

The development of earth pressure with displacement is illustrated in Fig. 24.2. In Fig. 24.2(a) a wall supported by a force P retains soil where the horizontal total stress is σ_h; obviously the stresses and the force must be in equilibrium. If P is increased the wall moves towards the soil with displacements δ_p and the horizontal stresses increase, as shown in Fig. 24.2(b); if P is decreased the wall moves away from the soil with displacements δ_a and the horizontal stresses decrease. If the movements are sufficient the horizontal stresses reach the limiting values of the passive pressure σ_p and the active pressure σ_a. If there is no movement the horizontal stress σ_0 is the earth pressure at rest, corresponding to K_0 (see Sec. 8.6).

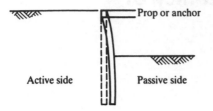

Figure 24.1 Characteristics of a retaining wall.

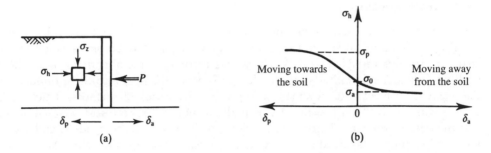

Figure 24.2 Development of active and passive pressures with displacement.

24.3   Types of retaining structure

There are a number of different types of retaining wall and the principal ones are illustrated in Fig. 24.3. Figure 24.3(a) shows a simple cantilever wall where all the support comes from the passive earth pressures. Figure 24.3(b) and (c) illustrates simple propped and anchored walls respectively. Figure 24.3(d) shows a gravity wall where the resistance comes from shear stresses between the ground and the base of the wall. In Fig. 24.3(e) the wall supports the sides of an excavation and in Fig. 24.3(f) the wall supports fill.

Permanent walls are used to support highway cuttings, bridge abutments, basements, dock and harbour walls and so on, while temporary retaining walls are used extensively during construction to support excavations and to provide dry working conditions in coffer dams. Gravity walls are usually of masonry or mass concrete but could also be made from gabions (wire baskets about 0.5 to 1 m cube filled with soil or rock). Slender walls are steel or reinforced concrete. Steel sheet piles are usually driven into the ground while slender concrete walls are usually cast *in situ* as rectangular diaphragm panels or as interlocking or touching cylindrical piles.

24.4   Failure of retaining walls

Retaining walls can fail in a number of different ways. Figure 24.4 illustrates typical failure in the soil where the wall itself remains intact and Fig. 24.5 illustrates typical

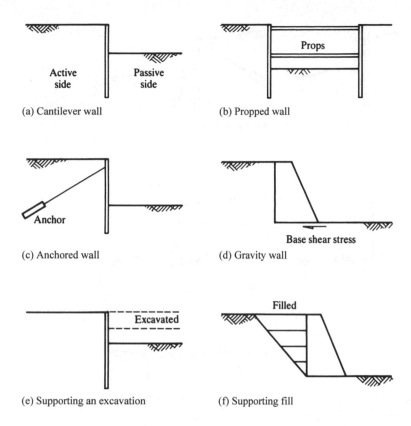

Figure 24.3 Principal types of retaining structure.

failures of the structural elements. The walls in Fig. 24.4(a) and (b) are failing because there are very large distortions in the soil in front of and behind the wall. In Fig. 24.4(c) and (d) a gravity wall may fail by sliding, overturning or by exceeding the limiting bearing pressure at the toe. In Fig. 24.4(e) any retaining wall may fail by slipping below the wall but this is really a problem in slope stability (see Chapter 21). In Fig. 24.4(f) the base of an excavation may fail by piping and erosion due to seepage or by movement of the soil. Figure 24.5 illustrates structural failures of the wall or an anchor or buckling of props.

24.5   Stress changes in soil near retaining walls

It is helpful to consider the total and effective stress paths in soil near retaining walls during and after construction to examine whether the undrained or long-term drained cases are most critical. For retaining walls it is necessary to separate those loaded by excavation from those loaded by filling. (Note that I am continuing to use loading to mean an increase of shear stress irrespective of what happens to the normal stresses.)

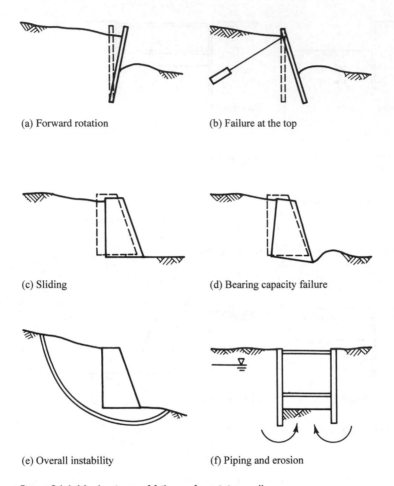

(a) Forward rotation

(b) Failure at the top

(c) Sliding

(d) Bearing capacity failure

(e) Overall instability

(f) Piping and erosion

Figure 24.4 Mechanisms of failure of retaining walls.

Figure 24.6(a) shows a retaining wall loaded by excavation. For convenience the excavation if kept full of water so the long term pore pressures after construction are the same as those before construction. If the excavation is kept dry by pumping, which would be the usual case, the long term pore pressures are governed by a steady state seepage flownet and will usually be less than those before construction. For both the elements shown on the critical slip surfaces, one on the active side and one on the passive side, the shear stresses increase while the mean normal total stresses decrease. The total and effective stress paths are shown in Fig. 24.6(b); these are like those for a slope, shown in Fig. 21.4. The effective stress path $A' \to B'$ corresponds to undrained loading: the exact path will depend on the characteristics of the soil and on its initial overconsolidation ratio, as discussed in Chapter 11.

As shown in Fig. 24.6(b), the pore pressure immediately after construction u_i is less than the final steady state pore pressure u_∞ and so there is an initial excess pore

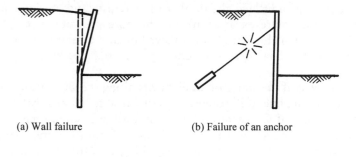

(a) Wall failure   (b) Failure of an anchor

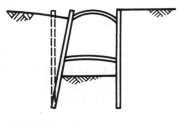

(c) Buckling of props

Figure 24.5 Structural failures of retaining walls.

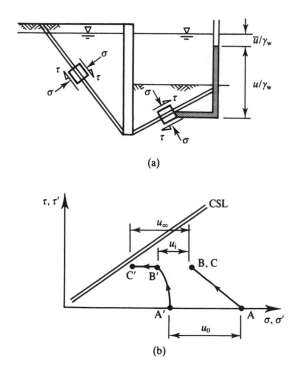

(a)

(b)

Figure 24.6 Changes of stress and pore pressure for a wall retaining an excavation.

pressure which is negative. As time passes the total stresses remain approximately unchanged at B (they will change a little as the total stresses redistribute during consolidation, although there is no more excavation) but the pore pressures rise. The effective stress path is B′ → C′, which corresponds to swelling and a reduction in the mean normal effective stress. The final state at C′ corresponds to a steady state pore pressure u_∞.

The wall will fail in some way if the states of all elements along the slip surfaces in Fig. 24.6(a) reach the critical state line; if B′ reaches the critical state line the wall fails during undrained excavation and if C′ reaches the line the wall fails some time after construction. The distance of the effective stress point B′ or C′ from the critical state line is a measure of the factor of safety against collapse and Fig. 24.6(b) demonstrates that the factor of safety of a retaining wall supporting an excavation will decrease with time. This is the same as for a slope, discussed in Sec. 21.4. We could also trace the state paths for failing walls as we did for failing slopes, but this is not really relevant as retaining walls should not be allowed to fail.

Figure 24.7(a) shows a wall embedded in soil and retaining coarse-grained fill. In this case the shear and normal stresses on typical elements on a slip surface both increase. Total and effective stress paths are shown in Fig. 24.7(b). The effective stress path for undrained loading is A′ → B′ and this is the same as that in Fig. 24.6(b), but the

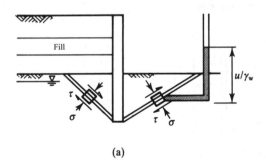

(a)

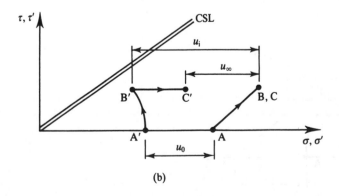

(b)

Figure 24.7 Changes of stress and pore pressure for a wall retaining fill.

total stress path A → B and the initial pore pressures are different. In particular, the initial pore pressure u_i is greater than the final steady state pore pressure u_∞, so the initial excess pore pressure is positive. As time passes the pore pressures decrease as the soil consolidates and the effective stress path is B′ → C′. The effective stress point is moving away from the critical state line so the factor of safety of a wall retaining fill increases with time.

The analyses and the stress paths shown in Figs. 24.6 and 24.7 are simplified and idealized and ignore a number of important aspects such as the installation of the wall into the ground. They do, however, illustrate the general features of the behaviour of retaining walls during and after construction. Notice particularly the fundamental difference between the long-term behaviour of walls supporting excavations and walls retaining fill: the one becomes less safe with time as the soil softens and weakens and the other becomes safer with time as the soil consolidates and strengthens.

24.6   Influence of water on retaining walls

Water influences the loading on retaining walls in a number of fundamentally different ways; the most important of these are illustrated in Fig. 24.8. Figure 24.8(a) shows a coffer dam wall embedded in soil and retaining water. The free water applies a total force P_w to the wall where

$$P_w = \frac{1}{2}\gamma_w H_w^2 \tag{24.1}$$

Figure 24.8(b) shows a wall retaining soil. There is water in the excavation which applies a total stress P_w and the wall is supported by a single prop with a load P_a. (It is assumed that the prop is placed so that the wall does not rotate.) The total stress applied to the soil arises from the sum of P_w and P_a: notice that this is the same whether the soil is drained or undrained and whether the wall is impermeable or leaky.

Figure 24.8(c) shows a wall supporting a coarse-grained soil which is loaded drained. The toe of the wall is embedded in relatively impermeable clay and the excavation is dry. If the wall is impermeable it acts as a dam and the pore pressures are everywhere hydrostatic. The pore pressures apply a force P_w to the wall in addition to the horizontal effective stresses. The strength of the soil on the slip surface shown is reduced by the influence of the pore pressures lowering the effective stresses. Figure 24.8(d) shows the same wall but with a drain near the toe and part of a flownet for steady state seepage. It is obvious that the force P_a required to support the wall has been significantly reduced: there are no water pressures acting directly on the wall and the effective stresses, and the strength, on the slip surfaces are greater because the pore pressures are less. The example illustrates the importance of providing adequate drainage for retaining walls.

Figure 24.8(e) shows steady state seepage into a pumped coffer dam. (The flownet is similar to the one shown in Fig. 24.7.) At the bottom of the coffer dam, along AB, there is upward seepage and the possibility of instability due to piping and erosion, discussed in Sec. 14.6. Piping will occur when the hydraulic gradient $i = \delta P/\delta s$ becomes close to unity. For the example illustrated, δP over the last element of the flownet is

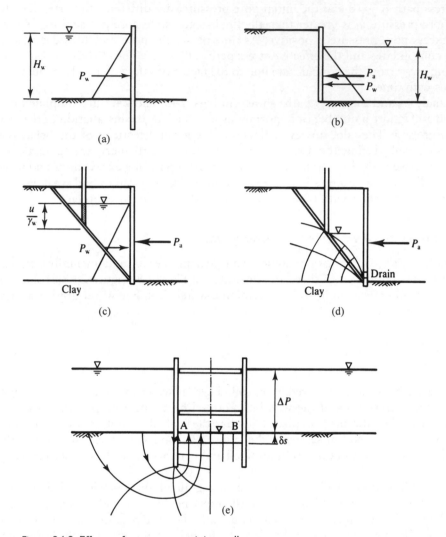

Figure 24.8 Effects of water on retaining walls.

$\Delta P/7$ (because there are seven equipotential drops in the flownet) and the size of the last element δs can be determined by measurement from a scaled diagram.

24.7   Calculation of earth pressures – drained loading

As a retaining wall moves the horizontal stresses change, as illustrated in Fig. 24.2, and when they reach the limiting active or passive pressures the soil has reached its critical state. The active and passive pressures can be calculated using upper and lower bound and limit equilibrium methods and, as always, it is necessary to distinguish between drained and undrained loading.

A limit equilibrium solution for the active force on a wall retaining dry soil was found in Sec. 20.3. The mechanism and the polygon of forces were shown in Fig. 20.6 and the solution is

$$P_a = \tfrac{1}{2}\gamma H^2 \tan^2\left(45° - \tfrac{1}{2}\phi'\right) \tag{24.2}$$

where ϕ' is the appropriate friction angle discussed in Sec. 24.10.

Assuming that the effective active pressure σ'_a increases linearly with depth the earth pressures corresponding to this limit equilibrium solution are

$$\sigma'_a = \sigma'_v \tan^2\left(45° - \tfrac{1}{2}\phi'\right) = K_a\sigma'_v \tag{24.3}$$

where σ'_v is the vertical effective stress and K_a is called the active earth pressure coefficient. It is quite easy to show that the solution for the passive pressure is

$$\sigma'_b = \sigma'_v \tan^2\left(45° + \tfrac{1}{2}\phi'\right) = K_p\sigma'_v \tag{24.4}$$

where K_p is called the passive earth pressure coefficient.

These solutions are for a smooth vertical wall with a level ground surface. A more general case is shown in Fig. 24.9 where the ground surface and the back of the wall are both inclined and the wall is rough. Shear stresses between the soil and the wall are given by

$$\tau'_s = \sigma'_n \tan \delta' \tag{24.5}$$

where σ'_n is the normal stress for the appropriate active or passive pressure and δ' is the critical angle of wall friction. Obviously $0 < \delta' < \phi'$ and a value commonly taken for design is $\delta' = \tfrac{2}{3}\phi'$. The general case was considered in Sec. 20.3 (see Fig. 20.7). Tables and charts are available giving values for K_a and K_p for various combinations of ϕ', δ', α and β.

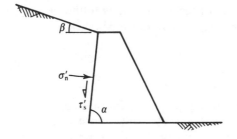

Figure 24.9 Earth pressures on a rough wall with a sloping face and with sloping ground.

24.8   Calculation of earth pressures – undrained loading

Active and passive pressures for undrained loading can be calculated using either the upper and lower bound methods or the limit equilibrium method. The procedures are similar to those described in the previous section for drained loading.

A limit equilibrium solution for the active pressures on a smooth wall was obtained in Sec. 20.3 from the limit equilibrium method using the Coulomb wedge analysis (see Fig. 20.4). The solution was

$$P_a = \tfrac{1}{2}\gamma H^2 - 2s_u H \tag{24.6}$$

where s_u is the undrained strength. Assuming that the stresses increase linearly with depth,

$$\sigma_a = \sigma_v - 2s_u \tag{24.7}$$

where σ_v is the total vertical stress. It is relatively simple to show that the passive pressure for undrained loading is given by

$$\sigma_p = \sigma_v + 2s_u \tag{24.8}$$

These expressions for active and passive earth pressures for undrained loading can be written as

$$\sigma_a = \sigma_v - K_{au}s_u \tag{24.9}$$

$$\sigma_p = \sigma_v + K_{pu}s_u \tag{24.10}$$

where K_{au} and K_{pu} are earth pressure coefficients for undrained loading.

The solutions with $K_{au} = K_{pu} = 2$ are for a smooth vertical wall with a level ground surface. Tables and charts are available giving values for K_{au} and K_{pu} for other cases including rough walls where the shear stress between the soil and the wall is s_w.

From Eq. (24.7) the active earth pressure for undrained loading appears to become negative (i.e. in tension) when

$$\sigma_v < 2s_u \tag{24.11}$$

This is impossible as the soil is not glued to the wall and a tension crack opens up as shown in Fig. 24.10(a). This is the same kind of tension crack as found near the top of slopes (see Sec. 20.8) and the critical depth H_c of a water-filled crack is

$$H_c = \frac{2s_u}{\gamma - \gamma_w} \tag{24.12}$$

If the crack is not filled with water put $\gamma_w = 0$ into Eq. (24.12). Notice that the position of the active force P_a has been lowered and if the crack is filled with water it is free water (not pore water) and applies a total stress to the wall. If there is a surface stress q as shown in Fig. 21.10(b), the tension crack will close entirely when $q > 2s_u$.

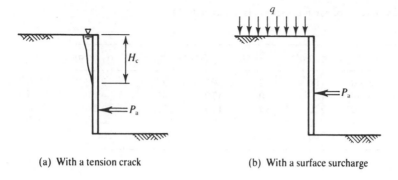

(a) With a tension crack   (b) With a surface surcharge

Figure 24.10 Active forces on walls – undrained loading.

Compare Eqs. (24.9) and (24.l0) for undrained loading with Eqs. (24.3) and (24.4) for drained loading. For undrained loading the earth pressure coefficients are expressed as a difference $(\sigma_h - \sigma_v)$ while for drained loading they are a ratio σ'_h/σ'_v. This is a consequence of the fundamental difference between the basic equations for drained and undrained strength.

24.9   Overall stability

The forces on a retaining wall arise from the active and passive earth pressures, from free water pressures and from loads in props and anchors. For overall stability the forces and moments arising from these pressures must be in equilibrium. For the simplified example shown in Fig. 24.11,

$$P + \int_0^{H_w} \sigma_w dz = \int_0^{H} \sigma_h dz \tag{24.13}$$

where the integrals are simply the areas under the pressure distribution diagrams. In order to take moments it is necessary to determine the moment arm of each force;

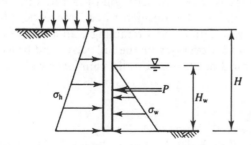

Figure 24.11 Influence of free water on the loads on a retaining wall.

Table 24.1 Calculation of earth pressures (see Example 24.1)

(a) Active side

Depth (m)	Soil	σ_z (kPa)	u (kPa)	σ_z' (kPa)	σ_a' (kPa)	σ_a (kPa)
0	Sand	80	0	80	27	27
2	Sand	120	0	120	40	40
7	Sand	220	50	170	57	107
7	Clay	220				140
10	Clay	280				200

(b) Passive side

Depth (m)	Soil	σ_z (kPa)	u (kPa)	σ_z' (kPa)	σ_p' (kPa)	σ_p (kPa)
2	Water	0	0	0	0	0
5	Water	30	30	0	0	30
5	Sand	30	30	0	0	30
7	Sand	70	50	20	60	110
7	Clay	70				150
10	Clay	130				210

the line of action of a force is through the centre of area of each pressure distribution diagram.

The best way to avoid making mistakes is to set up a table and draw the distribution of earth pressure with depth, as shown in Table 24.1 in Example 24.1 below. This gives calculations for the horizontal stresses on a wall bedded into sand over clay as shown in Fig. 24.17(a). The calculations use Eqs. (24.3) and (24.4) for the stresses in the sand and Eqs. (24.9) and (24.10) for the stresses in the clay: in the free water the horizontal and vertical total stresses are equal. Notice how the pore pressures come into the calculations in the drained sand but not in the undrained clay. There is a step in the earth pressures at the sand–clay junction, so it is necessary to calculate separately the stresses just in the sand and just in the clay.

Overall, a wall is considered to be stable if the forces and moments are in equilibrium and this is examined by resolving horizontally and taking moments about a convenient point. In most analyses the variable (or unknown) is the depth of embedment, which is increased until a suitable margin of safety is achieved. Selection of factors of safety for a retaining wall design is very difficult and will be considered in a later section: for the present I will simply consider the overall stability of a retaining wall at the point of collapse, such that the horizontal stresses are everywhere the full active and passive pressures. It is necessary to consider propped or anchored walls, cantilever walls and gravity walls separately.

(a) Anchored or propped walls

Figure 24.12 shows a simple propped wall with depth of embedment d. The active and passive pressures are as shown and from these the magnitudes P and depths z

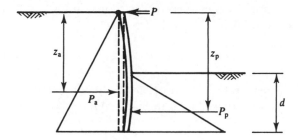

Figure 24.12 Forces on a propped wall.

of the active and passive forces are calculated as described in the previous section. Taking moments about P, the line of action of the prop forces, the wall is stable if

$$P_a z_a = P_p z_a \qquad (24.14)$$

Resolving horizontally, the prop or anchor force P is given by

$$P = P_a - P_p \qquad (24.15)$$

Notice that all the terms in Eq. (24.14) depend on the (unknown) depth of penetration d and solutions are most easily found by trial and error, adjusting d until Eq. (24.14) is satisfied. In Fig. 24.12 the toe of the wall rotates and translates and this is known as the free earth support condition. If the depth d is very large the toe of the wall will not translate or rotate; this is known as the fixed earth support condition.

(b) Cantilever walls

If there is no prop or anchor it is impossible to satisfy moment and force equilibrium at the same time with only the two forces P_a and P_p. Stiff cantilever walls fail by rotation about a point some way above the toe, as shown in Fig. 24.13(a), and this

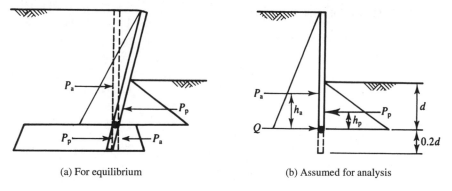

(a) For equilibrium   (b) Assumed for analysis

Figure 24.13 Forces on cantilever walls.

system of forces can satisfy moment and force equilibrium. It is convenient to replace the forces below the point of rotation by a single force Q, as shown in Fig. 24.13(b). Taking moments about Q the wall is stable if

$$P_a h_a = P_p h_p \tag{24.16}$$

which gives the unknown depth of penetration d. In order to allow the wall below the point of rotation to mobilize the pressures shown in Fig. 24.13(a), the depth d is usually increased by 20%.

(c)   Gravity walls

Gravity walls may fail by sliding, by overturning or by failure of the soil at the toe, as illustrated in Figs. 24.4 and 24.14. Figure 24.14(a) shows a wall failing by sliding along its base and $P_a = T$. For undrained loading,

$$T = s_w B \tag{24.17}$$

where s_w is the undrained shear strength between the soil and the base of the concrete wall. For drained loading,

$$T = (W - U) \tan \delta' \tag{24.18}$$

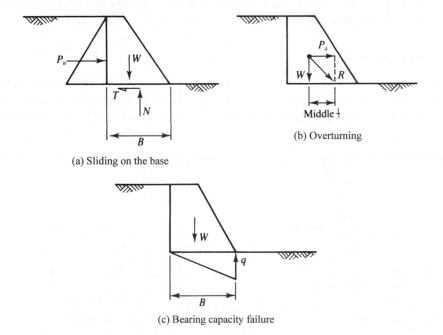

(a) Sliding on the base

(b) Overturning

(c) Bearing capacity failure

Figure 24.14 Equilibrium of gravity retaining walls.

where δ' is the angle of shearing resistance between the soil and the wall and U is the force due to pore pressures acting over the base area B. The wall cannot overturn provided that the normal stress at the upstream edge remains positive (i.e. in compression) and it can be shown by simple statics that this requires that the resultant R passes through the middle third of the base, as shown in Fig. 24.14(b). The resulting triangular distribution of normal stress shown in Fig. 21.14(c) implies that the maximum stress at the toe is given by

$$q = \frac{2W}{B}$$
(24.19)

The possibility of failure of the foundation due to excessive bearing pressure is really a problem of bearing capacity and is discussed in Chapter 22.

24.10 Soil strength and factors for design of retaining walls

So far I have described analyses for earth pressures, prop or anchor forces and overall stability of retaining walls using a friction angle ϕ' or an undrained strength s_u. As discussed in Sec. 24.4, if the wall is supporting an excavation and the soil is initially undrained it will become less safe with time and the drained case is most critical. However, if the wall is supporting a fill and the soil is initially undrained it will become safer with time and the undrained case is most critical. The question now is should the critical state strengths or the peak strengths be used to calculate the ultimate limit state and what factors should be applied.

Design of a retaining wall is something like a problem in slope stability where there must be an adequate margin against ultimate failure states and something like a problem in bearing capacity where it is necessary to limit movements. There are a number of standards, codes and advice notes which deal with selection of design parameters and factors for design of retaining walls and many of these give different designs. I am not going to deal with these; the issues are far too complicated for this simple book and you will have to consult books specializing in retaining walls for details. Instead I will describe some simple and logical procedures.

Firstly you should assume the wall is strong and stiff and demonstrate that it is in equilibrium with active and passive pressures calculated from the critical state strength and the worst credible groundwater and free water conditions. You can add partial factors to the critical state strength and water pressures to account for any uncertainties you may have in your estimates of them. This is the procedure for slope stability analyses described in Chapter 21.

Next you should repeat the stability analyses with active and passive pressures calculated from the peak strength with a load factor to limit ground movements. The load factor will probably be in the region of 2 to 3, depending on whether you took worst credible, moderately conservative or average values of measured peak strengths. This is the procedure described in Chapter 22 for design of shallow foundations.

You will then have to calculate the loads on props or anchors and shear forces and bending moments in the wall. A major difficulty here is that the distributions of stress on the active and passive sides of a wall found from simple earth pressure calculations with suitable factors of safety are very different from the true distributions of stress. This means you cannot determine the ground movements or the shear force and bending moments in the wall from simple earth pressure analyses. The point is illustrated in Fig. 24.15, which shows a simple cantilever wall that has been designed for drained loading with a substantial margin of safety against ultimate failure. The stresses in Fig. 24.15(a) are those calculated as described in Sec. 24.18 reduced by a factor of safety and are similar to those shown in Fig. 24.13(a). Figure 24.15(b) shows the stresses which would act on the wall if there were small movements. The stresses in the ground before construction correspond to K_0 and these are shown by the broken lines. Near the bottom of the wall there have been only very small movements and the stresses correspond to the K_0 on both sides of the wall. Near the ground level and the excavation the movements might be enough to develop full active and passive pressures which will be larger than those calculated with a factor of safety. The shear forces and bending moments in the wall and the ground movements calculated from the stresses in Fig. 24.15(a) will be different from those calculated from the stresses in Fig. 24.15(b).

Next you will have to calculate the ground movements and the deflections of the wall. If you have calculated the bending moments in the wall you can use simple structures analyses to calculate its deflections and these will have to be compatible with the ground movements. Matching wall movements and moments to ground movements and stresses is known as soil structure interaction and it is part of what makes analyses of retaining walls difficult.

There are a number of commercial computer programs for design of retaining walls. Like codes and standards these often produce different designs. Before you use one of these programs you should be sure that you understand the theories and assumptions in the analyses.

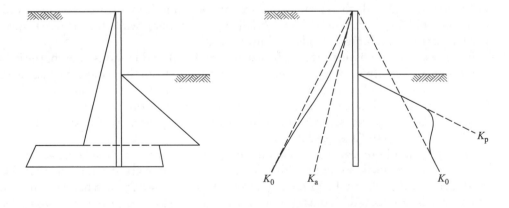

(a) From active and passive earth pressures   (b) Corresponding to wall movements

Figure 24.15 Distribution of stress on a simple cantilever wall.

24.11   Summary

1. Retaining walls are used to support slopes that are too high or too steep to remain stable if unsupported or to limit ground movements. There are a number of different kinds of retaining wall. They can fail in different ways including slipping in the soil, failure of the wall itself and failure of props or anchors.

2. As a wall moves away from the soil the horizontal stresses are active pressures and as it moves towards the soil they are passive pressures. For drained loading on smooth walls these are

$$\sigma_a' = \sigma_v' \tan^2\left(45° - \tfrac{1}{2}\phi'\right) = K_a\sigma_v' \tag{24.3}$$

$$\sigma_p' = \sigma_v' \tan^2\left(45° + \tfrac{1}{2}\phi'\right) = K_p\sigma_v' \tag{24.4}$$

where K_a is the active earth pressure coefficient and K_p is the passive earth pressure coefficient. For undrained loading on smooth walls active and passive pressures are

$$\sigma_a = \sigma_v - K_{au}s_u \tag{24.9}$$

$$\sigma_p = \sigma_v + K_{pu}s_u \tag{24.10}$$

where K_{au} and K_{pu} are earth pressure coefficients for undrained loading.

3. For walls retaining excavated slopes pore pressures rise with time and the safety deteriorates, but for walls retaining coarse grained fill the excess pore pressures developed in the foundations during construction will generally decrease with time.

4. The depth of the toe of a wall below the base of the excavation must be sufficient to ensure overall stability (with an appropriate margin of safety). Overall stability is examined by considering the statical equilibrium of the forces due to the active and passive earth pressures and the loads in props and anchors. Different calculations are required for cantilever and propped walls.

Worked examples

Example 24.1: Calculation of active and passive earth pressures   Figure 24.16 shows a 10 m high wall retaining layers of sand and clay. The active and passive total stresses in the drained sand are

$$\sigma_a = \sigma_a' + u = \sigma_z'K_a + u = (\sigma_z - u)\,K_a + u$$

$$\sigma_p = \sigma_p' + u = \sigma_z'K_p + u = (\sigma_z - u)\,K_p + u$$

where $K_a = \tan^2(45° - \tfrac{1}{2}\phi_c')$ and $K_p = \tan^2(45° + \tfrac{1}{2}\phi_c')$ and, for $\phi' = 30$, $K_p = 1/K_a = 3$. The total active and passive stresses in the undrained clay are

$$\sigma_a = \sigma_z - K_{au}s_u$$

$$\sigma_p = \sigma_z + K_{pu}s_u$$

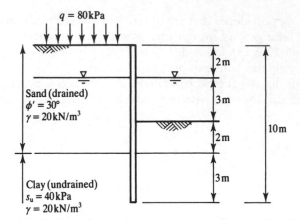

Figure 24.16 Cantilever retaining wall – Example 24.1.

where, for a smooth wall, $K_{au} = K_{pu} = 2$. The variations of σ_a and σ_p with depth are given in Table 24.l; to calculate active and passive pressures in layered soils and where there are pore pressures it is convenient to tabulate the calculations in this way. Notice that the stresses at the base of the sand are not the same as the stresses at the top of the clay. Figure 24.17 shows the variations of active and passive total pressures with depth.

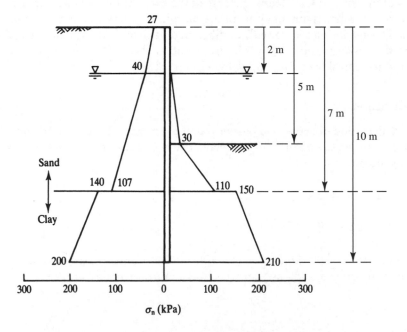

Figure 24.17 Active and passive pressures on a retaining wall – Example 24.1.

Example 24.2: Depth of a propped wall   Figure 24.18(a) shows a wall propped at the top retaining dry sand. The unknown depth of penetration is d. For a factor of safety $F_s = 1.6$ the safe angle of friction, given by $\tan\phi'_s = \tan\phi'_c/F_s$ is $\phi'_s = 20°$. Hence, from Eqs (24.3) and (24.4),

$$K_a = \tan^2\left(45° - \tfrac{1}{2}\phi'_a\right) = \tan^2 35° = 0.5$$

$$K_p = \tan^2\left(45° + \tfrac{1}{2}\phi'_a\right) = \tan^2 55° = 2.0$$

With the depth H measured from the ground level on either side of the wall

$$P_a = \tfrac{1}{2}\gamma H^2 K_a = \tfrac{1}{2} \times 20 \times (5+d)^2 \times \tfrac{1}{2} = 5(5+d)^2 \text{ kN}$$

$$P_p = \tfrac{1}{2}\gamma H^2 K_p = \tfrac{1}{2} \times 20 \times d^2 \times 2 = 20d^2 \text{ kN}$$

The distributions of active and passive earth pressures and the active and passive forces are shown in Fig. 24.18(b). Taking moments about the top of the wall and noting that the forces P_a and P_p act at the centres of the triangular areas (i.e. $\tfrac{1}{3}H$ above the base),

$$5(5+d)^2 \times \tfrac{2}{3}(5+d) = 20d^2 \times \left(5 + \tfrac{2}{3}d\right)$$

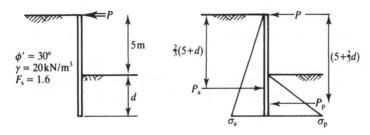

(a) Propped wall   (b) Earth pressures with prop

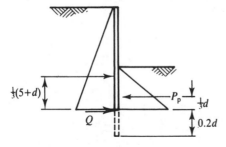

(c) Earth pressures without prop

Figure 24.18 Analysis of a propped retaining wall – Examples 24.2 and 24.3.

and, solving by trial and error, or otherwise,

$$d = 40 \text{ m}$$

With this value of d we have $P_a = 405$ kN and $P_p = 320$ kN. Hence, resolving horizontally, from Eq. (24.15) the force in the prop is,

$$P = 405 - 320 = 85 \text{ kN}$$

Example 24.3: Depth of a cantilever wall   If the wall in Fig. 24.18(a) is not propped it acts as a cantilever and the forces on the wall are shown in Fig. 24.18(c). From Eq. (24.16), taking moments about the toe where the force Q acts,

$$5(5 + d)^2 \times \tfrac{1}{3}(5 + d) = 20d^2 \times \tfrac{1}{3}d$$

and, solving that by trial and error, or otherwise,

$$d = 8.5 \text{ m}$$

To provide sufficient length to mobilize the force Q, the wall depth should be increased by 20% so the required depth of penetration is about 10 m.

Further reading

Atkinson, J. H. (1981) *Foundations and Slopes*, McGraw-Hill, London.
Clayton, C. R. I., J. Mililitsky and R. I. Woods (1993) *Earth Pressure and Earth Retaining Structures*, Spon Press, London.
Heyman, J. (1972) *Coulomb's Memoir on Statics*, Cambridge University Press, Cambridge.
Kerisel, J. and E. Absi (1990) *Active and Passive Pressure Tables*, Balkema, Rotterdam.
Padfield, C. J. and R. J. Mair (1984) *Design of Retaining Walls Embedded in Stiff Clays*, CIRIA, Report 104, London.

Chapter 25

Tunnelling in soft ground

25.1   Tunnel construction

Tunnels are built for transport of people, cars, trains and water, for storage and for mining. They may be deep or shallow, in rock or in soil, in urban or rural environments. They may be built by boring or by cut and cover methods or by sinking them into the bed of a river. In Fig. 25.1(a) the tunnel is advanced by mining the ground from inside the tunnel. In Fig. 25.1(b) the tunnel is constructed as a pair of retaining walls with a roof; the design is really design of retaining walls discussed in Chapter 24. In Fig. 25.1(c) tunnel sections are floated into place, sunk into a trench, connected together and covered by fill.

Some tunnels are deep and so they are mostly in rock and construction does not affect nearby buildings or other tunnels unless they leak and alter the groundwater conditions. Deep tunnels are for mining or road and rail connections through mountains. The construction problems are mostly excavation of strong rock and support of fractured rock in the roof of the tunnel. Other tunnels are relatively shallow: the ratio of their depth to diameter is less than about 10. These include tunnels for underground rail and road connections and water supplies in cities. Shallow tunnels are often built in soils and close to existing buildings and underground structures. The term soft ground is used in tunnelling to describe soils and weak rocks which require support to prevent collapse and damaging ground movements during construction and throughout their lives.

In this chapter I will discuss engineering design of shallow tunnels in soft ground. The criteria for design of a tunnel in soft ground are essentially the same as those for foundations described in Chapter 22 and retaining walls described in Chapter 24. Firstly it is necessary to investigate the ultimate limit state: there must be an adequate factor of safety against collapse. Secondly it is necessary to investigate the serviceability limit state: the ground movements caused by construction of the tunnel must not damage nearby infrastructure.

25.2   Construction of bored tunnels in soft ground

Tunnels in soft ground have stiff and strong permanent linings to prevent collapse and excessive movement. They are usually bored and often their cross-section is circular, as

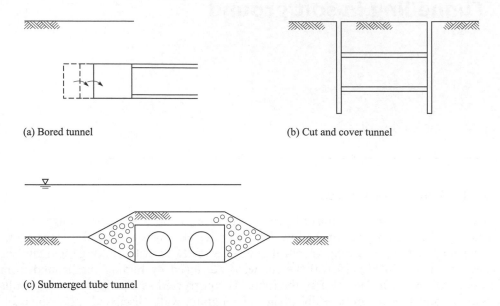

(a) Bored tunnel   (b) Cut and cover tunnel

(c) Submerged tube tunnel

Figure 25.1 Methods of construction of tunnels in soft ground.

shown in Fig. 25.2(a). The top of the tunnel is its crown, the bottom its invert and the dimensions are as shown where D is the tunnel diameter and C is known as the cover.

As the tunnel is excavated at the face the permanent lining cannot be constructed immediately and there is a short length P called the heading which requires temporary support. The temporary support is usually provided by a shield which is essentially a stiff and strong steel tube. In the front of the shield is an excavator: this may be a simple digger separate from the shield or a 'cheese-grater' rotating cutter which is part of the shield. The shield is pushed forward from the completed permanent lining and the face is excavated until there is enough space to build more permanent lining, often inside the tail-skin of the shield.

There are different types of shield. Figure 25.3(a) shows an open shield with an integral cutter head in which excavated ground falls directly onto a conveyor. The total

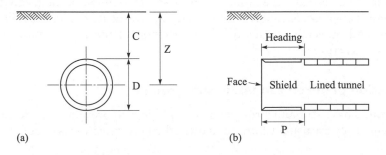

(a)   (b)

Figure 25.2 Dimensions of shallow tunnels.

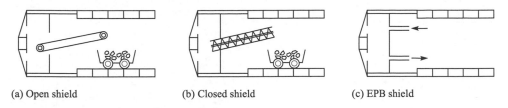

(a) Open shield   (b) Closed shield   (c) EPB shield

Figure 25.3 Types of shield for tunnelling in soft ground.

stress acting on the face is the internal pressure in the heading. This is usually atmospheric but sometimes there is an air-lock in the tunnel and the internal air pressure is raised. A closed face shield is shown in Fig. 25.3(b). The pressure of the excavated soil in the chamber is maintained by the relative speeds of the shield advance and the rotation of the screw conveyor. Figure 25.3(c) shows a simple form of earth pressure balance machine. The end of the shield is a closed chamber through which bentonite mud is circulated under pressure. This supports the face, prevents seepage of groundwater into the heading and transports the excavated material away. In both the earth pressure balance and the closed face shield the intention is to maintain the stresses in the ground ahead of the face to reduce inflow of ground water, to prevent collapse of the face and to reduce ground movements.

Tunnels are constructed relatively quickly. In coarse-grained soils the soil will be drained and in fine-grained soils it will be undrained but there will then be a period of consolidation or swelling as excess pore pressures dissipate. Most tunnels in soft ground are below the water table. The permanent lining may be waterproof or it may leak. If it leaks the tunnel acts as a drain and, in the fully drained condition, there will be steady state seepage from the groundwater into the tunnel. In coarse-grained soils uncontrolled seepage into the heading during construction causes erosion and instability like that near the toe of a slope described in Sec. 21.7.

25.3   Stress changes near tunnels

Figure 25.4(a) illustrates the ground ahead of an advancing tunnel. If the face collapses there would be a mechanism something like that shown. The total shear and normal stresses on an inclined slip plane are τ and σ and the pore pressure is given by the height of water in a standpipe as shown. During excavation of the face the total normal stresses decrease and the shear stresses increase because they are required to prevent collapse. The conditions are like those in the ground behind a slope as it is excavated, shown in Fig. 21.4 and near a wall retaining an excavation, shown in Fig. 24.6.

In Fig. 25.4(b) the total stress path for soil in front of the tunnel is $A \rightarrow B$ and the effective stress path $A' \rightarrow B'$ corresponds to undrained loading at constant water content, as shown in Fig. 25.4(c). The exact effective stress path $A' \rightarrow B'$ will depend on the characteristics of the soil and its initial state or overconsolidation ratio, as discussed in Chapter 11. The face will collapse immediately if the point B' representing the mean effective stress on all the slip planes in the mechanism in Fig. 25.4(a) reaches

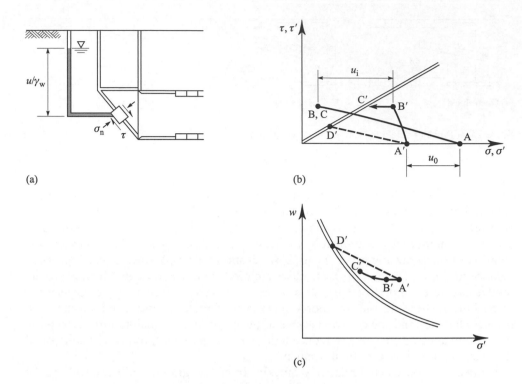

Figure 25.4 Stress and pore pressure changes during tunnelling.

the critical state line and the shear stress is then the undrained strength. The distance of the effective stress from the critical state line is a measure of the factor of safety.

The initial pore pressure immediately after excavation u_i is negative because the total normal stress at B is smaller than the effective normal stress at B'. If there is no further excavation the total stress remains at B but, as time passes, the pore pressures rise, the effective stress path is $B' \to C'$. The effective stress point C' moves towards the critical state line and the factor of safety decreases. The face open will collapse when C' reaches the critical state line.

The broken line $A' \to D'$ in Figs. 25.4(b) and (c) represents tunnelling in soil which is drained so the pore pressure remains constant. If the tunnel heading acts as a drain there will be steady state seepage and the pore pressures will be determined by a flownet. Notice that the shear stress at D' for the fully drained case is smaller than that at B' for the undrained case so the shield is required to provide more support for the drained case than for the undrained case. If the soil around the heading is drained the shield must also support pressures from the groundwater, as discussed in Sec. 25.5.

25.4   Stability of tunnel headings

The permanent lining of a tunnel in soft ground is usually made from cast iron or concrete segments which fit together to make a ring. You can often see these out of the window of an underground train. Usually the loads on the lining when it is being built and when the shield is pushed forward from it are considerably larger than the loads from the ground. As a result, once a circular lining has been built the tunnel is generally safe and the subsequent movements, which are due to dissipation of excess pore pressures, are relatively small. The critical stability condition is in the heading during construction and most of the ground movements caused by the tunnel occur during construction.

(a) Stability of tunnel headings for undrained loading

Figure 25.5(a) shows a simple tunnel heading. The soil has an undrained strength s_u and, because we are considering an ultimate limit state of collapse, this should be taken as the critical state strength, as discussed in Chapter 18. The total stress inside the heading is σ_t and the length of the heading is P. In an earth pressure balance shield like that shown in Fig. 25.3(b) σ_t is provided by the shield but in an open shield like that shown in Fig. 25.3(a) σ_t is generally zero. If the face is supported by slurry pressure or if the heading is pressurized then σ_t becomes the slurry or air pressure. There is a total stress q on the surface: this may arise from buildings, from free water (see Sec. 21.5) or from very soft ground whose strength can be neglected.

If the ground is a heavy fluid with no strength σ_t must equal the total stress in the ground which is $\gamma z + q$ at the axis level. (If $\sigma_t < \gamma z + q$ the tunnel will collapse inwards but if $\sigma_t > \gamma z + q$ the ground will heave and the tunnel will burst.) The undrained strength of the ground will have the effect of reducing the tunnel collapse pressure σ_{tc}, which can be written as

$$\sigma_{tc} = \gamma z + q - s_u T_c \tag{25.1}$$

where T_c is called the tunnel stability number. Notice that Eq. (25.1) which gives the collapse pressure for a tunnel for undrained loading is very like Eq. (22.6) which gave the bearing capacity of a foundation for undrained loading. The tunnel stability number T_c in Eq. (25.1) is comparable to the bearing capacity factor N_c in Eq. (22.6).

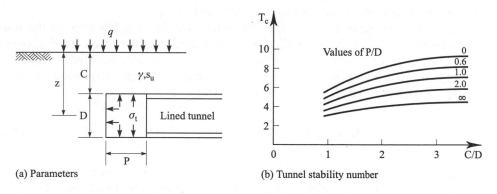

Figure 25.5 Stability of a tunnel heading for undrained tunnelling.

Values for T_c depend on the values of C/D and P/D. Some values have been obtained from upper and lower bound and limit equilibrium solutions of the kind described in Chapters 19 and 20 but these are in three dimensions and are quite complicated. The values of T_c commonly used in design are given by Atkinson and Mair (1981) and these are shown in Fig. 25.5(b). They were obtained from centrifuge model tests of the kind described in Chapter 27. This is a very good example of an instance where information obtained from centrifuge model tests has been applied with an equation derived from soil mechanics theory to give a design method.

To apply a single factor of safety, s_u can be replaced with a safe undrained strength s_{us}, as described in Sec. 19.5. Alternatively, partial factors can be applied to all the design parameters in Eq. (25.1.) Notice that the tunnel and its heading become safer as the support pressure σ_t is increased but may become less safe if the tunnel pressure becomes so large that it is close to causing a blow-out failure.

From Eq. (25.1) with $\sigma_{tc} = 0$ the face is self-supporting and can be safely excavated with an open shield if

$$s_u \geq \frac{1}{T_c}(\gamma z + q) \tag{25.2}$$

with a value of T_c obtained from Fig. 25.5(b). Notice that the stability of an unsupported tunnel face is similar to the stability of an unsupported vertical cut discussed in Sec. 21.8. In each case the undrained strength of the soil arises from negative pore pressures which are developed by the excavation. With time the pore pressures rise towards their steady state values, so the soil swells and weakens and the face becomes less stable and sooner or later both must collapse. The question is not whether an unsupported face or heading collapses but how long will it be before it does. This is a problem of consolidation time, discussed in Chapter 15.

(b)   Stability of tunnel headings for drained loading

Model tests on tunnels in dry sand show that the tunnel pressure at the collapse state σ_{tc} is always relatively small. It depends strongly on the tunnel diameter D and is almost independent of the depth of cover C. Figure 25.6(a) shows a circular tunnel section in dry soil. The stability is rather like that of an arch in a building: it is necessary only to maintain a ring of stable grains round the circumference.

Figure 25.6(b) shows a soil wedge behind the face which is like that on the active side of a retaining wall shown in Fig. 24.6. Due to arching the vertical stress at the top of the wedge is very small so the vertical stress in the dry soil near the invert is γD. The horizontal stress in the soil in the wedge corresponds to the active pressures so the collapse pressure on the face σ_{tc} is approximately

$$\sigma_{tc} = K_a \gamma D = \gamma D \tan^2\left(45° - \frac{1}{2}\phi'\right) \tag{25.3}$$

where ϕ' is the friction angle. As before, because we are considering an ultimate limit state of collapse this should be taken as the critical state strength friction angle ϕ'_c, as discussed in Chapter 18. To apply a single factor of safety ϕ'_c can be replaced with a

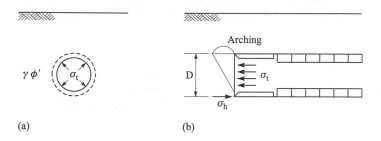

Figure 25.6 Stability of a tunnel heading for drained loading.

safe angle of friction ϕ_s', as described in Sec. 19.5. Alternatively you can apply partial factors to all the design parameters in Eq. (25.3).

The collapse pressure given by Eq. (25.3) is for a tunnel face in dry soil but many tunnels will be in ground which is below the water table in saturated soil. In this case it is necessary to take account of drainage and pore pressures, as discussed in Sec. 25.5.

25.5   Influence of water on tunnels

When a tunnel is driven through soft ground below the water table the groundwater influences both the loads on the completed lining and the stability of the heading and the face during construction. Tunnels in soft ground have to have a shield and face support to prevent collapse during construction and a structural lining to maintain long term stability. If the lining is fully waterproof then, after the excess pore pressures due to construction have dissipated, the groundwater outside the lining will be hydrostatic and there will be no seepage. In this case the pore pressures just outside the lining correspond to the original water table. The loads on the lining are the sum of the hydrostatic pore pressures and effective stresses. As discussed above the effective stresses acting on the lining are relatively small and the greatest proportion of the loading on a waterproof tunnel lining comes from the groundwater. The effective stresses acting on a supported face below the water table are given by Eq. (25.3) with $(\gamma - \gamma_w)$ instead of γ. With typical values for ϕ_c' in the range 30° to 37° values for K_a are in the range 1/3 to 1/4 so the loads on the face support from the groundwater are considerably larger than the loads from the soil.

Most tunnel linings are not completely waterproof and there is often some seepage of water through them. Figure 25.7(a) shows a section of a circular tunnel with a leaking lining. The water table is drawn down above the tunnel which is acting as a drain. There is part of a flownet for steady state seepage from the groundwater into the tunnel. Notice that the arrangement of flowlines and equipotentials near the tunnel is similar to that near a slope, with seepage outward from the ground shown in Figs. 21.8(c) and (d).The effect of the lining leaking is to reduce the water pressures on it while the stresses from the soil change very little.

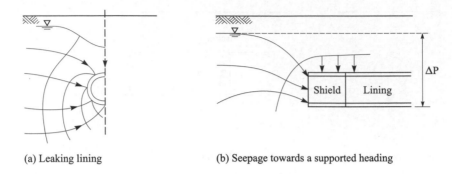

(a) Leaking lining   (b) Seepage towards a supported heading

Figure 25.7 Steady state seepage towards tunnels and headings.

Figure 25.7(b) shows part of a flownet for seepage towards a supported face. The shield and lining leak so there is radial seepage towards the tunnel like that shown in Fig. 25.7(a). The face has to be supported to maintain stability but the support is fully permeable. The seepage is three-dimensional and it is difficult to draw an accurate flownet in two dimensions; the one shown in Fig. 25.7(b) shows reasonably well the flowlines and equipotentials near the face. These are similar to those close to a slope with seepage out of the ground shown in Figs. 21.8(c) and (d). Taking the datum for potential at the invert the potential there is zero and the total drop in potential is ΔP as shown. The hydraulic gradient close to the face depends on the geometry of the flownet but there is a hydraulic gradient towards the face and there is the risk of piping and erosion like that described in Sec. 14.6. To prevent seepage towards the face and to maintain stability it is necessary to apply a total stress to the face at least equal to the hydrostatic pore pressures. This may be applied by air pressure inside the tunnel or, more usually, by use of a closed face or earth pressure balance shield like those shown in Figs. 25.3(b) and (c).

25.6   Ground movements due to tunnelling

The changes of stress and pore pressure in the ground around headings and tunnels during and after construction cause ground movements; this is inevitable. Figure 25.8 shows a tunnel excavated with a diameter D and the volume of soil excavated (per unit length along the tunnel) is $V_e (= \frac{1}{4}\pi D^2)$. Due to ground movements which occur during construction the diameter of the outside of the lining is a little smaller and the finished volume of the tunnel and lining is V_t. The ground loss v_l is the difference expressed as a percentage of the excavated volume

$$v_l = \left(\frac{V_e - V_t}{V_e} \right) \times 100\% \tag{25.4}$$

Values for volume loss in soft ground tunnels vary with the ground conditions and the method of tunnelling but are often 1% to 2%.

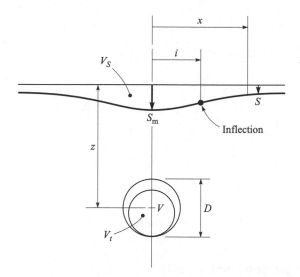

Figure 25.8 Settlement profile above a tunnel in soft ground.

Ground loss at the tunnel causes a settlement trough at the surface, as shown in Fig. 25.8. The settlements are greatest directly above the tunnel axis and become smaller away from it as shown. The shape of the settlement trough is often approximated by a normal probability or Gaussian distribution curve given by

$$S = S_{\mathrm{m}} e^{\frac{-x^2}{2i^2}} \tag{25.5}$$

where S_{m} is the maximum settlement above the tunnel axis, x is the distance measured away from the centre line and i is the distance to the point of inflection where the curvature changes and the slope of the ground surface is greatest. The volume of the settlement trough (per unit length along the tunnel) V_{s} is

$$V_{\mathrm{s}} = \sqrt{(2\pi)} i S_{\mathrm{m}} \tag{25.6}$$

If the soil is undrained then $V_{\mathrm{s}} \approx V_{\mathrm{e}} - V_{\mathrm{t}}$. The values of i depend principally on the depth of the tunnel and to a lesser extent on whether the soil is coarse-grained or fine-grained. Figures 25.9(a) and (b) show that i increases from i_1 to i_2 as the tunnel depth increases from z_1 to z_2 but, for a given ground loss the maximum settlement decreases from S_{m1} to S_{m2}. From Fig. 25.9(c) the value of i increases approximately linearly with z and

$$i \approx kz \tag{25.7}$$

For most cases you can take $k = 0.35$ for coarse-grained soils and $k = 0.5$ for fine-grained soils. If you make an assumption about the percentage ground loss you can use the above relationships to calculate the profile of surface settlement above a tunnel in soft ground.

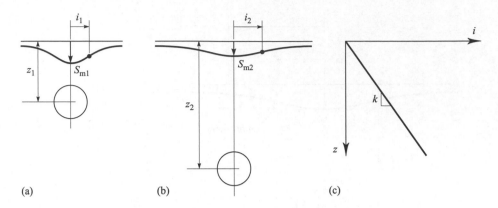

Figure 25.9 Influence of tunnel depth on surface settlement profile.

25.7   Load factors to limit ground movements

In Sec. 25.4 I investigated the ultimate limit state of tunnels and headings and calculated the tunnel support pressure at collapse σ_{tc} for undrained and drained tunnelling. However, if the tunnel support pressure approaches the ultimate limit state ground loss and settlements will be relatively large. In order to limit ground movements so nearby structures are not damaged it is convenient to apply a load factor to determine an allowable tunnel support pressure. This approach is the same as the one used to determine allowable bearing pressures for foundations in Chapter 22.

Figure 25.10(a) illustrates the volume loss V_l increasing as the tunnel support pressure σ_t decreases. If the tunnel support pressure is the same as the vertical stress γz in the ground at the level of the axis the settlements will be negligible. As the tunnel support pressure approaches the ultimate limit state σ_{tc} the settlements become very large as the tunnel and heading collapse. At the design point the allowable tunnel support pressure σ_{ta} causes an allowable volume loss V_{la} and this causes allowable

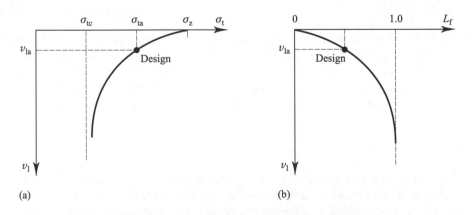

Figure 25.10 Relationship between tunnel support and ground settlements.

settlements S_a. Notice that ground loss and ground movements will decrease as the tunnel support pressure is raised. This is opposite to a foundation which will settle more as the bearing pressure is raised.

It is convenient to define a load factor as

$$L_f = \frac{\sigma_z - \sigma_t}{\sigma_z - \sigma_{tc}} \tag{25.8}$$

so $L_f = 1$ when $\sigma_t = \sigma_{tc}$ and as the tunnel support pressure σ_t is increased the load factor L_f and ground movements decrease. This definition of load factor is consistent with the definition of load factor for a foundation given in Sec. 22.5: in both cases settlements decrease as L_f reduces. Relationships between settlement and tunnel load factor like that shown in Fig. 25.10(b), based on observations made in centrifuge tests on model tunnels in soft clay, are given by Taylor (1984). From these data a load factor of 1/3 would give volume losses less than 1%.

25.8 Summary

1. Tunnels in soft ground have linings which are relatively strong and stiff so they are stable once the lining has been built. They require substantial support during construction.
2. Tunnelling in fine-grained soils is undrained and pore pressures near the heading are reduced. Dissipation of excess pore pressures decreases the effective stresses near the heading and the factor of safety against collapse will reduce with time.
3. There are simple analyses for undrained and drained stability of tunnels and headings which determine the collapse tunnel pressure σ_{tc}.
4. Groundwater has a major influence on lining loads and heading stability.
5. Surface settlements due to tunnelling take the shape of a normal probability curve and can be determined from the ground loss at the tunnel during construction.
6. Surface settlements are related to a tunnel load factor which is similar to the load factor used to limit settlement of a foundation.

References

Atkinson, J. H. and R. J. Mair (1981) Soil Mechanics Aspects of Soft Ground Tunnelling, *Ground Engineering*, July, pp. 20–26.

Taylor, R. N. (1984) Ground movements associated with tunnels and trenches. Ph.D Thesis, Cambridge University, Cambridge.

Further reading

Burland, J. B. R. J Mair, and J. R. Standing (2004) Ground performance and building response due to tunneling, *Proc. The Skempton Conference*, London, 2004. Ed. Jardine, R.J., D.M. Potts and K.G. Higgins, Thomas Telford, London, Vol. 1, pp. 291–342.

Mair, R. J. (2007) 46th Rankine Lecture. Tunnelling and geotechnics – new horizons, *Geotechnique* (to be published).

Mair, R. J. and R. N. Taylor (1997) Theme lecture: Bored tunnelling in the urban environment, *Proc. 14th Int. Conf. Soil Mech and Foundation Eng*, Hamburg, pp. 2353–2385.

Chapter 26

Behaviour of unsaturated soils

26.1 Introduction

Throughout most of this book I have considered the behaviour of soils which are either saturated or dry. In dry soils the pore spaces are filled with air and in saturated soils they are filled with water. In an unsaturated soil the pore spaces contain both water and a gas, which is usually air, but which may be water vapour or methane.

The behaviour of unsaturated soil is very complicated and there is, at present, no simple theory which adequately describes the engineering behaviour of unsaturated soils. A major difficulty is that the behaviour of the soil as it is drying and the proportion of air is increasing is different from the behaviour as it is wetting and the proportion of air is decreasing. All I can do in this short chapter is outline the basic features of unsaturated soils.

26.2 Occurrence of unsaturated soils

As illustrated in Fig. 6.3 soil below the water table and for some distance above it is saturated and soil above that is unsaturated. There may be dry soil at the surface but, in practice, this is rare. Soils which were initially saturated may become unsaturated when the pore pressure falls below a critical value and air is able to enter the pores. Alternatively, soils may be formed in an initially unsaturated condition. Unsaturated soils may become saturated if the water table rises.

Figure 26.1 illustrates different ways in which unsaturated soils are formed. In Fig. 26.1(a) desaturation of an initially saturated soil occurs as the water table falls naturally or by pumping. As discussed in Sec. 6.3 there will be a zone of saturated soil above the water table the height of which depends on the grain size of the soil. Water contents in soil near the surface may be reduced further by vegetation and evaporation. In Fig. 26.1(b) residual soils which are formed by weathering of rock *in situ* are often unsaturated from the start. In Fig. 26.1(c) excavated soil lumps, which themselves may be saturated or unsaturated, have been compacted into a fill.

26.3 Degree of saturation and specific volume

Figure 26.2 illustrates the volumes and weights of soil grains, water and air (or gas) in an unsaturated soil; it is similar to Fig. 5.3 which was for saturated soil.

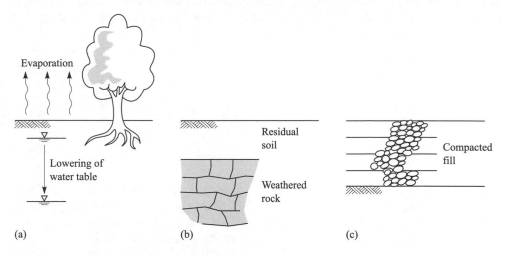

Figure 26.1 Formation of unsaturated soils.

	Volumes	Weights
Air	V_a	$W_a = 0$
Water	V_w	$W_w = \gamma_w V_w$
Grams	V_s	$W_s = \gamma_w g_s V_s$
Totals	V	W

Figure 26.2 Volumes and weights in unsaturated soils.

The degree of saturation S_r is defined as

$$S_r = \frac{V_w}{V_w + V_a} \tag{26.1}$$

so that for dry soil ($V_w = 0$) $S_r = 0$ and for saturated soil ($V_a = 0$) $S_r = 1$. The specific volume v, the water content w and the unit weight γ are defined in the same way as for saturated soil and they are given by

$$v = \frac{V}{V_s} \tag{26.2}$$

$$w = \frac{W_w}{W_s} \tag{26.3}$$

$$\gamma = \frac{W}{V} \tag{26.4}$$

Both unit weight and water content are obtained from direct measurements of dimensions and weights of samples before and after drying. Specific volume and degree of saturation can be calculated from these from

$$v = \frac{\gamma_w G_s (1 + w)}{\gamma} \tag{26.5}$$

and

$$S_r = \frac{w G_s}{(v - 1)} \tag{26.6}$$

The expressions in Eqs. (26.5) and (26.6) can be obtained from Eqs (26.1) to (26.4) making use of Fig. 26.2. Notice that from Eq. (26.6), the specific volume depends on both the water content and on the degree of saturation so, in an unsaturated soil, the water content can change without any change in the volume of the soil.

26.4   Distribution of air and water in unsaturated soil

The way in which water and air are distributed through unsaturated soil is important. Figure 26.3 illustrates an ideal soil. In Fig. 26.3(a) the water content and degree of saturation are small. Water collects at the points of contact forming meniscus water bridges. The air is continuous throughout the soil and air pressures are the same everywhere. The water is not continuous. Water pressures depend on the radii of the meniscuses, which may be different at different contact points and so water pressures may vary through the soil. The meniscus water bridges have the effect of bonding the grains together.

In Fig. 26.3(c) the degree of saturation is large. The water is continuous throughout the soil, the water pressure is the same at any horizon and it varies with depth. The gas, which is probably water vapour, is in bubbles and is not continuous. The pressure in the gas depends partly on the sizes of the bubbles, which may differ throughout the soil.

In Fig. 26.3(b) both the air and the water are continuous, in three dimensions, throughout the soil. The water pressure is governed by the radii of the meniscuses

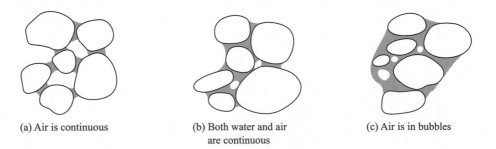

(a) Air is continuous   (b) Both water and air   (c) Air is in bubbles
                                    are continuous

Figure 26.3 Distributions of air and water in unsaturated soils.

which must adjust to maintain constant water pressure at any horizon in the ground. Since the air is continuous air pressures are constant throughout the soil. Pore air pressures and pore water pressures are not equal. The degree of saturation over which both water and air are continuous varies typically between about $S_r = 0.25$ to about $S_r = 0.85$ and this is the situation most common in practice.

26.5   Pore pressure and suction in unsaturated soil

The very important principle of effective stress was discussed in Chapter 6. This states that there is an effective stress which controls the behaviour of soil including strength and stiffness. The effective stress σ' in saturated soil is related to the total stress σ and the pore pressure u by

$$\sigma' = \sigma - u \tag{26.7}$$

In an unsaturated soil the simple effective stress principle and equation (Eq. (26.7)) does not work. There must be some stress which controls its behaviour and this ought to be some combination of total stress σ, pore water pressure u_w, pore air pressure u_a and degree of saturation.

Current practice is to consider the net stress and the matrix suction defined as:

$$\text{net stress} = (\sigma - u_a) \tag{26.8}$$

$$\text{matrix suction} = (u_a - u_w) \tag{26.9}$$

If the pore air pressure is atmospheric (i.e. $u_a = 0$) the net stress is the same as the total stress and a negative pore water pressure gives rise to a numerically equal positive suction. There is not, at present, a simple and satisfactory theory which can be used to determine soil behaviour from the net stress and the matrix suction or the degree of saturation and this is the subject of much current research.

26.6   Desaturation and water retention

If a saturated soil is subjected to an increasing suction there will be a critical suction at which the water cavitates or boils and water vapour forms bubbles. In normal circumstances water cavitates at room temperature at a suction of about 100 kPa but in fine grained soils the water in the very small pore spaces can sustain much larger suctions without cavitation.

If the suction is increased still further there will be a critical suction at which air is drawn into the pore spaces. The suction at which air can enter the soil depends on the size of the pore spaces. The analysis is similar to that in Sec. 6.4 for suctions in saturated soil. Figure 26.4 shows the surface of the soil. The pore water pressure is u_w, the pressure in the external air is u_a, the diameter of the pore spaces in the soil is $d_v = (v - 1)d_s$ where d_s is the mean grain diameter and T is the surface tension force between water and the material of the soil. For equilibrium

$$T\pi d_v = (u_a - u_w)\frac{\pi d_v^2}{4} \tag{26.10}$$

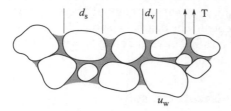

Figure 26.4 Meniscuses in unsaturated soils.

so air will enter the soil if

$$(u_a - u_w) > \frac{4T}{(v-1)d_s} \qquad (26.11)$$

Taking a value for T of about 7×10^{-5} kNm^{-1} air at atmospheric pressure will enter soil with a grain size of 0.001 mm and specific volume of 1.5 if the suction is about 600 kPa.

As air enters the soil the water content and degree of saturation reduce. There will be a relationship between suction and degree of saturation which depends principally on the grading of the soil. This relationship is known as the water retention curve or the soil water characteristic curve. Figure 26.5 shows typical water retention curves for a soil. There are different curves for drying and for wetting and there will be different curves for different soils. When the soil is saturated the degree of saturation is $S_r = 1$. At the air entry suction, which is related to grain size by Eq. (26.11), the degree of saturation starts to reduce. As the soil dries it will compress but the change in volume is not related to the change in water content as it was in saturated soil. There is a minimum degree of saturation where there are stable meniscus water bridges. At low water contents and low degrees of saturation suctions can be very large indeed, especially in fine grained soils.

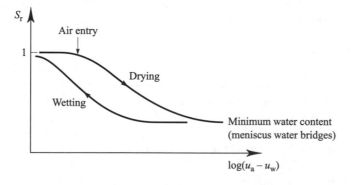

Figure 26.5 Water retention curves for unsaturated soils.

26.7   Undrained loading of unsaturated soil

A very important difference between the behaviours of saturated and unsaturated soil is the response to undrained loading. Strictly undrained means no drainage of water and no change of water content. In saturated soil this means there is no change in volume either because both soil grains and water can be assumed to be incompressible over changes of stress common in ground engineering. In unsaturated soil, however, the air or gas is highly compressible and so constant water content does not mean constant volume.

If soil is loaded isotropically and undrained at constant water content the change of pore water pressure δu_w which occurs during a change of total stress $\delta \sigma$ is given by

$$\delta u_w = B \delta \sigma \qquad (26.12)$$

where B is called a pore pressure parameter. For saturated soil there is no change of volume and so there is no change of effective stress, in which case $\delta u_w = \delta \sigma$ and $B = 1$. For unsaturated soil B is less than 1: the value of B decreases with decreasing degree of saturation and becomes $B = 0$ in dry soil when $S_r = 0$.

In Sec. 9.4 I showed that the strength of soil depends on its specific volume and for saturated soil this remains constant for undrained (constant water content) loading. This is the reason why we can use the undrained strength s_u for undrained loading of saturated soil. For unsaturated soil, however, constant water content does not mean constant specific volume and the concept of undrained strength cannot be used.

26.8   Strength of unsaturated soil in slope stability

The critical state strength of unsaturated soil cannot be described by an undrained strength or by an effective stress strength. The best we can do is relate the strength of unsaturated soil to the net stress $(\sigma - u_a)$ and the matrix suction $(u_a - u_w)$.

Figure 26.6 shows the relationship between the critical state strength τ_f of an unsaturated soil, the net stress and the matrix suction. Instead of matrix suction we could plot degree of saturation as these are related, as shown in Fig. 26.5, but the relationships between them are different for wetting and drying. If the soil is saturated with $S_r = 1$ and $u_w = 0$, total and effective stresses are equal and the strength envelope is defined by the critical friction angle ϕ'_c. As the degree of saturation reduces and the matrix suction increases the strength increases and there will be a surface, as shown in Fig. 26.6, which describes the strength at any net stress and matrix suction.

The lines on the surface represent the strength at constant matrix suction and if they are assumed to be linear each can be represented by the Mohr–Coulomb criterion given by

$$\tau_f = c + \sigma \tan \phi \qquad (26.13)$$

In Eq. 26.13 the cohesion intercept c depends on the matrix suction while the total stress friction angle ϕ may or may not be related to ϕ'_c.

The arrow on the surface marked wetting shows that the strength at a given net stress decreases with decreasing suction and increasing degree of saturation. The arrow

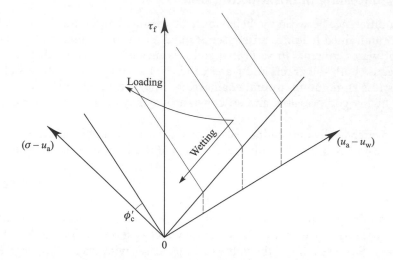

Figure 26.6 Strength of unsaturated soils.

marked loading shows what happens as the net stress is increased at a given water content. The gas compresses and the degree of saturation increases: the strength may increase or decrease depending on the relative contributions of c and $\sigma \tan \phi$ in Eq. 26.13. At very large total stress the initially unsaturated soil may become saturated, or nearly so, in which case the strength will converge on the envelope for saturated soil.

You can determine the strength surface shown in Fig. 26.6 from a set of triaxial or shear tests and you can use these to do total stress analyses for stability of slopes and retaining walls. But if the water content rises due to rainfall or changes in the groundwater the strength will reduce. The worst that can happen is the soil becomes saturated and then you can use the analyses given in Chapter 21.

26.9   Settlement of foundations on unsaturated soil

In Sec. 22.11 I discussed settlement and heave of shallow foundations on saturated soil due to changes of groundwater resulting from such things as dewatering, changes of vegetation and leaking drains. These apply also to settlement and heave of foundations on unsaturated soil but in this case there is another important mechanism for settlement.

In soil with a low degree of saturation the water forms matrix water bridges, which act like glue bonding the grains together. As a result the soil can be very loose but still relatively stiff and strong so shallow foundations can have large allowable bearing pressure and small settlements. If the soil becomes wetter the matrix water bridges are lost as the water changes from that shown in Fig. 26.3(a) to that shown in Fig. 26.3(b). The initially loose bonded soil now becomes unbonded and its strength and stiffness decrease. Its load factor and allowable bearing pressures decrease and there will be settlements which are known as wetting collapse settlements.

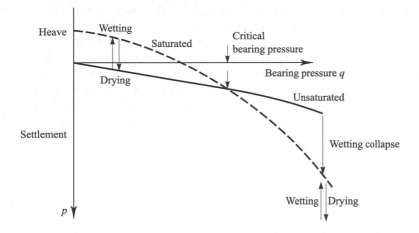

Figure 26.7 Volume changes in unsaturated soils.

Figure 26.7 shows bearing pressure settlement curves for unsaturated soil (the full line) and the same soil after it has been saturated at a small bearing pressure. These curves can be obtained from two oedometer tests: this is known as the double oedometer method. There is a critical bearing pressure at which the curves cross and wetting collapse can occur if the bearing pressure is greater than this critical value. Once wetting collapse has occurred, or for bearing pressures smaller than the critical value, wetting and drying at constant bearing pressure causes heave and settlement due to changes in matrix suction.

26.10   Compaction and compacted soils

Construction often involves excavation and use of soil in earthworks such as road and rail embankments, earthfill dams and land reclamation. When it is excavated soil comes out of the ground in lumps; you can see this happening when you dig the garden. When it is placed it must be compacted to make a stiff and strong engineering fill. The process is illustrated in Fig. 26.8.

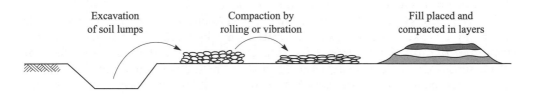

Figure 26.8 Earthworks.

Compaction of fine grained soils is usually by rolling or impact while compaction of coarse grained soil is usually by vibration. You can see this being done in roadworks and in back-filling trenches. The objective of compaction is to remove air trapped between the excavated lumps as they are first placed. This means that the lumps must be relatively weak to allow them to fail and deform plastically. However, if they are too weak the resulting fill will itself be weak. There is an optimum strength of the lumps which results in the best fill for a particular effort in compacting it. Since soil strength is related to its water content there is an optimum water content for placing and compacting soil.

Do not confuse the words compression, consolidation and compaction: they mean very different things and they describe very different processes. Compression was described in Chapter 8: it relates volume change to change of effective stress. Consolidation was described in Chapter 15: it relates volume change to time as excess pore pressures dissipate and water is squeezed from soil. In these chapters I have described compression and consolidation of saturated soil. Compaction is the process of removal of air from an assembly of saturated lumps by mechanical work: rolling, impact or vibration. The water content and the weight of soil grains do not change during compaction but the volume decreases and the soil becomes stiffer and stronger.

26.11   Compaction curves and behaviour of compacted soil

The degree of compaction of soil is measured by the dry density ρ_d which is given by

$$\rho_d = \frac{M_s}{V} \tag{26.14}$$

where M_s is the mass of dry soil in a volume V. (The usual units for dry density are Mg/m^3.)

The degree of compaction of soil depends on the effort put into compacting it, either in a laboratory test or in the ground, and on the water content. In a laboratory test the compactive effort is provided by a number of standard hammer blows and in the ground by passes of a roller or a vibratory compactor.

Figure 26.9 illustrates a typical compaction curve for a particular compactive effort as the dry density ρ_d related to the water content. The dry density reaches a maximum at the optimum water content. For greater or lesser compactive efforts the curve would be shifted but it should retain the same basic shape.

The chain dotted line is the relationship between ρ_d and water content for a saturated soil and the broken lines represent the relationships between ρ_d and water content as the degree of saturation becomes less. At high water contents the degree of saturation of a compacted soil become larger and approaches the fully saturated condition only at large water contents. At water contents below the optimum the degree of saturation, and the dry density, diminish rapidly. It is obviously best to compact soils at water contents as close to the optimum as possible.

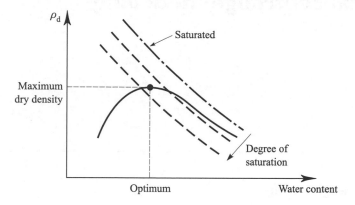

Figure 26.9 Compaction curve.

26.12   Summary

1. An unsaturated soil contains soil grains, water and a gas, usually air, and it is characterized by the degree of saturation S_r. The air may occur in discrete bubbles or it may be continuous, in which case the water collects at points the contact between the grains. For S_r in the range 0.25 to 0.85 both the water and the air are continuous in three dimensions.
2. The simple effective stress described in Chapter 6 does not work for unsaturated soils whose behaviour depends on the net stress $(\sigma - u_a)$ and the matrix suction $(u_a - u_w)$.
3. Matrix suctions are generated by surface tension. They are related to the grain size and for fine grained soils they can be very large. Suctions also vary with degree of saturation but the relationship is different when the soil is drying from when it is wetting.
4. The strength of unsaturated soil is related to the net stress and the matrix suction.
5. Foundations on unsaturated soils may settle or heave depending on whether the bearing pressure is greater or less than a critical bearing pressure.
6. Compacted soils are unsaturated and the degree of saturation decreases as the water content decreases. They should be compacted at a water content close to the optimum when the dry density will be maximum.

Further reading

Fredlund, D. G. and H. Rahardjo (1993) Soil Mechanics for Unsaturated Soils, Wiley, New York.
Mitchell, J. K. and K. Soga (2005) *Fundamentals of soil behaviour, 3rd edition*, Wiley, New York.

Chapter 27

Geotechnical centrifuge modelling

27.1 Modelling in engineering

Engineers frequently use scale models in conjunction with theoretical analyses. For example, wind tunnel modelling is used routinely by engineers to study the flow of air past vehicles, aircraft and buildings. Hydraulic engineers frequently use models to study the flow of water in river channels, tidal flow in estuaries and wave loading on structures. Scale modelling is used most often when the theoretical solutions contain major simplifications and approximations or when numerical solutions are very lengthy, as is often the case in geotechnical engineering.

A geotechnical model might be tested when it would be too difficult, expensive or dangerous to build and test a full-scale structure. For example, it would be very difficult to test the response of a large earth-fill dam to earthquake loading and it would be very dangerous to examine the collapse of a tunnel heading during construction. Usually a model will be smaller than the prototype (or full-scale) structure that it represents.

The principles for modelling fluid flows are well established and so too are the principles for geotechnical modelling. To achieve correct scaling in geotechnical models the unit weight of the soil is increased by accelerating the model in a geotechnical centrifuge.

At present modelling is used less frequently in geotechnical engineering than in other branches of civil engineering but it is an important and valuable technique and one that you should know about. Detailed discussion of geotechnical centrifuge modelling is obviously beyond the scope of this book and what I want to do in this chapter is simply to set out the basic principles and to describe the principal purposes of modelling.

27.2 Scaling laws and dimensional analysis

Normally a model and the prototype will be geometrically similar so that all the linear dimensions in a model will be scaled equally but, for various reasons, it is impossible to construct a model that behaves exactly like a large prototype in all respects. (You have probably noticed that the waves made by a model sailing boat are different from the waves made by a full-sized yacht.) Instead, the model should have similarity with the prototype in the aspect of behaviour under examination. For example, in a wind tunnel model of an aircraft wing the relationships between lift, drag and velocity should be similar while in a river model the relationships between water depths and velocities should be similar, but neither model need look very much like the prototype it represents. On the other hand, a model built by an architect or a railway enthusiast should look like the real thing.

The rules that govern the conditions for similarity between models and prototypes are well known and the simplest method for establishing scaling laws is by dimensional analysis. The basic principle is that any particular phenomenon can be described by a dimensionless group of the principal variables. Models are said to be similar when the dimensionless group has the same value and then the particular phenomenon will be correctly scaled. Often these dimensionless groups have names and the most familiar of these are for modelling fluid flow (e.g. the Reynolds number).

27.3   Scaling geotechnical models

In constructing a geotechnical model the objectives might be to study collapse, ground movements, loads on buried structures, consolidation or some other phenomenon during a construction or loading sequence. In earlier chapters of this book I showed that soil behaviour is governed to a very major extent by the current effective stresses (this is a consequence of the fundamental frictional nature of soil behaviour). Consequently, the stresses at a point in a model should be the same as the stresses at the corresponding point in the prototype.

Figure 27.1(a) shows the vertical total stress at a depth z_p in a prototype construction in the ground and Fig. 27.1(b) shows a similar point at a depth z_m in a model with a scale factor n (i.e. all the linear dimensions in the model have been reduced n times). In the prototype the vertical stress is

$$\sigma_p = g \rho z_p \tag{27.1}$$

where ρ is the density of the soil and $g = 9.81$ m/s^2 is the accleration due to Earth's gravity. If the model is placed in a centrifuge and accelerated to n times g the stress at a depth in the model $z_m = z_p/n$ is

$$\sigma_m = ng \rho z_m = \frac{ng \rho z_p}{n} \tag{27.2}$$

and $\sigma_m = \sigma_p$. Since the stresses at equivalent depths are the same the soil properties will also be the same (provided that the stress history in the model and prototype are the same) and the behaviour of the soil in the model will represent the behaviour of the soil in the prototype. Notice that you cannot reproduce the correct prototype stresses by applying a uniform surcharge to the surface of the model as, in this case, the stresses

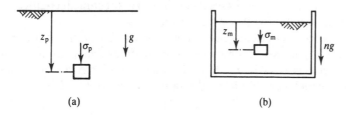

(a)   (b)

Figure 27.1 Stresses in the ground and in a centrifuge model.

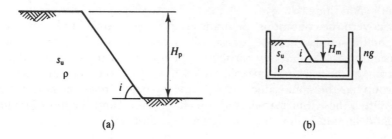

Figure 27.2 Scaling for the stability of a model slope.

in the model will be approximately constant with depth rather than increasing linearly with depth as in the ground.

Another way of looking at the requirements of geotechnical modelling is through dimensional analysis. The stability of a slope for undrained loading was described in Sec. 21.8. For the prototype slope in Fig. 27.2(a) with height H_p and slope angle i (which is itself dimensionless), the stability depends on the undrained strength s_u, the height H_p and the unit weight $\gamma = g\rho$. These can be arranged into a dimensionless group

$$N_s = \frac{g\rho H}{s_u} \tag{27.3}$$

where N_s is a stability number. Notice that this is exactly the same as the stability number in Eq. (21.42). A model and a prototype are similar (i.e. they will both collapse in the same way) if they both have the same value of N_s. If the scale factor is n so that the model height H_m and the prototype height H_p are related by $H_m = H_p/n$ the stability numbers can be made equal by accelerating the model in a centrifuge to ng so that

$$N_s = \frac{g\rho H_p}{s_u} = \frac{ng\rho H_m}{s_u} \tag{27.4}$$

Thus the stability of the model slope illustrated in Fig. 27.2(b) will be the same as the stability of the prototype slope in Fig. 27.2(a) and if the slopes fail they will both fail in the same way.

The stresses, and the basic soil properties, in a prototype and in an nth scale model will be the same if the model is accelerated in a centrifuge to ng, but time effects may require a different scaling. There are several aspects of time in geotechnical engineering, the most important being associated with consolidation.

Consolidation due to dissipation of excess pore pressures with constant total stresses was discussed in Chapter 15. The rate at which excess pore pressures dissipate during one-dimensional consolidation is given by Eq. (15.34) and for similarity the time factor T_v in the model and in the prototype should be the same. From Eq. (15.25),

$$T_v = \frac{c_v t_p}{H_p^2} = \frac{c_v t_m}{H_m^2} \tag{27.5}$$

In a model with the same soil and pore fluid as the prototype, c_v is the same and if the scale is n we have $H_m = H_p/n$. Hence, from Eq. (27.5), the times for consolidation in the model and prototype are related by

$$t_m = \frac{t_p}{n^2} \qquad (27.6)$$

so that consolidation will proceed much more rapidly in the model than in the prototype. For a typical scale factor $n = 100$, we have $t_m = 10^{-4} t_p$ so that 1 hour of model time represents approximately 1 year of prototype consolidation time.

The relationship between the rate at which excess pore pressures dissipate as drainage occurs and the rate of loading that generates additional excess pore pressures governs whether a particular construction event is drained, undrained or partly drained, as discussed in Sec. 6.9. Remember that for routine geotechnical calculations we have to assume either that the soil is fully drained or that it is fully undrained, in which case there will be subsequent consolidation. A model could, however, examine cases of partial drainage in which the rates of loading and consolidation were coupled.

If the accelerations in the prototype and in the model are related by the scale factor n and are given by

$$\frac{d^2 x_p}{dt_p^2} = a\omega^2 \sin(\omega t_p) \qquad (27.7)$$

$$\frac{d^2 x_m}{dt_m^2} = na\omega^2 \sin(n\omega t_m) \qquad (27.8)$$

then the displacements are given by

$$x_p = a \sin(\omega t_p) \qquad (27.9)$$

$$x_m = \frac{a}{n} \sin(n\omega t_m) \qquad (27.10)$$

and the times in the prototype and in the model are related by

$$t_p = nt_m \qquad (27.11)$$

Any motion can be represented by a Fourier series which is a summation of sine functions and so the time scaling rule given by Eq. (27.11) applies to any displacement or loading. Notice that the scaling requirement for the rate of loading is that the times should be related by n, which is not the same as the requirement for modelling consolidation where the times should be related by n^2. Therefore it is not generally possible to model coupled loading and consolidation in the same model. This problem can be avoided by using a pore fluid such as silicon oil with a viscosity n times greater than that of water. In this case the coefficient of consolidation and the rate of consolidation in the model are reduced by n times so that the scaling $t_p = nt_m$ is then the same for both the rate of loading and the rate of consolidation.

27.4   Purposes of modelling

It would be very convenient to be able to construct and test a scale model that reproduced all the significant features of the behaviour of a proposed construction. Unfortunately, however, this is not generally possible for a variety of reasons. The principal difficulties are rather like those associated with ground investigations and laboratory testing (see Chapters 7 and 17) and are due to test samples not being fully representative of the soil in the ground. It is also difficult to model geological history and complex construction sequences. Instead, geotechnical models are usually constructed and tested to meet specific objectives.

The principal purposes and categories of geotechical modelling were discussed by Taylor (1987) and these are as follows.

(a)   Mechanistic studies

The basic methodology of engineering design is that engineers imagine all the possible ways in which a proposed construction may fail or distort and they then carry out analyses that demonstrate that it will perform satisfactorily in any of these ways. Sometimes major failures occur when the construction finds some other way to fail or distort. For example, in the upper bound and limit equilibrium methods described in Chapters 19 and 20 it is necessary to define compatible mechanisms and the solutions depend on the mechanisms chosen. For relatively simple cases it is usually possible to choose the critical mechanisms from previous experience, but in novel and complex cases they may not be so obvious. In these cases relatively simple model tests may be carried out simply to observe qualitatively the way in which the structure distorts and fails, thus indicating the most appropriate analyses.

(b)   Validation of numerical analyses

Design of geotechnical structures often requires complex numerical analyses using finite element, or similar, methods with non-linear and inelastic soil behaviour (see Chapter 13). These analyses are highly complex and before they are applied in design studies they should be tested against exact analytical solutions or against observations of the real events. Observations from relatively simple model tests can be used to test numerical analyses. The models should be similar to the proposed construction but, since the models are used only to calibrate the analyses, they need not reproduce all the details of the prototype.

(c)   Parametric studies

Another important procedure in design studies involves examining alternative construction details and investigating the consequences of different design assumptions. Furthermore, standard design codes and charts rely on studies of many different alternatives. Normally parametric studies are carried out using analytical or numerical methods, but model studies have a role to play in parametric studies, either on their own or together with other methods.

(d) Site-specific studies

In this case the model is intended to represent a particular construction so that the behaviour of the model is used directly to assess the behaviour of the prototype. It is obviously not easy to model all the details of the ground conditions and the construction and loading sequence; these are the most difficult type of centrifuge models to construct and test satisfactorily.

Model studies may be carried out for more than one purpose, for example combining validation of analyses with parametric studies. In practice, designs are vary rarely completed on the basis of model tests alone and model tests are almost always used in conjunction with numerical analysis.

27.5   Geotechnical centrifuges

In a geotechnical centrifuge, a model in a strong container is rotated in a horizontal plane about a vertical axis, as shown in Fig. 27.3. At the model the centrifugal acceleration a is

$$a = ng = \omega^2 r \tag{27.12}$$

where r is the radius and ω is the angular velocity (in radians per second). To maintain a reasonably constant acceleration field through the model the radius r should be large compared with the size of the model.

The essential features of a geotechnical centrifuge are illustrated in Fig. 27.4. The motor drives a vertical shaft at constant speed. The arm has an adjustable counterweight for balance and the model sits on a swing. At rest the swing hangs down, but as the arm rotates it swings up to a nearly horizontal position as shown. The purpose of the swing is so that the self-weight of the model always acts towards the base of the container; if you put a strong bucket containing water on the swing and start the centrifuge the water will remain level in the bucket.

The selection of the dimensions and speed for design of a geotechnical centrifuge is a matter of optimization between a number of conflicting requirements. A given prototype size could be represented by a small model tested at high accelerations or by a larger model at smaller accelerations; a given acceleration, or scale factor, can be

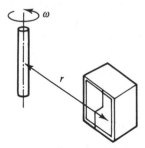

Figure 27.3 Centrifuge acceleration.

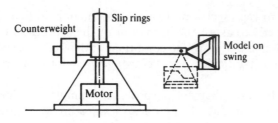

Figure 27.4 Characteristic features of a typical geotechnical centrifuge.

achieved by a high-speed machine with a relatively small radius or by a machine with a larger radius rotating more slowly.

From Eq. (27.12) the acceleration is given by $\omega^2 r$, so a small-radius, high-speed machine is more efficient than one with a larger radius and lower speeds. If, however, the radius is not large compared to the depth of the model there may be significant variations of acceleration with depth in the model. A small model, requiring large accelerations, will be relatively easy to manufacture and handle, but it will be possible to install only a limited number of instruments. On the other hand, a larger model which can be more easily instrumented will be heavy and more difficult to manufacture and handle on to the centrifuge.

The mass of the model, including the soil, the strong container and all the ancillary equipment for loading and observing the model, is called the payload. The capacity of a centrifuge is often quoted as the product (in g-tonnes) of the maximum acceleration (i.e. the scale factor) and the maximum payload at that acceleration.

There is a very great variation in the dimensions and capacities of geotechnical centrifuges. The optimization of size and capacity is determined largely by the resources of manpower available to the group who will run the facility, so that university groups tend to acquire machines requiring smaller and more easily managed models while commercial and government-run facilities tend to have machines able to test larger models that can accommodate more instrumentation.

As a very rough guide, about 50% of the payload could be soil, with the remainder required for the strong container and other equipment. For a medium sized geotechnical centrifuge with a working acceleration of 100 g the maximum payload of 400 kg could have about 200 kg of soil and this could be in a model (say) of 600 mm × 400 mm × 400 mm. At a scale factor of $n = 100$ (i.e. at an acceleration of 100 g) this represents a prototype volume of soil of 60 m × 40 m × 40 m or a 20 m thick plane strain slice 100 m wide and 50 m deep. A package up to 400 kg can be handled reasonably easily without expensive cranes and this represents an optimum size for a university facility.

27.6   Control and instrumentation in centrifuge models

During a typical geotechnical centrifuge model test the machine will be run at constant speed (i.e. at constant scale factor) while the model is loaded or unloaded and the behaviour observed. The requirements for control of loading and measurement of load and displacement in a model are broadly similar to those for laboratory tests described in Chapter 7.

Communication with the rotating model is through slip rings, as shown in Fig. 27.4. These may transmit fluids (e.g. water, gas or hydraulic oil) or power to operate motors or valves, and they will transmit signals from force, pressure and displacement transducers and from closed circuit television cameras set to observe critical points in the model.

Before conducting a test the model should be allowed to come into equilibrium under the increased self-weight stresses at constant centrifuge acceleration; larger models of fine-grained soils may require the centrifuge to be run continuously for several days to reach equilibrium. Often a small ground investigation will be carried out in flight using model cone penetration or shear vane tests similar to those discussed in Sec. 17.5.

A very large number of different events and construction activities can be modelled. Design and manufacture of model loading and construction devices taxes the ingenuity of the engineer and a number of sophisticated and novel examples can be found in the literature of centrifuge modelling. Some typical examples include: vertical and horizontal loading of foundations, piles and anchors; modelling excavation and tunnel construction by draining heavy fluids or by reducing pressures; embankment construction in stages by dropping sand from a hopper; earthquakes simulated by vibrating the base of the model; formation of craters and blast loading on buried structures simulated by detonating small explosive charges.

27.7   Summary

1. Modelling geotechnical structures can be used to examine mechanisms of deformation and collapse, to validate numerical analyses and for parametric studies. Models can occasionally be applied to site-specific cases, but this is usually very difficult.
2. For correct scaling of stresses and soil properties geotechnical models should be tested while under acceleration in a centrifuge. An n-scale model should be tested at an acceleration of ng, where g is the acceleration due to Earth's gravity.
3. At a scale factor of n, rates of loading should be raised by a factor of n and rates of consolidation will be increased n^2 times.

Reference

Taylor, R. N. (1987) Modelling in ground engineering, Chapter 58 in Geotechnical Engineers Reference Book, F. G. Bell (ed.) Butterworth, London.

Further reading

Craig, W. H., R. G. James, and A. N. Schofield (1988) *Centrifuges in Soil Mechanics*, Balkema, Rotterdam.

Ng, C. W. W., L. M. Zhang, and Y. H. Wang (eds) (2006) *Physical modelling in geotechnics*, Taylor and Francis.

Schofield, A. N. (1980) Cambridge geotechnical centrifuge operations, *Geotechnique*, 30, 227–268.

Springman, S. (ed.) (2002) *Constitutive and centrifuge modelling: two extremes*, Balkema.

Taylor, R. N. (ed.) (1995) *Geotechnical centrifuge technology*, Blackie Academic and Professional.

Chapter 28

Concluding remarks

My objective in writing this book was to set out the basic theories of soil mechanics and geotechnical engineering in a simple and understandable way. In common with introductory texts in other engineering subjects, I have dealt principally with simple idealization to construct a theoretical framework for soil behaviour. You should be aware, however, that this is only a part of the story and the behaviour of natural soils is often more complex.

I have tried to relate the basic principles of soil mechanics to the general theories of mechanics and materials to demonstrate that soil mechanics does have a sound theoretical basis linked to theories that will appear in other courses on structures and fluid mechanics. I have also tried to describe soil behaviour in the context of everyday experiences of the behaviour of soils and granular materials in the garden, on the beach and in the kitchen. I want readers to relate the simple theories of soil mechanics to their own observations. Broadly, the predictions of a theoretical calculation should be what you would reasonably expect to happen and the stability of a large excavation or foundation will be governed by the same theories that govern the behaviour of small holes in the beach.

If you have understood the simple theories in this book, you should be able to analyse a simple retaining wall or foundation and assess the stability of a slope in idealized soil. You should be able to say what soil parameters are required for a particular design, distinguishing between the total stress parameters for undrained loading and effective stress parameters which require knowledge of the pore pressures. You should also know how values of soil parameters for design are determined from ground investigations and laboratory and *in situ* tests and you should have some idea of what are reasonable values for different soils.

Of course, when you graduate you will not be a fully qualified and experienced engineer able to design major groundworks, and the next step in your career may take one of several directions. You might, for example, want to become an accountant, a manager or an inventor and you can do all these in civil engineering. Any construction enterprise is really a business and the engineers will need to manage their resources and account for income and expenditure. Any civil engineering design is really an invention because it is a unique creation and inventors must also be engineers because their inventions must be made to work.

The next step in your career as a civil engineer is to learn how to put theory into practice. You should start by working with experienced engineers and you will be trained through experience. Among other things you will learn how to do routine designs using standard methods. One of the most important things to learn is how to recognize when the problem has become so complex and difficult that you need to consult a specialist.

I hope that some of you will be sufficiently excited by the challenges of soil mechanics and geotechnical engineering to want to become a specialist called on to solve the difficult ground engineering problems. In this case you will probably want to take a higher degree in soil mechanics, geotechnical engineering, engineering geology or a related subject. You will need to know very much more about soil mechanics than I have been able to cover in this book, but it will provide an introduction to these more advanced studies.

The Mechanics of Soils and Foundations will have succeeded in its aims if it conveys to students and engineers the idea that there are relatively simple theories underlying engineering soil behaviour and that form the basis of engineering design. I hope that readers will be able to apply these theories to geotechnical design and use them to assess critically the conventional, routine design methods conventionally used in practice.

Author index

Subject index